Books by the author:
The Lost Cities Series:
Lost Cities of China, Central Asia & India
Lost Cities & Ancient Mysteries of Africa & Arabia
Lost Cities of Ancient Lemuria & the Pacific
Lost Cities & Ancient Mysteries of South America
Lost Cities of North & Central America
Lost Cities of Europe & the Mediterranean

The Lost Science Series:
The Anti-Gravity Handbook
Anti-Gravity & the World Grid
Anti-Gravity & the Unified Field

LOST CITIES & ANCIENT MYSTERIES OF AFRICA & ARABIA

DAVID HATCHER CHILDRESS

ADVENTURES UNLIMITED PRESS
STELLE, ILLINOIS

This book is dedicated to the ancient voyagers to Ophir, and the many nations that sailed the world in prehistory. I also dedicate this book to H. Rider Haggard, David Livingston, Sir Richard Burton, Howard Carter and Lord Carnarvon, James Churchward, Thor Heyerdahl, Barry Fell, Roy Mackal, Ivan T. Sanderson and all other researchers who have the courage to stand up for what they believe in.

Library of Congress Cataloging in
Publication Data.

Lost Cities & Ancient Mysteries of Africa & Arabia

Cover Photo by David Hatcher Childress

Copyright 1990 by David Hatcher Childress
All rights reserved.
Maps © Paul Barker
Printed in the United States of America.
First Edition
Second Printing July 1990
ISBN 0-932813-06-2

Portions of this book were first published in the book,
A Hitchhiker's Guide To Africa & Arabia
©1984 by David Hatcher Childress
Out of print since 1987.
Address all enquiries to:

Adventures Unlimited Press
Box 22
Stelle, IL 60919 USA

About the Author:

David Hatcher Childress was born in France, and raised in the mountains of Colorado and Montana. At nineteen, he left the United States on a six-year journey across Asia, Africa and the Pacific. An ardent student of history, archaeology, philosophy, and comparative religion, he has authored numerous articles, which have appeared in publications around the world. His many books include *Lost Cities & Ancient Mysteries of South America* , *Lost Cities of North & Central America, Anti-Gravity & the Unified Field,* and others. Many titles are now available in foreign language editions.

Currently, he travels the globe in search of lost cities and ancient mysteries. He also participates in the *World Explorer's Club's* expeditions to remote jungles, mountain, islands and deserts of our planet. For more information on the upcoming programs and or for a copy of our newsletter, please call our offices at: 815-253-6390 or write to:

Adventures Unlimited
Box 22
Stelle, IL 60919 USA

Our world headquarters is 75 miles south of Chicago.
Readers are invited to visit our bookstore at 303 Main Street,
Kempton, Illinois and visit the World Explorers Club.

Thanks to Linda Drane, Carole Gerardo, Paul Barker, John Moss, Pat Purcell, Dave Millman, Ted Martin, Harry Osoff, Anne Marie Bodo, Mom, Nadia, Alan, Aliza, Morry & the gang at Sunrise Hearth Bakery, everyone at the World Explorers Club and a host of other people who helped make this book a reality.

Every mystery solved brings us to the threshold of a greater one.

—Rachel Carson

The Lost Cities Series:
Lost Cities of China, Central Asia & India
Lost Cities & Ancient Mysteries of Africa & Arabia
Lost Cities of Ancient Lemuria & the Pacific
Lost Cities & Ancient Mysteries of South America
Lost Cities of North & Central America
Lost Cities of Europe & the Mediterranean

The Lost Science Series:
The Anti-Gravity Handbook
Anti-Gravity & the World Grid
Anti-Gravity & the Unified Field
Vimanas & Vailixi of Ancient Rama & Atlantis

LOST CITIES
& ANCIENT
MYSTERIES OF
AFRICA &
ARABIA

LOST CITIES
& ANCIENT
MYSTERIES OF
AFRICA &
ARABIA

AFRICA

TABLE OF
CONTENTS

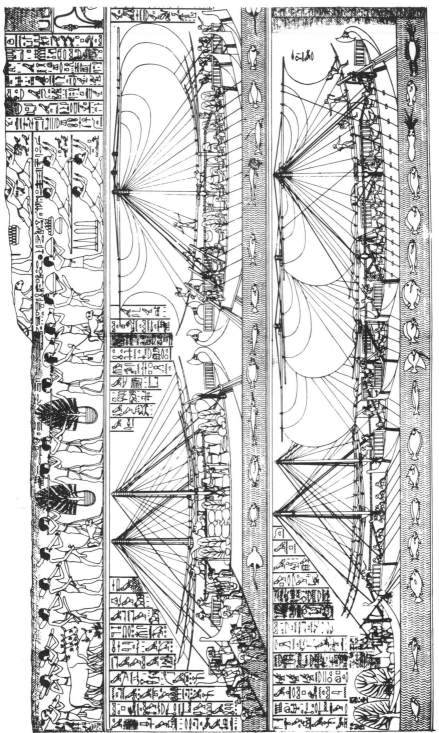

Egyptian seagoing vessels, ca. 1500 B.C.

From *Ships & Seamanship in the Ancient World*, Princeton University Press

Mappa Mundi Auctoris Incerti, Nuremberg, c. 1540

Reconstruction of an interior gateway to an Aton Sun Temple in Egypt. Notice the rays of the sun extending down, and the Key of Life (Ankh) being held in front of the Priestess' mouths. Behind them are devotees each holding a Djed column, used to illuminate the temple. Just what is happening in the center of the relief is not clear, bread may be being baked is a solar oven, or perhaps something is being levitated. From, *Akhenaten The Heretic King*, 1984, Princeton University Press.

Chapter 1

SYRIA & LEBANON: AT THE ANTI-DILUVIAN TEMPLE OF BAAL

What is it that pulls me away from what others
call happiness, home and loved ones?
Why does my love for them not hold me
down, root me?
Games. Adventures. The Unknown.
–Anais Nin

Once again, I was on the road. I downed the last bit of Turkish tea, half tea and half sugar, and headed for the bus station. Istanbul was a fun city, the gateway to the Middle East. Within a few minutes, I was leaving the Sultanahmet area of Istanbul, which is geographically in Europe, and was crossing the Bosphorus Strait into Asia.

I looked out the window and sighed. This was not my first time to the Middle East. I had only just come across from India, Afghanistan and Iran a month or so before. Now,with a new supply of traveler's checks and some dollars folded into my secret money belt, I was now on my way to Arabia and Africa. *Adventure Travel Fever* had seized me, and I was determined to embark on a long journey through remote lands that are reputed to be the most difficult in the world to travel through.

Taking trains and buses across Asia was easy compared to travel in Africa and Arabia. Generally speaking, there is very little public transport in most of these countries, and arranging rides with truck drivers is often necessary. Sometimes it is necessary to walk between borders or towns. In Africa and Arabia, transport, as well as other services, can be ridiculously expensive.

The distances in Africa are vast. It is the second largest continent with 20% of the earth's land mass. Africa is really far larger than it appears on most maps, because the typical Mercator projection distorts the southern hemisphere, making it seem smaller, and the northern hemisphere larger. Geologically, Arabia is considered part of the African continental plate. The great Rift Valley has cut the Arabian peninsula away from the African tectonic plate, creating the Jordan Valley, the Red Sea, and Lake Rudolf in East Africa. At Lake Rudolf the rift divides, the main branch forming Lake Victoria, the Ruwenzori Mountains, and Lake Tanganyika. The branches join again at Lake Malawi, and then runs down the Zambesi River and finally out to sea where it continues as an ocean valley. It extends over one fifth of the earth, over seventy degrees of latitude, and contains some of the deepest lakes in the world.[1] The ancient tectonic shift, of cataclysmic proportions that created the great Rift Valley will be explored in this book.

I wanted to do more than just see Africa and Arabia, I wanted to explore its mysteries. According to traditional history, mankind evolved in Africa millions of years ago, lived primitively in caves and trees for the same amount of time, and then, about ten thousand years ago, suddenly came out of the caves and started building ziggerats in Sumeria and Pyramids in Egypt. Yet, these civilizations were

15

still supposedly primitive, had little interest in the rest of the world except for the occasional conquering of a nearby country, and life today is far better by any standards.

Somehow, I didn't buy that version of history. The fertile crescent and the Nile civilizations were no doubt areas of ancient civilizations, but I wanted to find out the solutions to such riddles as what civilizations had inhabited the Sahara before it was a desert, where the Egyptians had inherited their extremely advanced sciences, where King Solomon had gotten his fantastic wealth, and what were massive, ancient cities doing in the wildest and remotest parts of central Africa? These were just a few of the mysteries that I intended to investigate, and I would have to do it in person. Reading about them in some book was not good enough for me; I wanted to see it for myself. Little did I know what I was getting myself into. Unbeknownst to me at the time, my trip was to last for two and a half years, and my life would be threatened on many an occasion. I would be arrested and thrown into jail. Wild animals would attack me. Crazed soldiers would try to kill me. The evil Idi Amin would get me into deep trouble. I would contract deadly diseases and see some of my best friends horribly murdered.

Perhaps if I had known all of the adventures that I was to have, I would have thought twice about my quest for lost cities and ancient mysteries. Yet, I was blissfully unaware of the many dangers, trials and tribulations that awaited me on the dusty roads I would trod. Lost treasure and riches are of interest to me, though I am not a particularly greedy person. My real thirst is for knowledge, not material wealth. The secrets of the past were my goal, no matter how strange the answers might be. I leave it to the reader to decide whether truth is indeed stranger than fiction!

§§§

"Who are you?" I was asked suddenly by a young European sitting across from me on the bus. I found out later that he was from Yugoslavia. Indeed, who was I? I had often asked myself that question. I was born in France, though my parents are both Americans. I was raised in the mountains of Colorado and Montana, and a deep curiosity about history, nature, the world, and just about everything had been instilled into my personality since I was a child. I ask a lot of questions; I did as child, I do as an adult. Weeks of backpacking and climbing in the Rocky Mountain as a teenager had given me the sort of confidence in myself that was necessary for sojourns into strange and unknown countries by myself. A year of traveling through Asia had further strengthened my self confidence.

I traveled light. Everything I needed was in my green, soft-frame backpack: a sleeping bag, a mosquito net, a nylon poncho, a large plastic water bottle and purification tablets, an assortment of shirts, pants, socks, underwear and other things to wear. With my first aid kit, camera and Swiss Army knife, what more did I need besides a good attitude, a friendly demeanor and a winning smile?

I am often described as happy-go-lucky fellow who is quick to smile and slow to get angry. Of medium height, build and looks, I hardly stand out in a crowd, except for the fact that my curly blonde hair and glasses make me a bit of an oddity in Africa. I find it difficult to disguise myself as a local, that's for sure.

I didn't tell all of this to the Yugoslavian student who was sitting across from me. That I was an American student of history and archaeology on my way to Syria and Jordan was sufficient for him. He too, with his friend who was sitting next to him, was on his way to Damascus, the capital of Syria. Their names were Uros and Dusan and they were students at a university in Ljublijana, a major city in northern Yugoslavia. Like me, they were young, and wouldn't pass very well as Arabs. They were friendly guys who spoke English well, liked American blues music, and were out to discover Arabia. We decided to travel together to Damascus.

It was late in the evening when our bus reached Adana, a large town in southern Turkey near the border of Syria. We found a hotel room with three beds for a few dollars and checked in. Then we went out for dinner at a local cafeteria. We were all up early the next morning, and caught a bus to the Syrian border. Border formalities were brief and within an hour we were off by bus to Alleppo, the largest town in northern Syria. As Uros had been to Syria once before, Dusan and I shouldered our packs and followed him to the Hotel Omara, where we got a triple room.

Moments after dumping our packs we out on the streets. It was obvious that we weren't in Turkey any more. The street was packed with something I had never seen before: Arabs! Unlike Turkey, where western dress is the norm, and it difficult to actually distinguish a Turk from other Europeans, Syrians wore the traditional Arab robe or "gellabia" with the typical "hata" or cloth, either white or checkered and "ocall", the black woven band that goes around the head, holding the hata to the skull. This typical Arab head covering is probably best known to Americans by photographs of Yasser Arafat, the head of the Palestinian Liberation Organization.

By the looks of all the people on the street, Yasser Arafat was everywhere. Except for camels, it was how one might imagine an Arab city: minarets towering above a bazaar; swarthy robed men walked in the streets, women in dark "chaderis" with their faces covered moved silently along; the smells of sesame bread baking and the sounds of guttural Arabic filled the street. This was Arabia, not Turkey or Iran. These were Arabs, not Turks or Persians.

Syria is at once an ancient and a new nation. Although full independence was secured from the French only after World War II, the capital city, Damascus, is on the oldest continuously inhabited cities in the world. The area now known as Syria was occupied by Canaanites, Phoenicians, Hebrews, Aramaeans, Assyrians, Babylonians, Persians, Greeks, Romans, Nabateans, and Byzantines. Each in turn left an imprint on the physical and social landscape. It was the conquest of Mohammed's armies in 636 A.D., however, that gives Syria the identity it has today.

Syria's modern existence is closely tied to that of Lebanon. In fact, Syria was split by the French in 1920 who drew up the present border around Lebanon in order to offset predominantly Moslem Syria from the largely (at that time) Christian population in Lebanon. Only slightly larger than North Dakota, Syria is a country that can be crossed in one day, but in ancient times, the eastern desert was a formidable barrier, and it was largely believed that it was virtually impossible to cross the desert from the Babylonian cities of the Euphrates to gain the inhabited areas of the mountains and coast in the west. As the powerful armies of the ancient world marched hither and yon across the Middle East, the Syrian desert was a major obstacle.

In 525 B.C., Egypt was in control of Syria and the major challenger to Egypt's power was Persia. The 26th Dynasty Pharaoh Aahmes II was aware that Persia intended to attack the eastern frontier of the Egyptian empire, but believed the Syrian desert to be such an impassable obstacle that there was little danger in such an attempt. Aahmes then brought his forces together at Pelusium, a port city on the Mediterranean coast of Syria to await a naval attack from Cambyses, the Persian king.

Indeed, Cambyses had intended to attack by sea, but then a bit of good fortune struck him: one of Aahmes' best Egyptian officers, Phanes, was aggrieved over his treatment by Aames and so angered that he set out across the desert to join the Persian army. Suspecting his purpose, Aahmes sent his favorite eunuch in pursuit, who overtook Phanes on the road. Yet Phanes escaped and reached the Persian camp and received a warm welcome by Cambyses.

Phanes not only told Cambyses all the secrets of his former master, but also told him how to cross the desert. As a first step, envoys were sent to the Bedouin sheiks, who were given presents and in return made treaties by which they promised to furnish the expedition with camels and water, and to guide them by the best and shortest route to Pelusium. Meanwhile, Aahmes died, and it was his inexperienced son Psamtek III who faced the Persian army Pelusium after they had crossed the Syrian desert. The Egyptians were defeated in the decisive battle and Psamtek fled to Memphis in the Nile Delta.

Cambyses, who was epileptic and slightly deranged, collected all the sacred cats of Pelusium and galloped by the main gate of the city flinging the cats high in the air to their deaths. After two hundred men on a ship sent to Memphis to demand surrender from Egypt were foolishly killed by the Egyptians, he laid siege to Memphis, took it and executed ten-fold: two thousand of the sons of the most respected citizens, among them the son of Psamtek, whose daughter and a number of leading young women were sold into slavery. Cambyses then became king of Egypt, founding the 27th Dynasty, and initiated the period of Persian rule. Thought to have been insane, he killed his own brother, committed the unpardonable sacrilege to the Egyptians by personally slaying the new bull during the festival of the Reincarnation of Apis and ultimately died of a wound which was self inflicted.[2] And so we see how the Syrian desert played an important part in the ultimate downfall of Egypt, as it was essentially under foreign domination from that point on.

§§§

After a good night's sleep after our weary travels, Uros, Dusan and I went out to explore Alleppo. It was a bright, sunny day, not unusual for Syria. The first thing that greeted us was the massive Citadel of Aleppo built on a hill overlooking the city.

"Wow, look at that!" exclaimed Uros. Indeed, it was an impressive sight, and we naturally headed in the direction of the ancient gate. While Aleppo itself is said to be over five thousand years old, the citadel is an Arab-Islamic military fort built about five hundred years ago. After a pleasant stroll around the fort, with its good views of the city below, Dusan wanted to go the post office.

While I was standing outside the post office waiting for Uros and Dusan, an Arab came up to me and asked me where I was from. His question was in English, and he was wearing a western suite, rather than an Arab robe. He was in his early thirties, I guessed, and was perhaps on his lunch break.

I thought about my answer for a moment, and then hesitatingly told him that I was from the United States of America. His eyes widened momentarily and then narrowed menacingly. He clenched his fists and fought back rage. "American! You are American! Why are you helping the Jews?" he demanded. "Is it true that Americans think that Arabs have tails? Is it true that Americans think that Arabs are animals and eat people? Is that what you think?"

I rolled my eyes and edged toward the door of the post office where Uros and Dusan were inside. "No. Why, uh, no, that's not what Americans think at all..." I stammered. His hostility increased, and he began shouting that America was evil, and that we should stop supporting the Jews (most Arabs will not use the name Israeli or Israel, but prefer just to refer to them as Jews). There was little that I could do or say, and I was quite relieved when Uros and Dusan emerged from the post office.

We left the ranting Syrian at the post office and headed for lunch at a local restaurant. While eating our shish kebabs, salad and bread, Dusan struck up a conversation in English with a man at the next table. He was in his late twenties, and also dressed in western clothes. He had short black hair and while he apparently already shaved that morning, his thick beard was already stubble on his face. When he found out that Uros and Dusan were from Yugoslavia, he then asked me where I was from.

Still reeling from my earlier confrontation at the post office, I wasn't sure whether I should tell him or not. Being Canadian in Syria seemed like a good idea for the rest of my trip, I thought. However, before I could say Hudson Bay, Dusan volunteered that I was American.

"American!" the man exclaimed, jumping to his feet. "You're American? Really!"

"Uh, no! No!" I protested, "I just live there. I mean..." I put my hands up to ward off any blows that might be forthcoming as the man approached me.

"Can I have your address?" he pleaded. "Please? I love America. I want to go there so bad! I need your address. Can you help me get a visa to America? It is my life's dream!"

I breathed a sigh of relief, an put my hands down. Well, certainly, being American in this country elicited some strong reactions from people, both positive and negative! He insisted on taking me out to dinner that night, though he did not extend the same invitation to Uros and Dusan. He was a bit too friendly, and I didn't feel like I was really in any position to help him get a visa to America.

"What hotel are you staying in?" he asked. "I'll come there tonight and we'll go to dinner!" I was reluctant to tell him, but Uros volunteered the name of our hotel.

After lunch, as we walked through the market back towards our hotel, I told them that I really didn't want to have dinner with that man.

"But what can you do?" asked Dusan. "He already knows where you are staying."

"We could leave today for Damascus," suggested Uros. "It's not even one o'clock yet, and Damascus is only a five hours away. Let's go back to the hotel, check out, and get a bus. They leave every half hour."

After a bit of discussion, we decided that this was the best solution. Aleppo was interesting, but it seemed like we had seen it all the night before and that morning. After packing up our bags, we were once again on the road, the road to Damascus.

§§§

The mystery of the Phoenicians & the Peoples of the Sea

One of the mysteries of coastal Syria and Lebanon, is its early conquest by a mysterious group of astonishingly sophisticated seafarers known historically as the Seapeoples. In the 12th century B.C. two ancient cities, Ras Shamra (ancient Ugarit) on the Syrian coast, and Byblos on the Lebanese coast, were both conquered and occupied by a mysterious group of seafarers who are believed to have been the ancestors of the Phoenicians and Carthaginians.

Ras Shamra was known in ancient times at Ugarit and has been dated to the 7th millennium B.C., which is an astonishing nine thousand years ago.[13] Considering that ancient Sumeria is only dated from about 3100 B.C.[14] and is still generally considered as the oldest known civilization, this is an astonishingly old date, making the city one of the oldest authenticated cities in history (along with Biblical Jericho and Catal Hyuk–see my book *Lost Cities of China, Central Asia & India*[15]). Any city which is is old as nine thousand years is likely to be associated with the ancient Osirian Civilization, about which we will shortly learn more.

The lost city of Ugarit was excavated by the French in the 1930s (who were controlling both Syria & Lebanon at the time by League of Nations Mandate) who made some interesting discoveries. It is situated only a few miles (12 km) north of the modern Syrian port of Latakia, and had its own ancient harbor, Minet-el-Beida. This prehistoric harbor was almost certainly linked the ancient and mysterious prehistoric harbor at Alexandria, which will be discussed in chapter 3. Ugarit was well placed for Mediterranean trade to the west, as well as for trading with the civilizations on the Tigris and Euphrates, and the Hittites to the north. In fact, while Ugarit was generally considered a Canaanite city, it was also, in some way, a vassal state of the powerful and mysterious Hittites to the north in modern day Turkey.

The French archaeologist Claude Schaeffer directed the excavations and published *Statigraphie comparée et chronologie de l'Asie occidentale* in 1948, which compared all the information collected from the Ugarit excavations with the surrounding area. Schaeffer was a respected archaeologist, but his beliefs about cataclysmic geology and the revelence of earthquakes to *absolute chronology* (dating according to our present calendar, counting backward; dates given in *absolute chronology* are given as Before Present or B.P. The problem of dating the ancient past, even with ancient texts that give us dates is difficult and controversial, as we are about to see) were not then and are not now accepted by mainstream archaeology.

Schaeffer' principal discovery at Ugarit was cuneiform texts written in Ugaritic. These give historians a better understanding of the development of the alphabet, and helps support the Greek tradition that the alphabet, as we know it, was invented by the Phoenicians.[18] Little is known about the early millenniums of Ugarit, but the later city of 2nd Millennium B.C. (just prior to its destruction by the Sea Peoples) was a magnificent city with stone blocks of massive proportions. There were two temples in the city, the temple of Ba'al on the west side of the city and the temple of Dagan at the summit of the central mound of the city. The temple of Dagan is built out of gigantic blocks of cut stone, much like Hittite cities to the north.

But, alas, after its destruction by the mysterious sea peoples in approximately 1170 B.C., the city was not to rise again but rather to await the French archaeologist in 1929. This time of destruction by the Sea Peoples is a strange and controversial period of history. The mysterious Sea Peoples were already laying siege to the Hittite cities to the north, which they totally destroyed with such fury and advanced weapons that the walls of many of the Hittite forts were vitrified! Vitrification is the fusing of brick or stone with a glass-like glaze and takes such an extreme heat that modern scholars are at a loss to explain the phenomenon and attempts to recreate vitrification have failed totally. Vitrified forts have been found in Scotland, Ireland, France, Turkey (the Hittite cities), India and even Peru.

The fusing of brick and stone takes such a tremendous amount of heat that a simple fire will just not suffice. Nuclear weapons do a very good job of vitrification, though that is not necessarily the answer in this case. A mysterious substance known in history as "Greek Fire" is reportedly responsible, a substance used in naval warfare that when flung from catapults on the ships would set the ships on fire, and could not be put out with water, as it would actually *burn underwater!* Obviously it was some chemical fire, and though we can make some guesses as to its composition, its actual contents remain a mystery. How such a substance might be used against brick and stone forts and actually vitrify the walls is even more of a mystery, and suggests that this was a different type of "Greek Fire" (for more information on the vitrification of Hittite forts see my book, *Lost*

20

Cities of China, Central Asia & India[15]).

When these same Sea Peoples were laying siege to the Hittite cities, they also laid siege to Ugarit, Byblos and other cities along the eastern Mediterranean coast. At Ugarit, the city army and navy was already up north attempting to help the Hittites. A tablet discovered by Schaeffer, still in ithe oven where it was baked. When deciphered, it proved to be from the last king of Ugarit, Ammurapi, to the king of Alashia (Cyprus). It said:

"My father, behold, the enemy's ships came here, my cities were burned and they did evil things in my country. Does not my father know that all my troops and chariots (?) are in the Hittite country and all my ships are in the Lukka land? Thus the country is abandoned to itself. May my father know it: the seven ships of the enemy that came here inflicted much damage upon us...."[18]

The identity of these attackers is highly confusing, and they appear to have actually been a number of groups working together who therefore had a number of names. Egyptian texts call them the Thekel, the Shekelesh, the Weshesh and Peleset. The Peleset are generally thought to have been the Philistines of the Bible. Goliath was a Philistine (or at least a mercenary working for them) and when he was slain by David in the famous Bible story, the Philistines became subjects of Israel. A race of men of unusually large stature, they became King David's Palace Guard and were known as the Cherethites and Pelethites. Martin Luther introduced them into the German language as the "Krethi and Plethi", a term which in German (and also obscurely in English) has the same general usage as the phrase "Tom, Dick and Harry".[16]

The Philistines, who occupied the Gaza Strip of Israel, have given their name to the country and also to the Arabs who still live there or are currently displaced: Palestine. It is generally believed that if the Peleset were Philistines, then the Cherethites were Minoans from Crete and the other Sea Peoples were from the Peloponnese and areas farther north in Greece. This assessment makes a great deal of sense, considering that the Sea Peoples were virtualy in total control of the Eastern Mediterrannean at the time and could also have made land excursions into present-day Turkey to wipe-out the Hittites. After all, they were destroyed by "Greek Fire".

Immanuel Velkovsky, the famous cataclysmic historian and author of *Worlds In Collision,* partially contests this identification, however. In his book *Peoples of the Sea,*[17] he contends that the Peleset were Persians, rather than Philistines, a major distinction, and that the entire episode of the invasion of the Sea Peoples happened 800 years later in the 4th century B.C., rather than the 12th century B.C. He does, however, agree that the other members of the Sea Peoples were indeed Greeks.

If Velikovsky is correct, he has singlehandedly demolished the dates assigned to the ruling dynasties of Egypt and to the most important events of ancient history as well. As he puts it himself in *Peoples of the Sea,* "We can stop here, perplexed by the evidently inadmissible thought that there could be a mistake of 800 years, or frightened at the sight of the perturbation into which this inquiry may lead us. But should we not make up our minds to try to probe a little further, and may we not perchance feel relieved if some new evidence should exonerate the centuries-old concept of ancient history? For this must be clearly understood: we cannot let Ramses III fight with the Persians and keep the hinges of world history in their former places. What a slide, what an avalanche must accompany such a disclosure: kingdoms must topple, empires must glide over centuries, descendants and ancestors must change places. And in addition to all this, how many books must become obsolete, how many scholarly pursuits must be restarted, how much inertia must be overcome? It is not merely an avalanche but a complete overturning of supposedly everlasting massifs."[17]

Velikovsky is a controversial and fascinating character. Perhaps no scholar in modern history has been more scorned and attacked by the establishment. While it is far beyond me to argue either side of this case, I think there are some very good lessons to be learned here. One is that ancient history is by no means established fact, and is extremely subjective. Another is that those who challenge the hallowed ground of the academic "experts" are heading for a battle with an extremely well-entrenched and vicious group of established scholars who have been known to take some very radical steps in silencing their opponents. The best that any "maverick" archaeologist or historian can hope for is to be ignored by the academics. I fortunately fall into this later category. When Velikovsky's first book came out, the academic community was so appalled, that they put a great deal of pressure on his publisher to drop the title. This they did, and it is perhaps the only known case in the history of book publishing when a current New York Times Top Ten bestselling book was actually dropped by a publisher! Fortunately, it was picked up by another publisher.

It seems however, that the invasion of the Sea Peoples did rather take place in the 12th century B.C. rather than the 4th century B.C. for one good reason: the result of the invasion by the Sea Peoples was the first step in the setting up of the Phoenician empire in the eastern Mediterranean and it was with the Phoenicians that King Solomon sailed through the Red Sea into the Indian Ocean to the fabulous gold mines of Ophir. King Solomon lived approximately from 961 to 921 B.C. and if Velikovsky was correct, he would rather date Solomon's reign at circa 300 B.C. which would be astonishing indeed. Velikovsky does not address this at all in his book, and I would surmise that were he to do so, he would adjust those dates accordingly, as he does with the rest of history.

The Phoenicians are as fascinating a group of people as ever lived, and they are key to many of the lost cities and ancient mysteries all over the world! The German historian Gerhard Herm in his book *The Phoenicians* (subtitled, *The Purple Empire of the Ancient World*)[16] says that the Phoenicians originated from Bedouin tribes of the Sinai desert who settled along the Levant, or mountains and coastal area of the Eastern Mediterranean and became seafarers. When they were conquered by the Sea Peoples (ancient Minoan Greeks, apparently) and merged with them, they became the seed of the Phoenician peoples and empire.

There was one particularly significant aspect of the Phoenician empire: it was located at one of the only sources of wood for good shipbuilding in the area. Most of the middle east was treeless desert, nor did Egypt have any ship building wood to speak of. The mighty cedars of Lebanon were the main source of wood, even for the Egyptians. Only in Greece and the forests of the Adriatic were there other abundant sources of wood nearby.

This source of wood, plus good trading skills and a bold, fearless attitude toward sea travel; eventually made the Phoenicians, with their partners the Carthaginians, masters of the Mediterranean and even the Atlantic. There are two main schools of anthropology, Isolationists and Diffusionists. Isolationists believe that ancient cultures were largely isolated and rarely traveled, other than occasionally migrating to conquer a nearby nation or if by sea, along some coast, probably stopping at nightfall on the shore. Diffusionists believe that early man was capable (he most certainly was) of traveling all over the world in ships, or via land, *and that he did so.*[19] The Isolationist theory is by far the prevailing view among academics, and essentially the only one taught in schools. While virtually any history or anthropology teacher in a university will be an Isolationist (often without even knowing it) it is the rare, and courageous investigator who will embrace the Diffusionist theory, as it is extremely perilous to his academic career (perhaps more so than actually sailing across vast oceans). The Norwegian anthropologist and explorer Thor Heyerdahl is perhaps the most famous of all Diffusionists, and has championed the theory for at least 40 years. Another famous Diffusionist is the American scholar (originally from New Zealand) Professor Barry Fell, author of the books *America B.C.*[20] and *Saga America.*[21]

The Phoenicians are a big problem for Isolationists, as they are known to have sailed around Africa in remote antiquity, as well as out into the Atlantic. With King Solomon, they are known to have sailed into the Indian Ocean. Their vast wanderings are a thorn in the side of Isolationists, and in fact, Barry Fell places Phoenicians all over the North American Coast, up the Mississippi to Illinois and Iowa, and into the deserts of the Southwest. There is evidence of Phoenician colonies in Namibia, of Phoenician exploration and gold mining cities in Brazil (see my book *Lost Cities & Ancient Mysteries of South America*[22] for more information on this fascinating city). In one fascinating theory, it was the Carthaginian/ Phoenicians that founded the Toltec civilization in the Yucatan after being defeated in the Punic Wars by the Romans, but that is another story...

§§§

"Passport! Let's see your passport!" said the man with the submachine gun, which was pointed in my face. A bit startled, I looked at him carefully. He was unshaven and wore a white shirt and black pants. Was I being kidnaped? My parents' worst fears were suddenly coming true! The bus had come to a stop, and this man was searching everyone on the bus. Another man with a machine gun stood menacingly at the front door. Neither of them had any uniform on, and except for their weapons they seemed like ordinary Syrians, though dressed in western clothing rather than robes.

"Show him your passport," urged Uros, nudging me in the side. "He is secret police!"

I fished into my money pouch and pulled out my American passport, not the best nationality to be in Syria, especially with irritable men with machine guns. Probably the only worse nationality was being Israeli. That would get dragged off the bus for sure. He glanced at the passport and then at me. Without a word he handed me my passport and went on to the next person. I breathed a sigh of relief and tucked my passport safely away. Damascus was still several hours away, and we were actually in Syrian-controlled Lebanon.

As the bus started to move again, I strained my eyes to the distance to see if I could possibly see ancient Ba'albek, one of the most incredible wonders of the world. Our bus would be passing near by, and though I would not get a chance to stop there, it has the world's largest carved stones. Ba'albek is 44 miles east of Beirut and consists of a number of ruins and catacombs. 2,500 feet long on each side, it is one of the largest stone structures in the world. Most agree that it is one of the most imposing sights in the world, a portion of it consisting of gigantic cut stone blocks from some ancient time, with a Roman Temple built on top of it. The Roman Temple to Jupiter and Venus was built on top of the earlier temples dedicated to the corresponding Semitic divinities–Ba'al and his partner, the Goddess Astarte, whose cult involved prostitution and sacred orgies.[23] However, it is quite likely that while Jupiter and Venus correspond to Ba'al and Astarte, the temple may actually have been built as a prehistoric Sun Temple, and even then on the ruins of an even more ancient structure, its purpose unknown.

The Roman architecture, largely destroyed by an earthquake in 1759, does not pose any archaeological problems, but the massive cut stone blocks beneath it certainly do! One part of the enclosure wall, called the Trilithon, is composed of three blocks of hewn stone that are the largest stone blocks ever used in construction; an engineering feat that has never been equaled in history. The blocks are so large that even their weight cannot be accurately calculated.

The temple was first excavated by German archaeologists just prior to World War One. In an article written in 1914 for the journal *Art and Archaeology* entitled "*The German Excavations at Baalbek*", the author Lewis Bayles Paton says," One of the chief results of excavation has been the disclosure of the original ground-plan of the temple-complex. The buildings stood upon a readied platform that was enclosed with a massive retaining wall. Instead of filling in this area with earth, arched vaults were constructed that were used as shops and storehouses. The sacred enclosure was reached by a grand staircase at the eastern end, at the head of which stood the Propylaea,consisting of a portico flanked by two towers. Within this was an hexagonal Forecourt from which three doors led into the great Court of the Altar. In the center of this court over the altar one sees the ground-plan of the later Basilica of Theodosius. On the west side of this court a flight of steps and a single portal led up to the Temple Court. Here, in line with the axis of the sanctuary, stood the Great Temple, traditionally called 'the Temple of the Sun,' but now known to have been dedicated to Jupiter and all the Gods of Heliopolis. south of this lay a smaller temple, traditionally called 'the Temple of Baal' but now known to have been dedicated to Bacchus.

"The lowest course of the outer retaining wall consisted of moderate-sized stones. Above these came three courses of stones, each about thirteen feet in length. The middle courses remain only on the western side, and there consist of three of the largest stones ever used by builders. One is sixty-four feet long, another sixty-three feet and one-half and the third sixty-two and one-half. All are about thirteen feet high and ten feet thick. The inner surface of one of these stones has been exposed by the excavations on the west side of the Temple Court. From these remarkable stones the temple probably received its ancient designation of Trilithon. A similar stone seventy by fourteen by thirteen feet and weighing at least one thousand tons remains in the quarry from which the other stones were obtained. How these gigantic blocks were transported and raised to a height of twenty-six feet in order to rest upon the lower courses, is an unsolved problem. The method of quarrying by which they were cut out like columns from the native rock and then split off may still be studied *in situ*."[27]

The above article is interesting, but the author fails to recognize several important points. One is that the structure is several structures in one, constructed over various ages, thousands of years apart. Therefore the different names are all correct, for the Roman Temple was dedicated to Roman gods, while the earlier construction was dedicated to earlier gods, and most probably, in prehistory it was a temple to the sun. According to an article by Jim Theisen in the *INFO Journal*,[26] the Greeks called the temple "Heliopolis" which means "Sun Temple" or "Sun City". Even so, the original purpose of the gigantic platform may have been something else entirely.

The weight, and even size of the stones, is open to controversy. According to the author Rene Noorbergen in his fascinating book *Secrets of the Lost Races*,[3] the individual stones are 82 feet long and 15 feet thick and are estimated to weigh between 1,200 and 1,500 tons each (a ton is two thousand pounds, which would make the blocks weigh an estimated 2,400,000 to 3,000,000 pounds each, slightly less in kilograms).[3] While Noorbergen's size may be incorrect, his weight is probably closer to the truth. Even conservative estimates say that the stones weigh at least 750 tons each, which would be over one million pounds (one million five hundred thousand, to be exact).[23]

The recent *INFO* article on the stones confirms the size and weight of the blocks, yet its statistics are the most incredible! According to the article, the blocks are 10 feet thick, 13 feet high and over 60 feet long. They are made of limestone, and knowing the density of limestone, the blocks are estimated to weigh over 1.2 million pounds![26]

It is an amazing feat of construction, as the blocks have been raised more than 20 feet in order to lie on top of smaller blocks. The colossal stones are fitted together perfectly, and not even a knife blade can be fitted between them.[3] Even the blocks on the level below the "Trilithons" are incredibly heavy. At 13 feet in length, they probably weigh about 50 tons each, an extremely large-sized bunch of stones by any other estimate, except when compared to the "Trilithons". Yet, even these are not the largest of the stones!

The largest hewn block, 13 feet by 14 feet and nearly 70 feet long and weighing at least 1,000 tons (both Noorbergen and Berlitz give the weight of this stone at 2,000 tons [3,4]), lies in the nearby quarry which is half a mile away. One thousand tons is an incredible two million pounds! The stone is called *Hadjar el Gouble*, Arabic for "Stone of the South". Noorbergen is correct in saying that there is no crane in the world that could lift any of these stones, no matter what their actual weight is. The largest cranes in the world are stationary cranes constructed at damns to lift huge concrete blocks into place. They can typically lift weights up to a 100 tons or so. 1,000 tons, and God forbid, 2,000 tons are far beyond their capacity. How these blocks were moved and raised into position is beyond the comprehension of engineers.

Large numbers of pilgrims came from Mesopotamia as well as the Nile Valley to the Temple of Ba'al–Astarte. It is mentioned in the Bible in the Book of Kings. There is a vast underground network of passages beneath the acropolis, which were possibly used to shelter pilgrims.

Who built the massive platform of Ba'albek? According to ancient Arab writings, the first Ba'al–Astarte temple, including the massive stone blocks, was built a short time after the Flood, at the order of the legendary King Nimrod, by a "tribe of giants".[23]

Naturally, Erich von Daniken and his adherents would tell us that it was built by extraterrestrials. Not such a wild conclusion, considering the facts, this makes more sense than saying that it was all built by the Romans a few thousand years ago. Interestingly, Charles Berlitz says that a Soviet scientist named Dr. Agrest suggests that the stones were originally part of a landing and takeoff platform for extraterrestrial spacecraft.[4]

I find a more natural conclusion is that Ba'albek is a remnant of the Osirian Empire. What is the Osirian Empire? According to esoteric tradition, about fifteen thousand years ago, there were a number of highly developed and sophisticated civilizations on our planet, each with a technology that largely surpasses our own. Among these fabled civilizations was Atlantis, a civilization which was, according to ancient Greeks and Egyptians, on a small continent in the Atlantic ocean outside the pillars of Hercules (Gibraltar). Another advanced civilization existed in the same time in India. It was known as the Rama Empire, and fortunately there are many ancient Indian texts still extant that relate the glory and technology of the Rama Empire. Like Atlanteans, the people of the Rama Empire (according to such well known ancient texts as the *Ramayana* and *Mahabharata*) flew around in airships known as vimanas and had weapons akin to laser weapons and nuclear bombs (see my book *Lost Cities of China, Central Asia & India*[15] for more information). Another civilization that is just as fascinating, but much less well known, is the ancient Osirian Empire.

It is said that at the time of Atlantis and Rama, the Mediterranean was a large and fertile valley, rather than a sea as it is today. The Nile river came out of Africa, as it does today, and was called the River Stix. However, instead of flowing into the Mediterranean Sea at the Nile Delta in northern Egypt, it continued into the valley, and then turned westward to flow in the deepest part of the Mediterranean Valley, just to the south of Crete, between Malta and Sicily, south of Sardina and then into the Atlantic at Gibraltar (the Pillars of Hercules). This huge, fertile valley, along with Sahara desert, a vast fertile plain, were known in ancient times as the Osirian Empire. The reason why we will see in a later chapter. Sometimes known also as "Pre-Dynastic Egypt", the Osirian Empire was closely allied with Atlantis in the devastating wars that raged throughout the world toward the end of the period of Atlantis's war-like imperial expansion.

According to esoteric information, stored (even today) in secret libraries in ancient Egypt, China, India, Tibet and other places, even today (according to certain sources) Atlantis was destroyed in a cataclysmic upheaval that was essentially isolated only to that mini-continent in the Atlantic. With this cataclysmic change in the Atlantic, the Osirian Civilization was slowly flooded as the Mediterranean Basin began to fill with water. Great cities were flooded, and Osirians began moving to higher ground. This theory helps explain the strange megalithic remains all over the Mediterranean, especially on the islands of Malta, Sardinia, Corsica, Sicily, Crete and the Baleric Islands of Spain. Sunken structures of megalithic proportions have been found off Morocco and Cadiz in Spain.[4]

It is an archaeological fact that there are more than two-hundred known sunken cities in the Mediterranean. Egyptian civilization, along with the Minoan and Mycenean in Crete and Greece are, in theory, remnants of this great, ancient culture. The subject of the Osirian Empire is a fascinating, little-known topic. As a potential solution to many of the mysteries of Africa and Arabia it will surface again and again in this book (also see the forthcoming book, *Lost Cities of Europe and the Mediterranean*, and *Lost Cities of China, Central Asia & India*[15] for more information on the theoretical war between Atlantis and the Rama Empire of India).

Suddenly, in view of an advanced and ancient civilization in the Mediterranean, the mystery of some of the awesome and inexplicable sites around the Mediterranean, such as Ba'albek, does not seem quite so mysterious after all. Yet, most researchers, even those who write books about Atlantis or ancient astronauts, have never even heard of the theory of the Osirian Empire. This theory of Ba'albek being some remnant of the Osirian Empire, along with some of the other megalithic sites in the Mediterranean, fits in well with the Arab legend mentioned previously: that the massive stone blocks were built a short time after the Flood, at the order of the legendary King Nimrod, by a "tribe of giants".[23]

King Nimrod, and Poseidon, are just other names for Osiris of Egyptian mythology. Just who Osiris really was, and the relevance of the Osirian myth to Atlantis, the Osirian Empire, and Egypt will seen in chapter 4. The Arab legend is definitely pointing to the Osirian civilization as the source of the stones.

Yet, even if Ba'albek is a remnant from the Osirian Civilization, how were such huge blocks transported and lifted? One clue is the massive block that still remains at the quarry a half mile away called *Hadjar el Gouble*. This stone was apparently meant to take its place on the platform with the other stones, but never made it for some reason. According to the INFO article,[26] the largest stones used in the Great Pyramid of Egypt only weigh about 400,000 pounds (these are several large granite blocks in the interior of the pyramid). They also point out that until NASA moved the gigantic Saturn V rocket to its launch pad on a huge tracked vehicle, no man had transported such a weight as the blocks at Ba'albek.

Says the *INFO* article, "one sees no evidence of a road connecting the quarry and Temple. Even if a road existed, logs employed as rollers would have been crushed to a pulp. But, obviously, someone way back then knew how to transport million-pound stones."[26]

There is not a contractor today that would attempt to move or lift these stones. It is simply beyond our technology – or is it? I find it interesting that there is no discernible road between the quarry and the massive Sun Temple. This indicates either or both of two things: the building of the lower platform occurred at such an ancient time in antiquity that the road is long gone, or a road was never needed for transporting the block. As the *INFO* article points out, a road would have been of little use anyway.

There is only one plausible and logical explanation for the transport and lifting of the blocks and that is that they were moved by levitation, some sort of anti-gravity, if you will. Throughout the world there are ancient legends, as well as authentic reports, of stones being levitated. On the remote Micronesian island of Pohnpei lies an immense canal city eleven square miles large built out of magnetized basalt crystals weighing from 5 to 50 tons. According to local tradition, the stones were made to "magically fly through the air from the other side of the island" (see my book *Lost Cities of Ancient Lemuria & the Pacific*[28] for more information on this mysterious site).

Tales of levitating stones can be found in nearly all cultures, all over the world. Stonehenge was said to have been built by Merlin, levitating the stones into place; Egyptians were said to have levitated stones, in India is found the levitating stone of Shivapur (see *Lost Cities of China, Central Asia & India*[15]) and in 1939 a Dr. Jarl from Sweden witnessed Tibetan monks levitation stones 250 meters up the side of a cliff to a monastery that they were building.[29,30]

While the skeptic may exclaim that "anti-gravity" is impossible, there is nothing inherently impossible about it at all. Gravity fields and waves are hardly understood by physicists. When Einstein died in 1955, he was working on the problem of the Unified Field, a mysterious force from which he theorized electricity, magnetism, and gravity all manifested. Therefore, the idea of creating an artificial gravity field or bubble around an object is not such a crazy idea at all.

Physicists tell us that all matter in the universe is really made up of space and vibration. According to this theory, gravity, too, has a frequency that can be duplicated. Incredibly, by using a sort of "ultra-sound" or special vibration directed at an object (preferably one that resonates easily, such as a crystal or hard stone) it will become weightless. In this way, using a device similar to a musical instrument, gigantic stones weighing millions of pounds, could become weightless, and easily be moved into place in some wall or temple. In this theory, we see parallels in such ancient traditions as Joshua blowing his trumpet and the walls of Jericho falling down, and in the mystical concepts of "music of the spheres". It is through some form of levitation, I believe, that the massive stones of Ba'albek were put into place. The subject is no doubt a very controversial one, and a full discussion of the theory is beyond the scope of this book (for more information on levitation and other theories on anti-gravity, see our popular books *The Anti-Gravity Handbook*,[28] *Anti-Gravity & the World Grid*,[29] and *The Bridge To Infinity*[30]).

And what was the purpose of Ba'albek? Ba'albek may well have been a sun temple, as the ancients universally worshiped the sun. According to ancient Nestorian Christian texts, Archangels live on the sun, and nothing is more important to all life than the energies that manifest from that wonderful ball in space. However, why should such massive stones have to be used to built a simple place of worship? Perhaps to impress others, though this seems like a great deal of trouble just to impress the Joneses next door. It appears that Ba'albek was constructed for a purpose that was beyond the imagination of your average person in antiquity, but yet, for a reason that was logical and necessary.

There are basically two practical reasons for constructing walls or platforms out of gigantic stones: one is so that when the cataclysmic earthquakes that occasionally rend the entire world occur, a platform or wall constructed thusly will not shift or crumble. The gigantic walls in the high Andes of Peru are a good example of this kind of construction; a construction that interlocks the gigantic blocks so that they shift as one piece, rather than separately. It is also notable that no mortar can be used in such a construction technique, as it would be useless. It is the sheer weight and interlocking of the stones that holds them together, not mortar.

The other reason is so that the gigantic platform itself can securely support even more weight on top of it, like more construction, or as was suggested by the Soviet scientist Dr. Agrest, some aircraft, possibly a rocket. This I find to be not such a ridiculous theory. According to the well-known ancient Indian Epics, the *Ramayana* and the *Mahabharata*, the ancient Indians of the Rama Empire flew around in zeppelin-type airships that were known in ancient Dravidian and Sanskrit as "vimanas". Ancient Indian sources on vimanas are numerous, and there are even whole flight manuals written for them which are still extant. While all scholars admit that ancient Indian texts do speak of airships called vimanas, those who subscribe to a more conservative view of history believe them to be merely the wild, prehistoric fantasies of a bunch of opium-eating neo-cavemen.

According to esoteric literature, Atlantis had their own version of vimana airships. These were called vailixi (one vailix, two vailixi). Vailixi and vimanas flew all over the world as cargo, passenger and military craft, and naturally, needed places to land. These long, cigar shaped craft (identical to many elongated UFOs that are commonly reported) were powered by a mysterious electro-magnetic-gravitional force that is analogous to Einstein's Unified Field. Here we see how the force that is theoretically powering vimanas is the same force that was theoretically levitating the stones at Ba'albek.

And finally, I come to the end of my point, far-out and crazy as it might be, which is: was the original platform of Ba'albek a vimana landing pad? If built by the "legendary King Nimrod of Atlantis" why could it not have been for the occasional vimana flight bringing cargo or passengers from Atlantis or Rama to "Sun City of Osiris"? Perhaps as a passenger walked toward the main platform, he handed an attractive attendant in a white robe embroidered in gold threads, his boarding pass. "Thank you," she might have said, adding, "I hope you enjoy your flight to Abydos, Malta, and Atlantis, sir. Please fly with us again sometime!"

With all this talk about Atlantis, vimanas, cataclysms, anti-gravity and prehistoric world exploration, the firm believer in Columbus discovering America, the Wright Brothers being the first to fly, Sumeria being the first civilization, and pyramids being just monumental tombs for the dead, should put this book down right now!

§§§

Ba'albek is one of the most fascinating and controversial structures on our planet, but because of the current problems in Lebanon it is difficult to get to the site and check it out for oneself. I definitely would not recommend it for Americans. At one time Lebanon was considered the Switzerland of the Middle East. Beirut was a wealthy and sophisticated city, and a banking center for all the Arab countries. Wealthy Lebanese, tourists and Arabs sat at fashionable cafes and drank wine, or they tanned in bikinis at the many beaches or skied at the snowclad mountains of cedars.

With the displacement of Palestinians, initially from British-held Palestine taken by Jewish National guerrillas in 1948 and then dispelled from Jordan in 1971 after attempting to take over the country with the help of Syria, the Palestinian problem was then focused on Lebanon, where the majority of the Palestinians fled from Jordan. The many different Palestinian factions (one group is named after the "Black Sunday" when Jordanian troops squashed the rebellion and attempted take-over of the government) now became entrenched in Lebanon.

Lebanon was already a factionalized country to begin with, created by the French to give Christian Maronites some independence from the overwhelming majority of Muslims in Syria. In 1860, Lebanon was part of the Ottoman Empire, and the Turkish Sultanate encouraged the mysterious Druse population to wage war against the Christians. After a massacre of 2,500 Christians, France intervened and occupied the country. It was turned back over to the Ottoman Empire in 1864 with a Christian military government running the country under nominal Turkish sovereignty. When Turkey was on the losing end of World War I, France received a League of Nations mandate over Lebanon & Syria in 1914.

Even though the government of Lebanon is largely controlled by Christians, the people still consider themselves Arabs, and Lebanon took part in the Arab invasion of Palestine on May 15, 1948. A civil war between Muslims and Christians broke out in 1958, and America intervened at the request of the Christian government ("From the halls of Montezuma to the shores of Tripoli..." sang the Marines as they landed at the Lebanese port of Tripoli.). Palestinian guerrillas using Lebanon as a base for operations drew Lebanon into conflict with Israel, which retaliated fiercely for Palestinian attacks in Israel. Israel invaded Lebanon on March 14, 1978 and then again 1981 and 1982.

On June 6, 1982, the Israeli army swept across the border and within four days had surrounded Beirut and destroyed the Palestinian strongholds in Tyre and Sidon, two ancient Phoenician ports just north of the Israeli border. The Israelis laid siege to Muslim West Beirut where some 5,000 to 6,000 PLO guerrillas were trapped, and pounded the city with bombs and artillery. Finally, after the dispersal of most of the PLO to other Arab nations, Israel pulled back, and an uneasy truce, interspersed with occasional attacks, has existed ever since.

One of the unusual new items reported during the 1982 invasion of Lebanon by Israel was the discovery of huge underground bases near the port city of Sidon. In several obscure news items that never made the national news magazines (for reasons unknown) the Israelis claimed to have discovered a vast underground submarine base that had allegedly been built by the Soviets and staffed by PLO guerrillas. Reportedly, Soviet submarines could entire this huge, secret underground base from the Mediterranean without ever surfacing. From the sea, the area around the base appeared to be only cliffs.

Photos from this base appeared in several publications, including an odd book written by a Baptist minister living in Israel who claimed he been given a tour of the strange facility by the Israelis. The Lebanese government claimed it had no knowledge of the existence of this secret base. The Israelis reportedly destroyed the base when they pulled out of Lebanon. The existence of such a base, a sort of "lost city" of sorts on the Lebanese coast, seems to confirm the stories of other such bases built into remote, strategic areas such as South Georgia Island in the South Atlantic (where part of the Falkland Island's War was fought), the South Coast of New Zealand, and other places, by a sort of conspiratorial "shadow government" made up of International Bankers and other super-wealthy power brokers. A secret submarine base like that found at Sidon was the subject of the late 60's spy film *Our Man Flint*. Did the movie's producers know more than they were willing say?

§§§

"Passport!" said the man with the machine pistol as he waved it in my face. I felt a sense of Deja Vu, hadn't this happened to me once before? I brought out my passport, the man glanced at it, and went on to the Arabs sitting behind me. I guess this was just how things were done in Syria; every few hours a secret policeman just boarded your bus, menacingly waved a gun in your face, and wanted to see your papers. It seemed that the Gestapo was alive and well in Syria, but this wasn't *Hogan's Heroes!*

It turned out we were just on the outskirts of Damascus, and within twenty minutes or so, we arrived in the city. Uros, Dusan and I took a pick-up taxi which ran a normal route through town that passed the Youth Hostel. However, since not even one of us had a Youth Hostel card, nor could we buy one at the hostel, we were not allowed to stay there.

Therefore, we walked down the hill towards the lights of the central city, and searched for a cheap hotel. For some odd reason, perhaps because we were backpackers, all the hotels were full. Finally, at our last chance, a kindly old man who hadn't shaven for days said that we could sleep on the roof of the hotel in some rope-strung beds for a dinar or two a night. Gratefully we accepted his offer, and exhausted from the many adventures and long journey, crashed out under the starry sky of Damascus.

I awoke the next morning early with the hot sun beating down on me and flies buzzing around my head. Though I would have liked to have slept later, it was just not possible in the bright sunshine and with flies crawling all over me. In the spirit of new discoveries and adventures awaiting me in this new city, I launched myself out of bed, and over to the edge of the building.

From the roof of the hotel, I had a pretty good look at the city, but I must confess it was not particularly appealing. A dull brown city sprawled among dull brown hills and everything the eye could see was dull brown in a scorching sun that washed out everything in its bright light. I struggled to find a bit of green or red, but all was brown. Damascus was not my dream city come true, yet I was here anyway, and I would make the most of it anyway.

Once Uros and Dushen were up, we tripped out into the streets of Damascus to see the city. Naturally, the first stop was the tourist office. According to the tourist literature, Damascus is said to be the world's oldest inhabited city, a dubious claim, considering that Jericho is at least 9,000 years old, and that Cuzco, Peru also makes a similar claim as the "Father Culture of the World". According to the tourist office, Damascus is "a very ancient town which is mentioned in Egyptian texts of the Eighteenth Dynasty and in Assyrian tablets and in the Book of Genesis." The material goes on to say that poets have called Damascus "The Eye of the East", "Legend of the Centuries", "The Inspiration of Poets", and "The Rival of Paradise".

Indeed, Damascus is an ancient city, many thousands of years old. It was not only important in ancient times, but is also important today. Three thousand years ago it was the capital of the Aramean Kingdom. The language of the Arameans, Aramaic, was probably the most important language of traders in the ancient Middle East, and most of the early books of the New Testament were written in this language. Saint Paul took refuge in Damascus at Saint Hanania's House and later escaped from a prison cell in the Damascus Wall in a basket.

To the east of Damascus in the Syrian Desert is the ancient city of Tadmor, now called Palmyra, founded by King Solomon. "And it came to pass at the end of twenty years, wherein Solomon had built the house of the Lord, and his own house,. . . he built Tadmor in the wilderness," says the Old Testament (II Chronicles, viii, 1,4).

Now ruins, Tadmor/Palmyra is in a desert oasis, and at one time challenged the power of Rome. The famous and beautiful Queen Zenobia inherited a great desert empire, but Rome crossed the desert and took her captive in 272 A.D. They destroyed the city and took her back to Rome, and that was the end of Tadmor.

Aside from the bazaars of Damascus, the two main tourist attractions in Damascus are the museum and the Omayyad Mosque. Uros, Dusan and I headed for the museum which has a large collection of Assyrian, Roman, Phoenician, and Arab artifacts. Of special note in the museum is the Canaan Alphabet of Ugarit tablet, which preserves the Phoenician alphabet, often said to be the world's first (though I think this is unlikely).

After the museum, we visited the Omayyad Mosque, one of the most grandiose in all of Islam. Built from 705 to 715 by Caliph Al-Walid mostly out of marble, it is indeed a beautiful building with many exquisite mosaics. We were met at the entrance by a young man who insisted on showing us the mosque.

"No thanks, we don't need a guide," I said.

"Oh, no, I am not a guide," said the young man. He was dressed in blue jeans and a polyester shirt. "Please, I am not a tour guide, I just want to show you our beautiful mosque!" With that, he led us inside and gave us the grand tour of the mosque. He pointed out the tomb of John the Baptist, where his head was located (or part of it, anyway – a portion of John the Baptist's skull can be seen at the Topkapi Museum in Istanbul, possibly taken from this mosque). John the Baptist's body is still in Palestine, our non-guide informed us.

We were then shown one of the tall towers called the Minaret of Jesus Christ. Here, according to Islamic tradition, Jesus and another Islamic prophet would descend from heaven to this minaret. This, our non-guide believed, would happen in about four hundred years.

As we departed the mosque, our friend asked for money. "But I thought you weren't a guide," I said.

"Of course I'm a guide," he protested. "What did you expect?" Dusan handed him a dinar, and we headed back for our hotel.

That night as I lay in my rope-strung bed underneath the clear night of Damascus I thought of the many adventures that lay before me. In my heart, I knew I was destined to travel many places in Africa and Arabia. In my mind, I wondered what trials and tribulations awaited me on those dusty roads to lost cities and ancient mysteries. Did I have what it took to find the answers to history? And more importantly, did I have what it took to find the answers to my own identity?

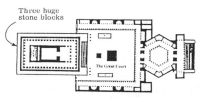

Ruins at Baalbek, Lebanon

Ba'albek. Colossal block of hewn stone.

Baalbek, Lebanon. Above: the three massive stones that are part of th
foundation of the Roman Temple to Jupiter. Placed there before the Romans, these
stones are among the largest in the world. Below, the largest block, still at the
quarries a few miles away.

Phoenician two banked warships and transports circa 700 B.C. Ships like these were perfectly capable of navigating the entire world. It was these same ships and even larger, that made the three year journey to the gold mines of Ophir. From *Ships & Seamanship In the Ancient World*, Princeton University Press, 1971.

Seirite-Sinaitic	Canaanite-Phoenician-Aramaic	Greek	Latin
		A	A
		B	B
		ГC	CG
		ΔD	D
		F E	E
		F Y	FVUWY
		I	Z
		H	H
		I	I J
		K	K
		ΛL	L
		M	M
		N	N
		+Ξ	X
		O	O
		ΠГ	P
		Q	Q
		PR	R
		Σ?	S
		T	T

Comparison of Alphabets Showing
Origins in Pictographs.

Chapter 2

JORDAN:
LOST CITY OF THE BLACK DESERT

These great works are those of the "Old Men of Arabia".
—*Bedouin tradition*

I was up early the next morning and caught a bus to Amman, the capital of Jordan. I said goodbye to Uros and Dusan, after swapping addresses. They were off to Iraq, of all places, a country Yugoslavs may enter easily, but where Americans have an extremely difficult time getting visas.

My visa for Jordan was not difficult to get, however, it was issued to me at the border. The immigration officer, a trim and friendly man with a thick mustache and well pressed uniform, greeted me with the traditional "salaam aleikum" and gave me two weeks to see the kingdom, which seemed plenty of time.

The Hashemite Kingdom of Jordan, in existence since 1946, is a largely uninhabited desert about the size of Indiana. Once the home of the biblical Samaritans, Jordan was part of the Roman and Greek Empires for centuries. Conquered by the Arabs in the seventh century, it was taken by the Turks in the sixteenth century and remained in their control until 1916. The League of Nations gave a mandate to Britain to govern the area that now comprises Jordan and Israel, then called Transjordan and Palestine. The Jordan River separated the two countries. In 1948 Jordan gained full independence as a kingdom ruled by the present king's grandfather, Abdullah.

When Britain withdrew from Palestine on May 15, 1948, the armies of Transjordan, Syria, Lebanon, Iraq and Egypt with Saudi Arabian and Yemeni increments simultaneously advanced into Palestine. They were met by an estimated twenty-five thousand Israeli troops who were European World War II veterans. The poorly trained Arab armies (with the exception of the British-trained Transjordanian army) were easily defeated by the Israelis after some early initial successes. By January of 1949, only the so-called West Bank was still in the hands of the Arab armies.

King Abdullah then annexed the West Bank into Jordan, much to the disapproval of other Arab countries. King Abdullah himself was assassinated in Jerusalem on July 20, 1951, by a Palestinian allegedly hired by relatives of Haj Amin al-Husseini, a former *mufti* of Jerusalem, who was a bitter enemy of Abdullah. The present King Hussein of Jordan, who was only sixteen at the time, was at Abdullah's side, also a target of the assassin's bullets.

Because Abdullah's oldest son, prince Talal, had a history of mental problems, his young son, the present King, was crowned King of Jordan on August 11, 1952, and is the head of state of Jordan to this day.

Arab-Israeli conflicts dominate Jordan's foreign policy, and Jordan, much to the government's dismay, remains the hot spot of Middle East negotiations. Jordan has become more and more involved in the conflicts of the volatile Middle East. Civil war broke out in Jordan in 1970 between Palestinian guerrillas and Jordanian forces, the Jordanian Army virtually eliminating all guerrilla bases in their country. Syria supported the Palestinian attempt to take over Jordan and then broke diplomatic relations with Hussein. During the Iran-Iraq war, Jordan was actively helping Iraq in its struggle to gain Iranian territory and allowed its Red Sea port of Aqaba to be used for the shipping of arms to Iraq.

After I crossed the border from Syria I headed for Jerash, the Pompeii of the Middle East and one of the best-preserved Greco-Roman provincial cities in the world. It is located just to the north of Amman, within easy access by a local bus for the morning or afternoon. Though many of the buildings were destroyed by earthquakes in the 8th century A.D., much remains, and was used by Crusaders as a temporary garrison and fort. The excavations have revealed a clear pattern of former magnificence: a massive oval Forum, baths, theaters, temples and a visible complex of streets, columns, fountains and shops.

In order to get to Amman before dark, I had to leave Jerash by mid afternoon, and take a bus heading south. It was an hour or so before dusk when I rolled into the city, and was dropped off in the town center. I stood on a sidewalk with my backpack on my feet, wondering what to do next. To my horror, I realized that it was Friday, the Muslim sabbath, and all the banks were closed! There was a money exchange shop nearby and I dragged my pack over to the window, to see if I could cash some traveler's cheques. They only dealt in cash, I was informed by the fat, balding man in stained white shirt. I dug into my money pouch to see if I had some cash dollars, and found only one slightly torn dollar bill.

It was worn and tatty, but it was all I had. I placed it under the thick, bullet proof window and said firmly, "I'd like to cash one dollar please."

With disdain, he picked up the bill by an edge and held it up to the window. "This?" he snorted. "You want me to change this?"

"That is correct, sir," I said politely.

"Humphh," he said, breathing out long and hard through his large nose. Silently he counted out a small amount of change for me, less than one Jordanian dinar, and pushed it under the window to me.

"Thank you, sir," I said cheerfully, taking my precious change. He said nothing, and I walked away.

It seemed that the cheapest place to stay in Amman was the youth hostel, and even though I had been turned away from the one in Damascus, I figured that I would give the one in Amman a shot. Therefore, I took a bus up one of the hills that surround Amman to the vicinity of the youth hostel. It took me an hour of walking around the neighborhood with my pack on my back to discover that the youth hostel no longer existed!

My day in Amman was not turning out too well, and as the first stars in the sky appeared, I asked a man who was dressed in a nice suit the way back into the center of town. He told me to take a certain bus, but then I explained to him that I needed to walk, because I didn't have the money for a bus, having spent my last few *fils* on a feta cheese sandwich.

He looked at me with genuine concern and said, "Here, take this money. This is plenty for your bus back to the town center. You will even have change for a *shwarma* sandwich."

With tears in my eyes I accepted the money. It was painful being reduced to a beggar here in Jordan. "Well, thank you very much, sir! That is very kind of you!" I said.

"Think nothing of it, young man," said the gentleman in excellent English. And with that he continued on his way, to his home presumably. This was to be my first experience with one of the pillars of Islam: the giving of alms.

Back down in the center of town I found a hotel that would allow me to sleep on a mattress on the roof for half a dinar a night, which was about two dollars at the time. Exhausted from my sojourn around Amman trying to find a cheap place to stay, I collapsed onto the mattress and gazed up at the cloudless sky sparkling with a million stars. Though I had intended to go out that evening, I fell asleep on that mattress, only to awake feeling cold late in the night. I pulled my trusty sleeping bag out of my pack, climbed inside it, and once again fell sound asleep.

§§§

Known in Biblical times as Rabbath Ammon and Philadelphia (City of Brotherly Love), Amman, the capital of Jordan, is built on seven hills like Rome. There are many things to do, including walking around the market, visiting the museum, or going to Kerak, south along the King's Highway. High on a plateau 3,400 feet above sea-level, is the ancient fortress town of Kerak. While the ancient ruins date back more than 2,000 years, the fortress that stands today in an old Crusader castle known as "the rock of the desert".

Things were going much better for me the next morning. I was able to cash some traveler's cheques at the bank, and after paying my hotel bill for the night before, I headed for the Jordanian Ministry of the Interior to apply for permission to go to the West Bank of Jordan, essentially Israeli-occupied territory. It would take several days to process my application, they said, so I decided to hitchhike south out of Amman, and search for the lost cities of ancient Samaria.

Jordan has one of the most awesome archaeological sites in the world, the ancient caravan city of Petra, built during the first millennium B.C. to protect the caravan routes from Mecca to Damascus. Petra is cut into solid red sandstone cliffs in a remote canyon in southern Jordan. This remote and little-known canyon was my main destination while in Jordan.

I decided that I would leave for Petra early the next morning, but had the afternoon to kill in Amman. After a brief visit to the museum, I caught an unusual film entitled *The Message*. *The Message* is a film about Mohammed and the beginnings of the Muslim religion. Paid for by the now deceased Shah of Iran, it was a multi-million dollar production about the life of Mohammed, filmed in English, largely meant to educate the ignorant masses in Western Society about the whys, and hows of Islamic religion, and to give Westerners at least a small glimpse of Muslim thought, history and religion. It was a good idea, I thought, as I bought my ticket, though it seemed this film was going to be mainly viewed by Muslims themselves.

One of the problems in the filming of the movies is the Muslim admonition that icons of prophets, gods, angels, deities never be shown, including any representation of Mohammed himself. When it was first announced that a film was going to be made about the life of Mohammed, this sparked widespread riots throughout the Middle East, because it was expected that any film about Mohammed would naturally have him in it. The film was able to get away from this paradox by filming the movie in such a way as if the action being seen from the eyes of Mohammed. Therefore, the camera was Mohammed, and actors such as Anthony Quinn, one of Mohammed's friends in the film, would walk up to the camera, stare at the audience, and address Mohammed, who was never seen in the movie.

It was a fascinating film, full of desert battles and interesting bits of historical tidbits from the 7th century. As Mohammed and his pal Anthony Quinn rallied their followers to attack the evil, pagan merchants of Mecca who had banished them to Medina and had vainly sought to kill Mohammed, the crowd of Jordanians in the theater let loose a deafening yell throughout the theater that did not die down until the very end of the movie. I left the the theater far better informed than I had been before, and with the distinct impression that this was a movie which your average Arab could get very excited about.

The next day, I left some of my luggage at the hotel and caught a bus out of Amman to the crossroads of the Desert Highway and the King's Highway, both of which stretched south through the desert to Aqaba on the Red Sea. The King's Highway cuts through virtually uninhabited desert along the eastern shore of the Dead Sea. I decided to take the Desert Highway, thinking that there would be more traffic, walked a short way down the two-lane blacktop and took my stand in an attempt to catch a ride through the Jordanian desert.

Shortly, a Bedouin woman appeared and walked up to me, hoping to get a ride when a truck stopped for me. She wore a long black dress, embroidered in pink and green threads with squares, crosses, and other geometrical designs. Her face was brown and wrinkled, and she appeared to be very old. After talking with her, I noticed that both of her eyes were cloudy with cataracts.

"Salaam Aleikum," I greeted her as she stood silently next to me.

"Aleikum a Salaam," she answered, looking vaguely at me through her cloudy eyes. I gave her a piece of pineapple candy, and she tried to flag down a public bus, but it wouldn't stop.

Shortly after that an old Mercedes dump truck stopped. He indicated he wanted me to get in, but I let the old lady in first, knowing she would have troubles getting a ride on her own. "Where are you going?" asked the driver in English, a large, full-bellied and unshaven trucker with a white shirt, brown polyester slacks and a white knit Muslim cap covering a bald spot on his head.

"Petra!" was my enthusiastic answer and with that we roared off down the empty highway. It wasn't long before we let off the Bedouin lady in the middle of the desert where a small village was just visible in the distance. After this the driver began talking to me. He was a Palestinian, his name was Mustaffa and he had learned English in Cairo; indeed, he spoke it fairly well. He now drove this dump truck around and was taking it to some quarry to get a load.

"You know about these Israelis?" Mustaffa asked.

"Well, uh, I guess I have heard of them..." It wasn't a subject I wanted to appear too knowledgeable about, as Arabs can get pretty worked up over it, understandably.

"I kill them!" he said, and he made a pistol out of his right hand and pretended to fire shots out of the windshield. "I kill them all!" I remained quiet and merely nodded. This volatile issue, one that encompasses religion, politics and even sex in the same breath...well, I decided it was better just to look out the window...

"Why do you Americans support Israelis?" he wanted to know.

I could think of several "official" reasons, but decided not to give them. It is, after all, a very complicated situation.

"I don't know," I said.

"I like America," he said, "but I just don't see why you support our enemies." We were cruising down a typical Arabian desert road, flat, dry, a few shrubs and a long straight black ribbon of tar lending some reality to an otherwise dream-like world of mirages and hallucinations. Suddenly Mustaffa stopped the truck, pulling off the side of the road.

"Just a minute," he said, jumping out and running into the desert. I figured he was going to move his bowels or something and decided to urinate myself. There were no villages or trees in sight, a few hills and dry shrubs were all I could see.

After a minute he came back and we both jumped into the truck without a word. He started it up and drove straight off the road and into the desert toward the hill he just had come from.

"Hey, where are you going?" I wanted to know. This guy was a bit strange and talked violently. Why was he driving off into the desert? I braced myself to jump out of the truck and run, if necessary.

"Just a minute," he said. "Five minutes." And he drove around the low hill where there were two tents pitched, tall enough to stand in and ten feet square. Two men were standing around a truck watching us as we pulled up. My driver stopped his truck and we piled out, but first he grabbed a couple of Jordanian dinar notes out of the glove compartment. He went straight into one of the tents without saying anything to me.

"Hmm," I thought, "Is this some kind of desert tea house?" I tried to follow my driver but an older, grey-haired man with a stubbled beard came up to me and aggressively pulled on my shirt sleeves saying, "shuk-shuk" over and over, and rubbing his thumb and fingers together in a way that indicated money. I pulled away from him and went back to the truck. He seemed a rather distasteful person and his aggressiveness made me a bit worried. For the life of me, I couldn't figure out what was going on.

Then I heard a female voice and a woman, pregnant and in her late thirties, came out of the other tent in a pink nightgown. It suddenly struck me: to my utter astonishment, this was a desert whorehouse! I banged the dashboard in my realization. "Of course!" I said aloud. "I would never have expected...!"

Mustaffa had only been in the tent a few minutes when he emerged, zipping up his pants. "OK," he said, "let's go." I was still looking around in wide-eyed amazement. I didn't think they had prostitutes in Arab countries! This certainly exposed my naiveté.

"No good," he said. "Let's go! No good!" As we pulled out back toward the highway, a woman came out of the tent, younger and though less attractive than the pregnant woman. Her lonely face was strained. It was a hard life in the desert, I thought, and knowing Islamic law, things could get harder.

We roared off down the Desert Highway and soon the driver stopped for a beer at a small roadside store. I began to realize that this guy was a little crazy. He drove like a maniac, smoked cigarettes one after the other, guzzled beer and Arabian whiskey alternately and spoke of sex outrageously.

As we blasted down the highway in the early afternoon sun, we found ourselves behind a large semi-truck and trailer with Kuwaiti license plates, the Desert Highway being the main road into Saudi Arabia and then to Kuwait and the other Gulf States. After a slug of whiskey from a bottle he kept under the seat, he tried to pass the truck and trailer. The other truck wouldn't let him pass, and we played games for a while until my driver finally passed the truck, forcing it off the road for a moment.

After a few more miles, a car was stopped in the road and Mustaffa slowed down as a cautionary measure. The Kuwaiti truck then passed him and pulled across the road, blocking all the traffic. The driver of this truck jumped out, absolutely seething with rage. His face was beet red and his lips curled back over his thin mustache. He was young, but powerfully built like a professional wrestler. Except for his black hair, he was a dead ringer for Hulk Hogan.

I guess it was just one of those normal King's Highway traffic disputes, though they were quite a new thing for me. Mustaffa reached down beneath his seat and grabbed a sledgehammer. "Don't worry," he said to me. He opened his door and brandished the hammer menacingly at the other driver who was on his way to our truck. When the younger driver saw the sledgehammer, he ran back to his cab and reemerged with the biggest monkey wrench I had ever seen, plus a pair of Chinese Kung-fu *nenchakas* (two large wooden pegs with a heavy chain between them which can be very lethal when used by someone who knows how).

These guys were far too macho and crazy for me. I grabbed my pack and jumped out of my side of the truck, heading out into the desert. There was no sense getting caught in this fight, I figured, especially if Mustaffa should be the loser.

The two men face off, shouting obscenities in Arabic at each other. Mustaffa brandished his sledgehammer menacingly, while his opponent held the gigantic wrench high in one hand and swung the *nenchakas* with the other. With one hand steadying my pack, and prepared to run into the desert should Mustaffa lose, I watched in horror as this Arab truck driving ritual unfolded before my eyes.

Fortunately, the armed confrontation was stopped just as the two were about to attack each other. Two policemen who happened to be coming up behind us stepped in and the whole fight was reduced to a lot of name calling and abuse — words I'd never read in the Koran. Eventually, after some serious negotiations and brave mediation on the part of the Jordanian policemen we got back into our trucks. "My," I thought, "tempers do flare out here in the heat! With little things like this causing such trouble, it was easy to see why Arabs couldn't get along with each other, much less Arabs and Israelis."

Once back in the cab of the dump truck, Mustaffa said that he had told the other driver that he was a boxer and a sergeant in the army who didn't want to hurt kids, and he claimed that that eventually settled the argument. Somehow, I figured that the other driver, my Arab "Hulk Hogan" had another story to tell.

Just a short way up the road, Mustaffa let me off at another crossroads in the desert where the road forked west toward Petra. I waved goodbye to my temporary bodyguard, who gave me a toothy grin as he drove away. Probably a nice guy at heart, I thought as I waved and called, "Goodbye, King of the Road!" With that, I started walking toward Petra, which was still about thirty miles away.

My next ride was with an off-duty police officer in his thirties, friendly and well-educated. He wore a neatly pressed shirt and slacks, and was clean shaven and good-looking. Also in the car was a businessman from Amman, also in a suit and in his thirties, although he didn't speak English.

"I would like to go to Canada. I have a relative there," the policeman told me. "How much does a wife cost in Canada?" He was speaking of the Arab custom of · paying a dowry to a wife's family, usually many thousands of dollars.

"Free," I said. "No cost at all."

He sat in silence for some time and then talked in Arabic with the other man for a while. Then in his best English, to be sure I understood his question, he asked again, "In Canada, how much cost one wife?"

"Canada—wife free!" I repeated. "No pay!"

This occupied the policeman for some time, contemplating a country where wives were free. If they didn't cost anything, I imagined he thought, perhaps he'll take three or four...

§§§

They let me off at Petra, where there is a small visitor's center. Located near the foot of Mount Hor, Petra was the capital of the Edomites and the Nabateans. It declined near the end of Roman times and was lost to the world between the twelfth century when it was briefly occupied by the crusaders and 1812, when it was rediscovered by the Swiss explorer John L. Burckhardt.

Burckhardt was the son of a Swiss Colonel in the French Army who volunteered to carry out exploration work for the "Association for Promoting the Discovery of the Interior Parts of Africa". In preparation for the dangerous journey into Arabia, Burckhardt spent time in London and Cambridge attending lectures in chemistry, astronomy, medicine and Arabic. He also took long walks, slept on the ground and ate only vegetables in order to harden himself to the rigors of desert travel. Burckhardt was an explorer and adventurer of high caliber. Yet his journey, like that of many adventurers before and after him, was to end in tragic death. Fortunately, the rigors and dangers of Burckhardt's time are much lessened, though others have taken their place. At least now it is not necessary to sleep on the floor and walk twenty miles every day for a trip to Petra, but it helps!

After spending seven weeks in Malta honing his Arabic, Burckhardt was ready to enter unknown Arabia under the guise of a Moslem trader from India. For two years his visited ancient sites in Syria (which included Lebanon) and studied Koranic law, of which he was later deemed an expert, even by the Arabs, who often consulted him as to Koranic doctrine. When asked about his strange accent, Burckhardt would say that Hindustani was his mother tongue and launch into guttural Swiss-German. His critics would then nod in recognition of Hindustani, satisfied at his explanation. Burckhardt was well aware that his discovery as a spy would mean certain death at the hands of the Arabs or Turks.

After three years in Syria, he then headed into the interior of Arabia, which was his mission, after all. When he arrived at Amman and continued southward, he was told by Arabs of a wonderful ancient city buried in the heart of a seemingly impenetrable mountain. Naturally, he wanted to investigate, and proclaimed his desire to local Bedouins of wanting to sacrifice a goat to the Biblical Prophet Aaron, whose tomb he had been told could be found in city.

He was led through a narrow *wadi* (Arabic for a canyon) that suddenly emerged into the spectacular sight of the so-called treasury, a huge cliff cut into a magnificent building. Burckhardt nearly blew his cover when he went inside to investigate the beautiful structure, as this was all work of a great magician, the "Pharaoh", to the Arabs, and should be of no concern to a devout Muslim. Burckhardt made sketches in a notebook that he kept concealed beneath his robes. If his hosts ever saw it, they would know him to be a spy, and he would surely be tortured and killed.

After several diversions, the disguised Swiss and his guides reached the foot of the stairs that were cut into the solid rock and led to an altar on top of a peak. Because it was getting late, the goat was sacrificed at the foot of the stairs, and the men returned from the ancient city back through the narrow *wadi*. Burckhardt realized that he seen the mysterious city mentioned in the Bible and by ancient historians, known as Petra, which means rock in Greek. Burckhardt was to go down in history for his discovery, which he recounted in his book, *Travels In Syria and the Holy Land*, published in 1922. Unfortunately, Burckhardt was never saw his own book, for in 1917 in Egypt, he was to die of dysentery, a disease which still takes its toll of travelers to this day.[6,32]

> It seems no work of Man's creative hand,
> By labor wrought as wavering fancy planned;
> But from the rock as if by magic grown,
> Eternal, silent, beautiful, alone!
> ...rose-red as if the blush of dawn

That first beheld them were not yet withdrawn;
The hues of youth upon a brow of woe,
Which Man deemed old two thousand years ago,
Match me such a marvel save in Eastern clime,
A rose-red city half as old as Time.

– *Petra* by Dean Burgon

With the romance of Victorian discovery still fresh in my mind, I replenished my canteen at the visitor's center and I began the mile walk down the narrow canyon, Wadi Musa, toward the city. A dry riverbed that has cut and twisted its way through the sandstone is the only entrance into Petra, making it a highly defensible city and especially safe as a caravanserai for the spice caravans coming north from the Arabian peninsula, loaded with frankincense, myrrh, amber, and exotic goods from the far east. Suddenly, the narrow wadi opens up into a box canyon, and standing opposite the entrance is the Treasury Building, a two-story, colossal stone building carved straight into the red sandstone walls of the canyon. With pillars more than fifty feet high and two stories of Roman-sculpted bits of architecture, it is a most impressive site, as awe-inspiring today as it was for Burckhardt. Throughout the canyon are temples, amphitheaters, courts, tombs, apartments and other buildings carved directly into the stone. It took more than 500 years to build Petra, and now it is a ghost city with a few archaeologists and Bedouins.

As I walked through the remains of this ancient capital, I was once again reminded of the magnificent past and the great civilizations that had come and gone throughout history. Who had built this amazing complex and where did the builders go?

Generally, Petra is said to have been built by a mysterious desert people known as Nabateans. Yet, architectural historians found Petra a fascinating puzzle, filled as it is with buildings showing how the influences of Assyria, Egypt, Greece, and Rome have been adapted and transformed by these desert people who controlled the caravan routes not only north-south from southern Arabia, but east-west from Mesopotamia. Excavations have indicated that the area around Petra was inhabited at least as far back as 7,000 B.C. Yet, Petra enters history in the first book of the Bible, Genesis, where we are told that the area of Mount Seir (Petra) is the home troglodytes (cave dwellers) called Horites or "Nuzians". Later, Abraham's son Esau destroys the Horite troglodytes and establishes his own kingdom, known as Edom. When Moses sought to bring the Israelites to the promised land from Egypt circa 1453 B.C., he was refused. That Edom, with its capital at Petra was a powerful nation at the time is evidenced in that Moses and his followers did not attempt to cross Edom, but probably went to the east, further into the desert.

The Edomites were constantly at war with Israel and Judah, and there are many verses and prophesies directed against the Edomites. King Solomon subdued them during his reign, but when the Israelites were carried into Babylonian captivity in 587 B.C. and Jerusalem destroyed, the Edomites rejoiced and sang, "Down with it, down with it, even to the ground," as recorded in the 137th psalm. At this point, the Edomites began moving into Palestine themselves, while a mysterious group of "nomads" began moving into Petra. These were the Nabateans.[6,32]

By the late 4th century B.C., the Nabateans had established themselves at Petra and controlled the caravans that passed on the way to Damascus or Gaza. Just as important as the famous frankincense route from Yemen and the Hadramout, were the caravans that arrived from Busra in Mesopotamia through Maan in the

Arabian desert. The Nabateans became extremely wealthy from the caravans, and the city grew to an estimated 30,000 people. This was the golden era of Petra, and most of the amazing structures that can be seen today date from this time. The Nabateans managed to repel an attack by Alexander the Great, and it is believed that the Greek influence in the architecture of Petra stems from this time. Later, the Romans did manage to conquer Petra in 196 A.D., when it became a Roman province. The amphitheater at Petra was probably built at this time, but Petra began to decline, and it is likely that it was uninhabited by the time of the Islamic conquest. Then the crusaders occupied it briefly, only for it to lost again and then finally rediscovered by Burckhardt.

Petra has its mysteries. It was believed to hold a great treasure at one point. This was (and possibly still is) the rumor among the Bedouins. Both the inner and outer facade of the Treasury is pockmarked with countless bullet holes, fired by Bedouins on the belief that a lucky hit would release a shower of gold.

Also, no one is sure just what the purpose of the buildings carved into the solid cliffs was,. They are often said to have been tombs, but this may not be the case. The absence of any Nabatean written history makes it difficult to know what the purpose of any of the structures was. One mysterious structure is the staircase that ascends to the top of a steep hill. The path suddenly becomes a wide, smooth ramp flanked by a high stone wall. The size of the passageway and its careful construction suggest that it must have been used for important processions. The peak was probably an important sanctuary, for there are apparent ruins of a temple at the top. Perhaps the ramp was used to lead animals to the top for sacrifice, and some historians believe that the Nabateans practiced human sacrifice.

The origin of the Nabateans is also a mystery. They are often said to have been desert nomads that suddenly settled down and became very industrious and wealthy. This seems unlikely, though they were probably somehow linked to the early caravan trade, and did come from somewhere in the interior of Arabia. Another mysterious city, almost identical to Petra, can be found in the harsh interior desert of Saudi Arabia. It is even more mysterious than Petra, and will be discussed in a later chapter. It seems that the Nabateans are linked to the lost cities of the Empty Quarter, a land still dreaded and unexplored even to this day.

I inquired about staying in the Guest House in the heart of Petra itself, which provides tents and cave homes for its guests. However, it was full of an American archaeological team. After spending the afternoon wandering around the city and talking with the archaeologists, I watched the sunset from the amphitheater and then walked back through the *wadi*, looking for a good spot to camp. One of the American archaeologists had told me that scorpions, spiders with a deadly bite, and heat-seeking poisonous vipers roamed the sands of Petra and the *wadi*, so it was best to sleep as high up in the cliffs as possible. After looking around a while, I climbed up on the rocks to a small ledge on a rock. I rolled my sleeping bag out on the solid rock, keeping a watchful eye for spiders, scorpions and snakes.

As I drifted to sleep beneath the myriad of stars that shone down on ancient Petra, I thought of the ancient caravans that used to tread the sands of the *wadi* to the city. Suddenly, I heard the sound of what appeared to be a galloping camel or horse. I sat up from my ledge and looked around. A half moon was just rising, and it cast a fair amount light into the canyon. I had a good view of the entire area, but as I looked around I could see nothing. Who would be galloping a camel through the *wadi* at night like that, I thought? Perhaps it was the ghost of some long-forgotten desert Sheik, or a Nabatean scout coming to warn the city of an impending attack. While the stars shone down on me, I drifted into a restless sleep. Did the ghosts of Petra have more tales to tell?

§§§

It was dawn when I awoke. After packing up my things, and checking my shoes for possible scorpions, I walked back to the visitor's center. From there I was able to catch a local bus back towards the King's Highway. Once again I was standing at the lonely crossroads in the bleak and windswept desert that is most of Jordan, especially in the east. As I gazed out into the trackless wilderness, I wondered about the lost cities that might be there. How could anyone but a nomad with a camel have lived out there?

Yet, the deserts of Arabia have their share of lost cities and ancient mysteries. Because of the harsh nature of this desert, these lost cities have generally been discovered from the air. In 1948, it was reported by the aviator Alec Kirkbride in *Antiquity* magazine[33] that a stone wall ran for about 30 kilometers to the east of Petra and about 12 kilometers west of the desert town of Ma'an. The wall ran north and south and was thick at 2 meters wide. It ran over ridges, had breaks or gates when it ran over important tracks, so as not to impede traffic, and at one point ran over a sheer cliff some 15 meters in height!

Local inhabitants know the wall as "Khat Shebib" or "Shebib's wall". They claim that a prince of the Himyarite dynasty named Amire Shedbib el Tubba'i el Himyari built the wall before the advent of Islam. Kirkbride believes that the wall was not a defensive structure, like the Great Wall of China, but rather a boundary marker, separating the arable land with permanent water to the west, from the barren desert to the east. In this way nomads would know where they could legally graze their flocks without encroaching on someone's property. As Kirkwood points out, the most impressive thing about the wall is the large amount of physical effort to build it, no little feat indeed!

Of more interest are the reports by early ex-RAF aviators flying the Cairo-to-Baghdad mail route who claimed to have seen strange ground drawings and rock formations of clearly human manufacture on the ground. These strange rock ruins, often running for miles across the desert, were dubbed "kites" by the aviators because their shapes were rather like that of a common kite as flown by children. It was also reported that in the vicinity of these mysterious walls was a lost city of gigantic proportions.[33,34] What was a huge city doing out in the middle of that waterless desert? And what was the purpose of these strange "kites"?

Before the bizarre markings on the Nazca plain in Peru caught the fancy of archaeologists and anthropologists, the "kites" of the Arabian desert, only seen from the air, were a topic of heated debate. Unlike Nazca, it was never suggested that they were built by extraterrestials or such. Rather, it was believed that these walls, which had a narrowing opening at one end, ran for miles, and opened into a walled enclosure with points, were somehow part of either a fortification or were pens for cattle, or were traps for wild gazelles, or a combination of all three. Yet, certain questions were brought up by these theories, such as, how could cattle graze in such a barren desert and who would attack a city that was many miles from nowhere?

Because of the remoteness of the sites, and the difficulty in spotting them from the ground, no one was able to confirm the existence of the city until the early 50's. Even the famous Israeli archaeologist (and spy) Nelson Glueck believed that the story of the mysterious lost city of the Black Desert, known as Jawa, was a mere myth. In 1948, Glueck passed within a few hundred meters of the city, but saw nothing. He later reported, "Jawa marks the location of a small, filthy spring, below the west end of which a crude birkeh (pool), about 20 m in circumference, has been dug....There was probably never more than a small police post at Jawa."[34]

Glueck was somehow missing a lost city with an area more than 100,000 square meters! The city was finally reached in 1951 by an epigraphical expedition but it was not until Svend Helms and his co-workers from the London University Institute of Archaeology that serious excavation of the city really took place. What they found was a huge city complex that was at least five thousand years old. There were massive walls of basalt, ancient damns and canals, the mysterious desert "kites", Safaitic inscriptions (an early, pre-Islamic Arab script), and petroglyphs of domestic animals, as well as wild gazelles.[34]

When the Bedouin of the desert are asked who built the ancient structures: the "kites", the city, the canals; they say these things were built by "the Old Men of Arabia". Who were these "Old Men of Arabia"? Were the inhabitants of the lost city of Jawa (known as Jawaites) the same people who built the desert "kites"? Sven Helms in his book, *Jawa, Lost City of the Black Desert*[34] says that Jawa was probably built around 3,000 B.C., though his dates are through a process of elimination, rather than carbon dating. There is precious little to date at Jawa: basalt cannot be dated (as with any stone structure, a fact often lost on many people), the few inscriptions are difficult Safaitic, an archaic form of Arabic going back to the mists of time, any carbon datable material such as wood or bones is non-existant and the pottery found at Jawa is, according Helms, of a more recent date than the city.

There are a number of definite enigmas associated with Jawa, and Helms is well aware of this. First of all are the numerous petroglyphs of long horned cattle and other domesticated animals. One carving is especially interesting as it depicts a long horned bull being vaulted in the same manner as Minoan bull vaulters! This is highly unusual, to say the least, and Helms only comment on this scene is, "This animal, like cattle, is similar to some Arabian rock art which is rather dubiously attributed to the third millennium B.C."[34] Helms actually believes the rock drawing to be of an older date, in which I would probably agree with him.

Secondly, just what civilization was occupying this bleak area still remains a mystery, and thirdly, how they could have possibly scratched out an existence in the "Black Desert" is startling. Because of the sophistication of the city, Helm cannot believe that ancient nomadic Arabian Bedouin built the city. Therefore, it must have been built by refugees from Mesopotamia or Palestine. Yet, even that doesn't make sense. Says Helm, "Jawa is quite far away into the Black Desert and there seems no compelling reason why these people should have gone there."[34] He believes that the Jawaites, whoever they were, were expelled with their belongings from some civilized center. Says Helm, "But why choose a desert in which to rebuild urban life lost?"

Helm's book is a thorough, but rather dry, academic treatment of the city, and its mysteries. However, some of his conclusions are, to my mind, skirting the obvious. Helms bases his conclusions on several very shaky foundations. These foundations are for the typical isolationist anthropologist and uniformitarian geologist (these two dogmas generally go hand-in-hand) seem only natural foundations to base conclusions on, yet, for the maverick archaeologist, they are patently without foundation! Firstly, Helms notes that the city is not only ancient, but it is located near several extinct volcanoes (hence the basalt) and it "lies in a region of frequent tectonic disturbance that can cause wells to shift....The ancient settlement could therefore have been abandoned because of a failing water supply."[34]

45

What genuinely puzzles Helm is why anyone would have settled out here in the middle of this horrible desert, which is called, of all things the *Bilad esh-Shaytan* in Arabic, or the *Land of Satan* or the *Land of the Devil!* Similarly, Helms does not believe that the "Old Men of Arabia" who built the desert "kites" are the same people as the Jawaites. The rock inscriptions of cattle must also belong to some other earlier period (more than 5,000 years ago).

What doesn't seem to occur to Helms is that the climate of Arabia may have changed from the time that Jawa was built and settled. In cataclysmic geology, climate changes can occur suddenly and rapidly, much like the extinct volcanoes that Helms' observes. That Arabia was once a green, fertile pasture-land with plenty of water, cities and animals does not enter into his theories about Jawa at all. Instead, he racks his brain over why and how anyone could live out in this barren desert. Certainly, there is no way that herds of cattle as shown in the rock inscriptions could graze the barren "Land of the Devil"!

Also, his refusal to believe that prehistoric Safaitic Arabs could have built and inhabited Jawa, even though Safaitic inscriptions are found, is part of this myopic conception of prehistory: Arabia has always been a desert, and Arabs have always been nomads. This, I will argue, is not the case!

Similarly, we have the solution to the desert "kites". Early in their discovery they were thought to have been cattle pens or defensive walls, or a combination of both. Helms rejects these two hypotheses because no one would attack a desert city and there could have been very little cattle to have put into pens. He maintains, not illogically, that they were trapping wild gazelles. Probably, the answer is a combination of all three, although the use of the "kites" as gazelle traps would be their best function. But they are undoubtedly useful as cattle pens, and with their towers, also useful in defensive combat.

What emerges out of study of the fascinating and mysterious city of the Black Desert, a city that must have house thousands of people, is evidence that Arabia was once a vast fertile land. This was not so long ago. Like the Sahara desert, the Arabian desert was once inhabited by huge herds of wild animals, had plenty of water and food, and is the home of lost cities and advanced, ancient civilizations. Devastating climatic changes forced the animals to other areas, and the cities that once thrived became ruins known only to few Bedouin. Here we also see a reversal in logic: Arabs were forced through climatic changes into a nomadic life. From their dying cities, they had to move seasonally to find food and water. Eventually, such magnificent cities as Jawa fell to the dust of the ages.

To add a more fantastic touch to Jawa, the Lost City of the Black Desert, it would be fun to speculate on two subjects. One is the fact that Jawa is built out of magnetized basalt. Basalt is a crystal that is naturally magnetized, and it lends itself extremely well to advanced forms of levitation. The gigantic basalt crystal city on Pohnpei in Micronesia is apparently built in this way (see *Lost Cities of Ancient Lemuria & the Pacific* [35] for more information on this fascinating city). Were the massive basalt blocks of Jawa also levitated into place as legend says the basalt logs of Nan Modal were moved? Not necessarily, yet, when we read of Joshua making the walls of ancient Jericho fall down by blowing a "trumpet", the science of levitating stones is not so far-fetched! However, I easily concede that Jawa could easily been built by far more primitive means.

Another fascinating question is why the Black Desert is known as the *Bilad esh-Shaytan?* Was some ancient battle between Atlantis and Rama fought here? Perhaps the reason for the sudden desertification of the area is due to more than geological changes! Considering the history chronicled in the ancient Dravidian texts of the *Ramayana* and *Mahabharata*, this is hardly so farfetched. Furthermore, the Minoan-like bull-vaulter gives a hint of ancient Osiris and the lost empire of the Mediterranean. History, contrary to popular belief, need not be stuffy!

§§§

I waved my arms wildly at a rusty, rickety Jordanian bus heading toward me. It coasted to a stop at those desert crossroads and I eagerly jumped on board. Moments later and a few Jordanian fils less, I was on my way back to Amman, the Black Desert and its lost city fading away in my mind.

Back in Amman, I went back to my hotel near the central mosque, and checked in again. They still had the portion of my luggage that I had left there, and after having a cup of sweet tea with the manager, a friendly old man with a scraggly beard and a twinkle in his eye, I took off to the Ministry of the Interior to get my permission to visit the West Bank. After I had gotten permission to enter the West Bank, I headed out for a chickpea sandwich dinner, and a tea in one of the shops. I was tired from my trip to Petra and my restless night among the scorpions. I crashed out on a mattress on the roof of the hotel and fell asleep while humming to myself, "When the stars fell on Jordan".

The next day I was off by bus to the Jordanian/Israeli checkpost on the River Jordan, only a few miles from Amman. On the bus, I met a Dutch traveler named Han, who was also traveling to the West Bank and Israel, just like myself. It was only natural that we decided to travel together. On a bus full of displaced Palestinians, now returning to their "occupied" homeland, the two of us with our blond hair seemed out of place. This was the second trip to Israel and the West Bank for Han, so he knew all the steps, and I decided that it was wise to stick with him.

Like many people before them, the Israelis for instance, the Palestinians are displaced people. I talked with one young Palestinian on the bus as we neared the Allenby Bridge checkpost. He was young, friendly and handsome. His English was excellent, and he was obviously well educated. He lived on the West Bank, and was coming from Amman, just like me.

"Which side are you on?" he asked me.

This was certainly an awkward question. "I'm not on any side," I answered truthfully, "I'm just an observer."

As the Israeli officials began separating the Palestinian nationals from the tourists (there were only two of us) the Palestinian said to me, "This is very painful for us. In 20 hours I can travel to the United States. But it takes me 8 days to go to my own country." Ironically, Palestinians have always felt bitterness at Jordan for having annexed the West Bank. Still, that is nothing to the bitterness they feel at having the Israelis control their country. The history of the world is one of migration, displacement, and unfortunately, genocide. In the ancient world, this was a common occurrence. Things have changed little since the great migrations of man across our globe. Perhaps the only difference is the political power that can now be mustered on a global scale. With such remote world leaders as the Ayatollah Khomeini sending death squads out to assassinate authors and bomb bookstores in far away countries, suddenly the political machinations of tribal, cultural, religious and racial movements have taken on a new importance.

After our luggage was separated from us, Han and I went through immigration, getting a stamp on a card that was separate from our passports. This was important to travelers who intended to continue visiting other countries in Africa and Arabia, because an Israeli stamp in one's passport can mean imprisonment in some countries, and extradition or refusal of entry at the least.

After we were reunited with our luggage, customs officials went through our packs with a finetooth comb. When everything had been taken out and examined, our empty packs were X-rayed. Not finding anything incriminating, our possessions were given back to us, and we were officially inside the Jewish State of Israel. It was seeing the reincarnation of an ancient Biblical nation come to life. Here, in the wake of the incredible extermination of World War II, a new country had been born, the fulfillment of Biblical prophecy thousands of years old! The thought was as bizarre as the experience of crossing the border.

Jordan is still the legitimate government of the "West Bank" and the eastern half of Jerusalem, both of which are now occupied by Israel. It is possible to get a permit in Amman to visit the "West Bank," which is in fact a permit to go to Israel. Jerusalem is only a few hours from Amman, and is a fascinating city with a history that dates back beyond the first millennium B.C. Historically, Jerusalem has been the center of some of the most important events in history, many of them wars of some kind or another, though its name means, ironically, "City of Peace."

Once Han, the Dutch traveler, and I had cleared customs on the border, we were off via "share-taxi" to Jericho. Here we would transfer to another taxi that would take us to Jerusalem. Jericho, presently a city largely of Palestinians under Israeli rule, with a few controversial Israeli settlements, is something of a lost city itself. Jericho is one of the most startling ancient cities presently known to scholars. Long before Joshua "blew his horn, and the walls fell down," a remarkable feat by anyone's standards, Jericho was a city of small, domed, round houses built of mud-brick inside a secure stone wall of 8 to 10 acres (3-4 hectares). At least one very large tower (probably more) was built into the walls. This was an astonishing 11,000 years ago (9th millennium B.C.), making Jericho one of the oldest cities in the world by official academic estimation.[13]

No less than 17 successive phases of building have been discovered at Jericho, the town being repeatedly destroyed through wars and rebuilt throughout history and so-called prehistory. Even an innovative architectural device for rendering battering rams useless against the walls was employed.

When the wandering Israelites finally made it to their "Promised Land", they needed no battering ram to destroy the walls of Jericho. It would have been of little use against the ditches and sloped high walls. Rather, Joshua blew his "horn" and the walls came tumbling down. This may be but a cute analogy for something else – but what? What sort of primitive method might have used by the Israelites to destroy the walls? I can think of none. If they had simply stormed the walls and overwhelmed the city, why doesn't the Bible just say so? Other battles are not told of in such enigmatic speech.

Unfortunately, there is very little archaeological evidence from this time period (circa 2000 B.C.) discovered at Jericho. We therefore have little to correlate the Biblical text.[13] Yet, Joshua using a "horn" to destroy the walls of the city is not so far out. Sound waves, and wave forms of any kind, can be highly destructive! Ultrasounds and high frequency resonation can indeed make walls fall down, and much more! Did Joshua have some sort of high-tech gizmo that allowed him to defeat Jericho? As the Israelites were coming from Egypt (after 40 years of wandering the desert (or deserts, as the Israelites could have literally circumnavigated the world during that time), they may have had all kinds of advanced scientific devices, as we shall see in coming chapters.

Another unusual discovery at Jericho is preserved skulls of the dead buried beneath the floors of homes. The extreme age is remarkable: many thousands of years beyond the date generally given for the oldest civilization (Sumeria: 5,000

B.C. versus Jericho: 9,000 B.C.) Discoveries at Jericho themselves totally destroy the notion that Sumeria was the first civilization, yet, the history taught to students in our schools today has not changed at all. Facts are one thing, changing the school text books is something else.

It is also interesting to note that not only is Jericho an extremely ancient city, but it is located in an extremely hot and barren zone in the rain shadow of the Judaean hills, 650 feet below (200 m) below sea level in the Rift Valley that cuts up from East Africa and the Red Sea, a geological zone of high disturbance. The only water at Jericho issues from a powerful spring which maintains Jericho's oasis environment. We can assume that this spring has been at Jericho since its existence...or has it?

Perhaps Jericho, and the surrounding area, like the Arabian desert, was not always the parched desert it is today. Was there once a cataclysmic change in Arabia and the Holy Land that changed the geography and climate? According to the Bible there was. Not only do we have the flood of Noah, but we also have the destruction of Sodom and Gomorrah at the southern end of the Dead Sea, which is also part of the Rift Valley. Perhaps the only thing that allowed for the resurrection of the Jericho was the sudden occurrence of a powerful spring.

There wasn't much left of the walls of Jericho, I saw as I glanced out the rear view mirror of a share taxi on my way to Jerusalem. Recently there have been a number of demonstrations and violent clashes in Jericho. Things haven't changed that much in 11,000 years, I guess. But, I wondered, was the military technology that was being used today more, or less, primitive?

The "Treasury Building" at Petra.

OLD ENGRAVING OF THE SIQ LEADING INTO PETRA

David Roberts 1839 print of the Treasury. Notice the stair-
way in the background going to the summit of a steep hill.

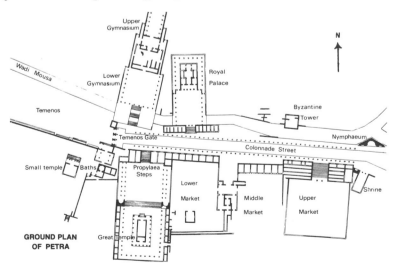

GROUND PLAN
OF PETRA

PETRA'S CARVED MOUNTAINS

Father Poidebard's bi-plane over the Black Desert in the 1930s ... the first photographic record of Jawa

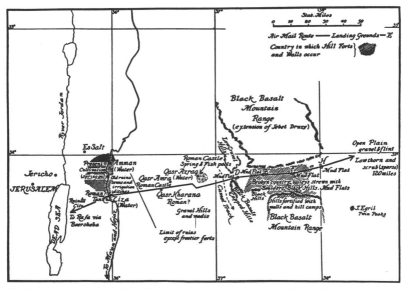

Flight Lieutenant Maitland's map: the airmail route from Cairo to Baghdad along which ex-RAF pilots discovered the mysterious 'desert kites' and other enigmatic signs of man's past in the Black Desert

The volcanic landscape of the Black Desert. Photograph of the eastern Qurma Gap. Jawa is on the summit of one of the hills in the distance. From *Jawa, Lost City of the Black Desert.*

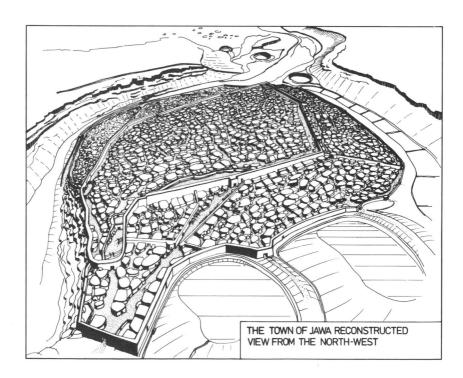

THE TOWN OF JAWA RECONSTRUCTED
VIEW FROM THE NORTH-WEST

Massive basalt formation at Jawa, with ancient petroglyphs of cattle. Though the area is now bleak desert, it is believed that the area was once a fertile plain.

Jawa seen from the south. One time this must have been ferti[l]
pasture. From *Jawa, Lost City of the Black Desert.*

The massive walls of Fortress Jawa

Petroglyphs of cattle at Jawa. Though the actual date of these petroglyphs cannot be determined, they are believed to be from 5,000 to 8,000 years old. In the top illustration cattle can be seen in a pen. In the center illustration, we see an acrobat vaulting over a longhorned bull in the style of Minoan Crete. This draws strong connections with the ancient Osirian Civilization. In the bottom illustration, a man is spearing a bull in a manner similar to the bullfights of today. From *Jawa, Lost City of the Black Desert.*

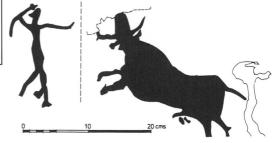

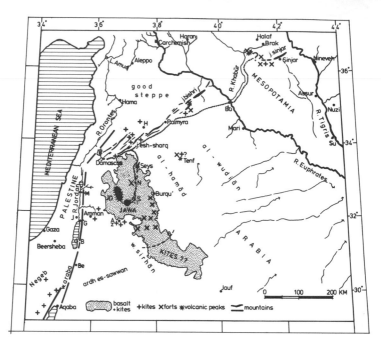

Road of the Rising Sun. Az el-Azraq, B Bab edh-Dhra', Be
Beidha, Bu Bouqras, G Tuleilat Ghassul, H Qasr el-Ḥair el-Gharbi, J Jericho,
JD Jebel Druze, M Munhata, N Numeirah, R Tell Ramad, S Khirbet
Shubeiqa, Su Tell es-Sawwan

A cave beneath the upper town of Jawa. Now a sheep pen with
recent wallsto keep in sheep. Thousands of years ago this cave
was used for other purposes. From *Jawa, Lost City of the
Black Desert.*

Chapter 3

ISRAEL:
RAIDERS OF THE LOST ARK

There is nothing new under the sun.
–Attributed to Solomon

It is a short ride from Jericho in the deep Rift Valley below sea level to Jerusalem, though it is all uphill. The "share taxi", driven by tight-lipped Arab with a red checkered scarf, let us off at Damascus Gate, the northern entrance to the old city of Jerusalem. Han, the Dutch traveler I had met in Jordan, and I grabbed our packs, and headed through the large, restored stone gate, and into the ancient city.

We walked down narrow, stone-paved streets that wound downward past vendors who sold sesame rolls and Arabs playing backgammon in tea shops. Tourist curios of all shapes, sizes and prices were thrust at us we walked by. "Here, special price for you," said one young hawker, recognizing a tourist when he saw one.

"Come on!" shouted Han from in front of me. "Our hotel is down this way."

I pressed on, nodding cordially at the young salesman. Jerusalem was a wonderful and exotic city, and attracted a lot of tourists, that was obvious. Since this was Han's third time to Israel, he knew the cheapest place to stay in all of the city. It was a small, hidden Turkish Steam Bath near the Lion's Gate in the Arab Quarter, now defunct.

A grizzly old man with as many as three or four teeth welcomed us inside. We sat down on a bench in the central room of the baths and took off our packs. "Do you want to take a bath?" he asked us in English, and then coughed.

"No, we want a double room. With two beds," said Han. The old man coughed again and went to the old white refrigerator by the spiral stairs that wound up to the second floor. Taking out a bottle of "Arak", Arabian spirits flavored with anise seed, giving it a licorice taste, he took a swig.

"That'll be two dollars each," he said, wiping his lips against his sleeve. We took it. There were only four rooms in the place, and we seemed to be the only travelers staying there. It was really a bathhouse, and not a very popular one at that. Our room was a few feet square with two single beds a table and a chair. We showered in the baths, and spent as little time in the room as possible. We weren't complaining though, not at that price.

There are more things to do in Jerusalem than even the people who live there have time to do. In the old city alone, there is The Dome of The Rock, one of the most beautiful and famous mosques in all Islam, and just below it is the Wailing Wall, the last remaining vestige of Solomon's Temple. Jerusalem is one of the three holiest cities to Muslims; Mecca and Medina in Saudi Arabia being the other two. The Wailing Wall is the Jews' most sacred spot and you will often see Orthodox Jews praying there, especially at sunset on Fridays. Jews will often write a prayer or wish on a piece of paper and insert it into the wall. For Christians there is the Holy Sepulcher where Jesus "rose from the dead" after his crucifixion. These and many other religious shrines, the winding streets and bazaars of the old city, and many old buildings make Jerusalem especially

59

interesting and popular.

For several days, Han and I wandered the old city. He had called another Dutch friend in Tel Aviv, who was to help him get a job on an oil rig in the Mediterranean somewhere. His friend would call him back in a few days, and so we killed time around the city.

One day, we decided to go out to a movie. It was one that Han had never seen before: *Raiders of the Lost Ark*. This movie, one of the most popular films ever made, is about an archaeologist, Indiana Jones, who enlisted by the U.S. Government just prior to World War II to find the Ark of the Covenant which may be found inside an Egyptian Temple at the lost city of Tanis in the Nile Delta. After many adventures, Indiana Jones and his friends discover the resting place of the Ark, but the Nazis who are also after the Ark take it away from them.

Indiana gets the Ark back, and attempts to take it via ship back to the United States, but the ship is captured by the Nazis, and the Ark is taken to an island in the Mediterranean where it opened after a Qabalistic ritual. The opening of the Ark unleashes an unholy power that then destroys all who look upon the Ark, and then reseals itself. At the end of the movie, the Ark is taken to a gigantic government warehouse in Washington D.C. where it is stored and presumably lost forever.

It was an exciting movie, to say the least, and the audience in that Jerusalem theater yelled and clapped in satisfaction at the end. We all felt that it was a few shekels well spent. As we left the theater and started for the walls of the Old City, I couldn't help pondering the fascinating subject of the Ark of the Covenant.

Just what was the Ark of the Covenant, anyway? The Ark of the Covenant first appears in the Old Testament story of the Exodus. Moses is said to have symbolically placed a copy of the Ten Commandments inside the Ark, which was a nesting of three boxes one inside the other. Descriptions of the Ark in the Bible are brief and scanty, but it seems that the box, or "Ark" was something between four and five feet long and two to three feet in both breadth and width. The three boxes were a sandwiching of gold, a conducting metal, and acacia wood, a non-conductor. There are reports in the Bible of persons touching the box, and being struck dead. This is probably quite true, because such a sandwiching of a conductor and a non-conductor creates what is known as an electrical condenser. A condenser such as the Ark would accumulate static electricity over a period of days (or years) until it suddenly discharged onto a person, or grounded by means of a conductor, like a wire or metal rod touching the ground. If the Ark had not been grounded for some long period of time, the electrical charge built up in it could give a very nasty and fatal shock to someone who touched it. If the shock itself was not fatal, then the surprise of the shock could well be. After the Ark had been discharged, however, it would be quite safe to touch as many of the Temple Priests did.

Another part of the Ark of the Covenant was a golden statue, whose importance is often missed. Indeed, in esoteric literature, it is the most significant part of the Ark. It is described in the Bible as the Holy of Holies. It was a solid gold statue of two cherubim (angels) facing each other, their wing tips touching above them. They hold between them a shallow dish with their outstretched arms. This was known as the *Mercy Seat*.

It is upon this *Mercy Seat* that an esoteric flame called in Hebrew the *Shekinah Glory* rests. The *Shekinah Glory* is supposedly a kind of "spirit fire" which was maintained from a distance, originally by Moses and later by an Adept of the Temple. If the person viewing the *Holy of Holies* was able to detect the *Shekinah Glory*, that showed pychic talent, as it could only be perceived by a person with clairvoyant ability.

This statue, along with the Ark, is suddenly referred to in the Biblical book of *Exodus*, and it is usually believed that the Hebrews manufactured them, including the gold statue, while they were out in the desert. This seems unlikely, especially the manufacturing of the solid gold statue of the Angels. Rather, it is more likely that the Holy of Holies, and the Ark, were relics from an earlier time, and were being taken out of Egypt by the fleeing Israelites. Indeed, it is quite possible that for this very reason the Egyptian army decided to pursue the fleeing Israelites, even after they had given them permission to depart.

According to an obscure esoteric order known as The Lemurian Fellowship, the Holy of Holies was a statue created many tens of thousands of years ago on the lost continent of the Pacific generally known as Mu or sometimes Lemuria (a term coined by geologists in the late 19th century). The statue was created to test a person's clairvoyant ability, which was shown by whether they could see the *Shekinah Glory* on the Mercy Seat. Persons of sufficient psychic ability were then offered the chance to take citizenship training and join the Commonwealth of Mukulia, the name which The Lemurian Fellowship attributes to this civilization which reportedly covered the entire Pacific Basin, including Australia (for more information on the belief that the Holy of Holies is from Mu/Lemuria, see my book *Lost Cities of Ancient Lemuria & the Pacific*[35]). This lost continent, a controversial subject among geologists and mystics, then reportedly sank in a cataclysmic pole shift circa 22,000 B.C.[36,37]

According to The Lemurian Fellowship, with the downfall of the Mukulian civilization, the Holy of Holies and plans for rebuilding the Tabernacle were removed to Atlantis where they were kept in a gigantic pyramidal building called the Incalathon, which was sort of the government headquarters and museum at the same time. Just prior to the supposed destruction of Atlantis circa 10,000 B.C., the Holy of Holies was taken to Egypt, which was part of the Osirian Empire at that time.[36] According to the *Ultimate Frontier*,[37] the relic was first kept in the Temple of Isis and then secreted in the large stone crypt which occupies the King's Chamber of the Great Pyramid at Giza. For 3,400 years it remained there, until the birth of Moses.

The box, or Ark, within which the Holy of Holies was kept, was probably constructed in Egypt. Electricity was used by the Egyptians, as evidenced by electroplated gold objects, electrical lighting reportedly being used in the temples, and the use of the Djed column as an electrical generator (more on the sciences of Egypt in the next two chapters). Because many persons still knew the significance of the gold statue, it was important that the Holy of Holies and the Ark be kept away from the evil Amon priests who fostered mummification in Egypt and controlled the country for thousands of years. Therefore, secret Mystery Schools operated in Egypt that kept the ancient traditions of Atlantis and "Mu" alive. The Holy of Holies and possibly the Ark were sealed in the so-called King's Chamber of the Great Pyramid, and the entrance to the inside was a carefully guarded secret, known only to a select few.

The story of Moses is a familiar and fascinating tale. Abandoned child of a Hebrew family living in Egypt, he was raised by the inner court of the Royal Family and taught the ancient mysteries in the secret Mystery Schools, a heritage from the days of Atlantis and Osiris. With Egypt slipping deeper and deeper into the grip of the evil and powerful priesthoods, the Mystery Schools decided that Moses should take the Holy of Holies to his "Promised Land" where a Temple would be built for it, and mankind would again have access to the important relic from the dawn of civilization.

In theory, Moses and a companion were able to gain entry into the Great Pyramid, which had been sealed thousands of years before in early Egyptian history. Opening the main door from the inside, Moses and his companion removed the Holy of Holies from the so-called "King's Chamber" of the pyramid and took it with them during their Exodus from Egypt to their "Promised Land."

The 25th chapter of Exodus says that while on Mt. Sinai, Moses received instruction as to the rebuilding of the Tabernacle and its furnishing. The Ark of the Covenant was built to carry the Holy of Holies at this time, according to some. This would be an alternative theory to the Ark having been built in Egypt hundreds or thousands of years before. Certainly, being constructed partly out of wood, it would tend to disintegrate after a long period of time, although the preservative energies inside the pyramid would have halted any decay.

The Ark was then housed in what the Bible calls the Tabernacle in the Wilderness. The Biblical Old Testament speaks a great deal about the Tabernacle in the Wilderness. There are several publications about the Tabernacle which describe the actual structure of the exhibit and how it was used. These include *Christ in the Tabernacle* by A. B. Simpson, *The Tabernacle of Moses* by Kevin J. Conner, *The Tabernacle, the Priesthood and the Offerings* by Henry W. Soltau, and *The Tabernacle, Priesthood and Offerings* by I. Haldman.

According to The Lemurian Fellowship, the Tabernacle was also used in ancient Lemuria the same way the Hebrews used it. The Tabernacle was moved from place to place so that as many persons as possible would be able to visit it. The site where the Tabernacle stood was enclosed with a high fence constructed of heavy cloth. The area enclosed by the fence and just outside the Tabernacle was known as the Outer Court and contained an alter as well as a Laver used for washing the hands and feet. Inside the Tabernacle were two rooms, the East Room and the West Room, which were separated by a veil. The East Room was the only room accessible to the visitor unless he was of sufficient advancement.

Based on the combined information of several sources the use of the Tabernacle seems to have been as follows: The visitor entered into the Outer Court through an opening on the eastern side of the fence. The opening was covered by three curtains which had a symbolic meaning though sources do not agree as to what. (A portion of the Great Pyramid at Giza also has an entrance known as the *Triple Veil.*)

Once inside the Outer Court, the visitor viewed an altar called the *Brazen Altar* upon which burned an animal sacrifice. This seems to have symbolized man's endurance of suffering as long as he lives in a state which is out of harmony with the principles of God and the universe.

The visitor was given the opportunity to leave the site at this point and allowed to come back the next day. Upon his return he had to satisfy his guide that he understood the significance of the Brazen Altar before he was allowed to continue. He was then given permission to bathe his hands and feet in the Laver. This washing, apparently, symbolized the cleansing process one must pass through before he can transcend the personal problems he must face in this world.

The visitor then went into the Tabernacle entering by way of the East Room. The room was without windows and was illuminated solely by light created by what was known as the Golden Candlestick. The Candlestick had six branches and a shaft upon which rested seven lamps. The purest olive oil was used in the lamps and they burned continuously.

Inside the East Room to the right stood a table upon which were arranged twelve loaves of Shewbread. There were two piles, each containing six loaves of the unleavened bread. The bread was representative of the service of mankind, which culminated in the planting and harvesting of this grain.

The visitor faced the veil which separated him from the West Room. At this point he was told that when he fitted himself to become a better person, he would be able to enter this room, which, he was told, contained the Holy of Holies and God. Until such time as he developed inner illumination, he was permitted only to view the room from the doorway.

The Ark of the Covenant within the Tabernacle was kept in Jerusalem until, during Solomon's reign, a permanent home was built for it in Solomon's Temple.

Also placed in the Ark is a copy of the ten commandments, a golden pot of manna, and Aaron's rod that budded. That the statue of the angels was sometimes kept in the box, rather than being merely part of the lid, is evidenced by the fact that the statue was not always seen by persons who viewed the Ark.

§§§

By the rivers of Babylon,
where we sat down,
and there we wept,
when we remembered Zion.
—*popular reggae song based on the Biblical book of Lamentations*

There is no question that the Ark of the Covenant, and its contents, were the most important articles kept inside Solomon's Temple. Today, there is probably no more valuable or important artifact in the world. Yet, what happened to the Ark? Where is it now? In 586 B.C. the huge Babylonian army of Nebuchadnezzar attacked Jerusalem. The Babylonians were jealous of Israel's wealth and power, and had for many generations sought to destroy the country, which was at the crossroads of most of the major trade routes of the day.

Jerusalem could not withstand the siege, and finally the city fell. The Temple was looted and destroyed, and the Israelites were taken into captivity in Babylon. Yet, where was the Ark? Was it destroyed or taken back to Babylon? This seems unlikely, as such an important artifact as the Ark would have been mentioned as part of the spoils, which it was not (Jeremiah 52).

In Maccabees 2:1-8, we read that Jeremiah the prophet, "being warned of God, commanded the Tabernacle and the Ark to go with him, as he went forth into the mountain, where Moses climbed up, and saw the heritage of God. And when Jeremiah came thither, he found an hollow cave, wherein he laid the tabernacle, and the ark, and the altar of incense, and so stopped the door."

What the Bible is speaking about is Jeremiah removing the Ark to Mount Nebo (now in Jordan, across from the Dead Sea), hiding it in a cave, and then concealing the entrance. This is almost certainly the case. The question that now remains is, was the Ark then reinstalled in the Temple at Jerusalem when the Jews returned from exile in 538 B.C. and rebuilt the temple between the years 520 and 515 B.C.? Although the Bible makes no mention of it, it is sometimes believed that the Ark was returned to the Temple, in the possession of the Tribe of Judah. Then, at the time of Christ's crucifixion, an earthquake and a hurricane hit the city. The Ark of the Covenant vanished forever.

In accordance with the theory that the Ark of the Covenant and the Holy of Holies were returned to the Temple and then vanished at the time of Christ's crucifixion, it has been said that the Ark was taken via ship to the Phoenician port of Catalunya, where it is allegedly kept in secret by the Essene Brotherhood. At the monastery of Montserrat in the Pyranees mountains outside of Barcelona, there is a secret tradition that the Ark of Covenant is kept in a secret cave. Indeed, the symbol for the monastery is three mountain peaks with a box on the top of the center peak. Is this the Ark of the Covenant?

Another tradition says that the Ark of the Covenant was taken to Ireland by Jeremiah himself. He fled to Egypt when the Babylonians invaded Israel, and then to Ireland with Queen Tamar Tephi, who married King Eochaidh of Ireland. She died only a short time after her marriage and is buried in county Meath, at Tara, north of Dublin. Buried with her was a great chest said to contain relics from Palestine. Some believe that this chest was the Ark of the Covenant! It is also interesting to note that Jeremiah is buried on Devenish Island in Loch Erne in western Ireland, according to local tradition.

Otherwise, it believed by Biblical scholars and professional "Ark Raiders", that the Ark of the Covenant is still in the Middle East somewhere, possibly on Mount Nebo, or somewhere nearby in the vicinity of the Dead Sea. Another possibility is that the Ark is hidden in secret chambers beneath Solomon's Temple Mount, upon which stands today the Dome of the Rock, third holiest mosque in Islam.

Maimonides, also called Rambam, a Jewish Torah scholar who lived in Cordoba, Spain, from 1135 to 1204 during the period of Moorish rule, quotes an earlier Jewish scholar named Abaraita as saying: "...when Solomon built the temple he foresaw its destruction and built a deep secret cave where Josiah ordered the Ark to be hidden." This theory may be partly responsible for the ill-fated expedition to uncover the hidden Ark from beneath the temple mount in 1908. The so-called Parker Expedition began at the Topkapi Museum/Library in Istanbul where a certain Swedish Biblical scholar named Valter H. Juvelius accidentally stumbled on a sacred code in the book of Ezekiel while studying it in ancient manuscript form. This code, he claimed, described the exact location of the long-lost treasures hidden within a tunnel system underneath the temple mount.

Juvelius teamed up with Captain Montague Parker, who got, among other backers, the duchess of Marlborough to put up $125,000 to search for the elusive treasure. They bribed their way through the red tape of the Ottoman Empire and worked beneath the city of Jerusalem from 1909 to 1911. They tunneled beneath the city, uncovered hidden passageways, but their search came to a sudden halt on April 17, 1911, when Captain Parker and his men attempted to enter a natural cavern that they had discovered was beneath the surface of the Sacred Rock on the temple mount itself.

The Sacred Rock is where Abraham was suppose to have offered his son Isaac to God, and where Mohammed is said to have ascended to heaven on his horse Borek. Local tradition said that evil spirits in this cavern guarded an ancient treasure vault.[38]

Captain Parker and his crew lowered themselves down into the cavern and began to break the stones that closed off the entrance to an ancient tunnel. But, unfortunately, one of the temple attendants had chosen to spend the night on the temple mount, and heard the sounds of the expedition. He followed the noises to the sacred stone, one the holiest in all of Islam, and, to his utter horror, found the sacred shrine occupied by a group of strangely clothed foreigners!

With mad shrieks, he fled into the city, spreading cries of how the temple was being desecrated. Within the hour, the entire city was in a tumult. There were riots in the streets as the rumors spread that Englishmen had discovered and stolen the Crown and magical ring of Solomon, the Ark of the Covenant, and the Sword of Mohammed! For this they must pay with their lives! Parker and his men fled the city and headed for Jaffa Harbor on the Mediterranean Coast where Parker's yacht allowed them to escape. He was further banned from ever entering Jerusalem again, and the local Turkish governor and commissioners were replaced.[38]

It was a close call for those lucky "Ark Raiders", but the exploration did not stop there, though it took a different turn. In the 1920s an American explorer named Antonia Frederick Futterer searched for the Ark on Mount Nebo in Jordan (then Trans-Jordan) where the Bible says that Jeremiah had hid it more than 2,500 years before. Futterer published a pamphlet in 1927 entitled *Search Is On For Lost Ark of the Covenant*, (Los Angeles, 1927) and claimed that while either on Mount Nebo or Mount Pisgah, in the Nebo Range, he squeezed into a cave leading to a long vault or corridor with "hieroglyphics" on the walls. At the end of the corridor he found two locked doors. Futterer took notes of the "hieroglyphics" and when he returned to Jerusalem, "a Hebrew scholar" deciphered his "hieroglyphic" signs "numerically". The numerical value of the signs totaled 1927, claimed Futterer. If there is any truth to this, it would seem that the signs were ancient Hebrew, which do indeed have numerical values, as well as phonetic meanings, rather than being genuine hieroglyphics in the Egyptian sense. Futterer interpreted this to mean that he would discover the Ark of the Covenant in 1927. After uncovering the Ark he planned "to build a Tourist Resort here out of these already prepared stones of old ruins." His pamphlet solicited funds for the project, asking, "What will you give to see the lost Ark restored to Jerusalem? Will you help us materially?"[39] What became of the funds Futterer collected, we do not know. That the tourist resort was never built, can be considered fact.

Yet, to his credit, Futterer never claimed to have actually discovered the Ark, though in his second book entitled *Palestine Speaks*, published in 1931, (pp. 536-537) he stated that he still believed that the Ark was to be found there. A close friend of Futterer's, a minister named Clinton Locy, fell heir to Futterer's papers. Then, in 1981, a Kansas resident named Tom Crotser visited the aging Reverend Locy and obtained a copy of the inscription that Futterer had taken from the wall on Mount Nebo. According to Crotser, the inscription read, "Herein lies the Ark of the Covenant." From Locy, Crotser also obtained a sketch from Futterer showing where the cave was located.

Crotser and three associates then proceeded to Jordan to discover for themselves The Ark of the Covenant, perhaps the most precious relic in the world. In October of 1981, he and his companions spent four days in the area, sleeping in sleeping bags. On Mount Pisgah, they found a depression, or crevice, which they believed to be the cave-opening identified in Futterer's sketch. Without any permits from the Jordanian government, the "Ark Raiders" removed a tin sheet covering the opening and proceeded into a passageway at 2:00 A.M. on October 31, 1981, the night of their third stay.

Crotser estimates that the initial passageway was 600 feet long, 4 to 6 feet wide and about 7 feet high. It led through several room-like enlargements with numerous tomb openings on both sides containing two or three levels of tombs. In the course of their exploration, Crotser and his associates illegally broke though two walls. The walls were made of mud and rock mixed together, sort of like cement, according to Crotser. He believed that someone had been there not long before, and had plugged up the wall.

At the end of these passageways, Crotser and his two friends encountered a third wall, more substantial than the ones he had already broken through. They found no inscriptions, as Futterer had described, though they broke through the wall anyway with hand picks. They cut a 4 feet by 4 feet opening in the wall, which led them into a rock hewn chamber about 7 feet by 7 feet. Crotser estimates that this chamber is directly beneath an old Byzantine church that is built on the very summit of the mountain and was connected to the church by a vertical shaft.

It was here in this chamber that Crotser claims to have seen the Ark of the Covenant. He describes it a gold-covered, rectangular box measuring 62 inches long, 37 inches wide, and 37 inches high. Wisely, they did not touch the box, though they took photographs, measured it and took notes. The golden cherubim were not on the lid, as often depicted. In the corner were some gauze-covered packages that Crotser took to be the golden statues of the Angels. He did not touch the packages, so he was unable to confirm his suspicion. He also noticed that poles to carry the Ark lay beside it, and that gold rings for holding the poles were fastened to the sides.[39]

Crotser and his companions then departed for Amman, where they unsuccessfully attempted to interest the Jordanian authorities in their important discovery. They then flew back to Kansas and reported their find. The UPI bureau in Kansas was quite interested in Crotser unusual announcement that he had discovered the Ark of the Covenant, and released it to the world's newspapers the next day, creating something of a scandal. Crotser refused to allow his photos printed, but later published them in the book, strangely entitled "Elijah, Rothschilds and the Ark of the Covenant".[40]

One person to view Crotser's photo of the supposed Ark of the Covenant is the well-known archaeologist Siegfried H. Horn, who was asked to check into Crotser's claims by a number of interested groups. Only one of the slides was clear enough to give Horn a good view of the box. From memory, Horn drew a description of the box from the slide, and noted a few tell-tale details. Horn decided that the box in Crotser's slides was quite modern, noting such details as the heads of nails, the regularity of machine-produced decorative strips.[40] Said Mr. Horn, "I became convinced that the object he had found is a comparatively modern box covered with metal sheets and strips."

The ramifications of Crotser's wild expedition in search of the Ark of the Covenant are quite far reaching. For one thing, the Jordanian government canceled all archaeological expeditions to that country as a direct result of Crotser's claim! Then, a staff member of Biblical Archaeological Review resigned because of the article, which the member stated was poor journalism and had slandered Jordan because of Jordan's refusal to allow any more archaeological expeditions.

More importantly, what of Crotser's claims, and what did he really find? It is quite possible that Crotser has fabricated the entire story as well as the "modern Ark" of which he took photos. This possibility is not even addressed in the *Biblical Archaeological Review (BAR)* article, perhaps out of respect and politeness to Crotser. Rather, *BAR* asks, what was it that Crotser and friends photographed in the sealed room that they illegally entered on that night of October 31, 1981? No one has an answer, but one suspects that whatever it was, it was known to the Byzantine Church directly above it. Was it a copy of the Ark, or a box containing some other sacred relics? We may never know. If there really was something of great importance in that chamber, you can bet that the Jordanian government has it now. It may have been more than embarrassment that made them cancel all archaeological expeditions to the countries.

If the Jordanian government had the Ark of the Covenant, would they acknowledge the fact? Probably not, for political reasons. The Israelis would gain a great boost by its discovery, though the fantastic wealth of the object would make it difficult to merely store it in some basement. While Jordan could always use a few billion dollars in revenue, Saudi Arabia would have the money and desire to purchase the Ark and keep it a secret.

Yet, was this really the Ark of the Covenant? I think not. The real Ark is much too important an artifact to be kept in such an accessible place, as evidenced by the shaft between the rock-cut chamber and the church. Plus, the golden statue of the Cherubim with the *Shekinah Glory* held between them, would not be wrapped in gauze and kept on the floor. It is more important than the Ark itself though, admittedly, it may have been kept inside the box.

Even more bizarre, perhaps, were Crotser's claims that God told him to only release the photos of the Ark to London Banker David Rothschild. Crotser claims that Rothschild is a direct descendant of Jesus and is to rebuild the Solomon's Temple.[39,40] The Ark would then be put in Rothschild's restored temple. Rothschild, part of the dynastic French- British- Jewish banking family that has controlled Europe's monetary wealth for hundreds of years, will have nothing to do with Crotser.[39] It is interesting to note here that the Rothschilds are a key element in nearly every banking conspiracy book every written and have strong links to the so-called Illuminati.[41] It is also worth noting that according the New

Testament book, *Revelations*, the final world battle of Armageddon is to occur after the third temple is built, which is exactly what Crotser is claiming Rothschild should now do.

Crotser is not the last "Ark Raider" to believe that he had discovered the Ark of the Covenant. Larry Blaser, a Seventh Day Adventist from Englewood, Colorado, also believed that he had discovered the resting place of the Ark of the Covenant. Blaser, however, concluded that the Ark could not have been hidden on Mount Nebo because Mount Nebo was too far from Jerusalem, beyond the borders of Judah, across the Jordan River. Blaser also contends that the book of Maccabees, in the Apocrypha, which tells of Jeremiah hiding the Ark on Mount Nebo, is not accurate. He came to the belief that the Ark was hidden near the Dead Sea in "David's Cave" where David had hidden from King Saul and his army with his own six hundred soldiers.

Near En-Gedi, the ancient Essene retreat on the Dead Sea, is a cave called "David's Cave" but Blaser believed that this cave could not be the actual cave that David and his army had hid in; it was far too small and afforded little in the way of protection or shelter. Therefore, the real cave was still to be found somewhere in the hills, and inside it, Blaser reasoned, lay the Ark of the Covenant.[38,42]

After a preliminary scouting trip in 1976-1976, Blaser returned in 1977 with Frank Ruskey, a geophysical engineer, and Richard Budick, an engineering geology technician, in order to conduct a thorough geophysical investigation for a hidden cave on the En-Gedi nature reserve. From the resistivity work and the seismic survey, combined with visual observations of the area, the scientists concluded that there was indeed a cave-like void, possibly twenty feet high, fifteen to twenty feet wide, and several hundred feet deep, with tunnels branching out like a two-pronged fork. Further visual investigation confirmed the initial impression that the cave had two possible entrances–both blocked–about ten to fifteen meters (30-45 feet) apart.[38]

This was a large cave indeed. In it was found what appeared to be a man-made wall and man-made works around the cave, including a system designed to divert seasonal run-off water over the entrance of the cave, thereby concealing it and calcifying the entrance at the same time. Blaser, Ruskey and Burdick then returned in 1979 with the author Rene Noorbergen and scholarly archaeologists, like Dr. James F. Strange of the University of South Florida, to attempt to open the cave. Dr. Strange did not believe that the Ark could be found in the cave, but felt that any cave in the area was worth exploring, especially with the important discovery of the Essene "Dead Sea Scrolls" in a cave nearby in 1948.

Unfortunately, it proved impossible to enter the cave. A huge boulder, said to be natural in the *BAR* article[42] and placed there in Noorbergen's book,[38] blocked the entrance to the cave system. Without the use of dynamite, expressly forbidden by Israeli authorities, as the area is a nature reserve, there was no way to enter the cave system, which further soundings indicated was genuinely there.[38,42] The expedition returned to the United States, and another group of "Ark Raiders" became history.

What is interesting in reading the stories of those who would be "Raiders of the Lost Ark" in the scholarly journal *Biblical Archaeology Review (BAR)*, is the obvious disdain that the academic scholars seem to have for such pursuits. Admittedly, Crotser's quest for the Ark led to Jordan to cancel all foreign archaeological research in country, yet for what reason? *BAR* itself states that "It is well known, however, that the Jordanians do not want any Biblical discoveries made in Jordan."[39] Similarly, they point out that no mention of discovering the Ark was given to the Israeli authorities by Blaser when they sought permission to enter the cave. What is obvious here, as *BAR* alludes, is that anyone seeking permission to excavate for the Ark of the Covenant would never get a permit either by the Israeli Government, or by the Jordanian. Therefore, only "Raiders" would be able to search for the Ark. An other attempt would be doomed by red tape.

Is the Ark of the Covenant inside the strange cave system near En-Gedi? We may never know. Perhaps there is another entrance, still cleverly concealed. *BAR* says that no person, not even animals, ever entered those caves.[42] How do they know? There is an interesting correlation to Blaser's search. Noorbergen relates a fascinating story in his book *Treasures of the Lost Races*[38]: In 1944, says Noorbergen, part of Rommel's North Afrika Corps split away from Rommel's army as it disintegrated. They stabbed north east and attempted to reach the Balkan states by going around Jerusalem to the east and into Syria and Turkey. Aware of their plans, the Allied High Command dispatched roving armored units to the area east of Jerusalem to intercept them.

On one night, a small American unit was camped in a narrow valley east of Jerusalem when it was strafed and bombed by German dive bomber. When one of the explosives hit the side of a cliff, it opened a small hole in the rock, exposing a cave. Scrambling for shelter, several men clawed their way through the opening and into a cave. As their eyes grew accustomed to the darkness, they saw "a coffin with what looked like two angels with outstretched wings on top. It had been covered with cloth which had disintegrated and was now hanging down like torn cobwebs..."[38]

Investigating the story, Noorbergen discovered that an army chaplain by the name of Captain Diefenbach had told the same story and had been assigned to the 28th Field Hospital in Palestine in 1944. Noorbergen attempted to find Diefenbach during his research into the Ark in the 1970s, but unfortunately learned that Captain Diefenbach had died on June 10, 1957. No other information was to be had on him, not even a list of relatives. Even his army records were accidentally destroyed in a fire. Noorbergen's quest for the Ark via Diefenbach came to a sudden end.[38]

Tales of the lost Ark of the Covenant are many and contradictory. In the film, *Raiders of the Lost Ark*, the Ark is taken by the Egyptians and placed in the city of Tanis, which is then buried in a sand storm, until uncovered by the Nazis and Indiana Jones. This is mere fanciful storytelling, though, there is some truth to the story, as the Nazis were indeed after quite a few ancient, mystical relics, and the Ark of the Covenant was probably one them. It is known to historians (though rarely related by them in more mundane histories of the Third Reich) that the Nazis were after the Spear of Destiny (which they actually obtained, it being in the National Museum in Vienna), the Chintamani Stone taken to Tibet by Nicholas Roerich (see my book, *Lost Cities of China, Central Asia & India*[15]) and other sacred relics reputed to have magical powers.

In personal correspondence, I have been shown a photograph of the "real Ark of the Covenant" as it was photographed in a cave "somewhere in the Middle East". I have been sworn to secrecy on this subject, and am true to my word. Let it suffice to say that the claim, though fantastic, has been given to me with evidence to back it up, and a good deal of sincerity. I attempt to be an objective and openminded person on the subject, so I am not convinced either way.

Where is the Ark today? It may be in a cave in the Middle East, probably somewhere in Israel or Jordan. It may have been taken to Ireland or Catalunya, Spain. It has also been said that the Ark of the Covenant will be moved to an island in the Pacific sometime before the end of this century. I have even been told that the Ark of the Covenant is kept in a secret cave in Arkansas! There is one other place where the Ark may be kept, and that is in Ethiopia (to be discussed in chapter 10).

The New Testament, describes a time when the need for the Tabernacle and its physical wonders along with the need for a human priesthood, will cease to exist. Under a "new covenant" people will come to know for themselves the truth of spiritual matters without outward inspiration and the use of "miraculous" signs:

"See, the days are coming—it is the Lord who speaks—when I will establish a new covenant with the House of Israel and the House of Judah, but not a covenant like the one I made with their ancestors on the day I took them by the hand to bring them out of the land of Egypt. They abandoned that covenant of mine, and so I on my side deserted them. No, this is the covenant I will make with the House of Israel when those days arrive. I will put my laws into their minds and write them on their hearts. Then I will be their God and they shall be my people. There will be no further need for neighbor to try to teach neighbor, or brother to say to brother, 'Learn to know the Lord.' No, they will all know me, the least no less than the greatest, since I will forgive their iniquities and never call their sins to mind." (Hebrews 8:9-11, *Jerusalem Bible*).

§§§

Little did I know that once I left Jerusalem, I would not return for six months. Han and I left and went to Tel Aviv, where he was planning to get a job on an oil rig. As we turned the corner one afternoon, after coming from the office of a kibbutz placement agency, he suddenly ran into an old friend.

"Han!" cried the man, a middle-aged Dutch man with thinning hair and a deep tan.

"Hank!" said Han, genuinely surprised. "Wow, it's great to see you here!"

Chance meetings in strange countries usually call for sitting down and having a beer, so we headed for a sidewalk cafe on the beach. Hank was a geologist who worked for Shell Oil Company. He would also help Han get a job on one of the oil rigs. We sat and had a few Gold Star beers, and my mind began wandering to the sight of what appeared to be several prostitutes standing on the corner in extremely short dresses and heavy make-up.

Hank saw me looking at them, with mere curiosity rather than genuine interest. A bit embarrassed I told him that I was just coming from the Middle East and was not used to the sight of ladies of the evening so flagrantly displaying their wares.

"Those aren't women!" laughed Hank, just as one walked up to our table. Her blouse was unbuttoned to give us a good view of her pendulous breasts.

The prostitute then stood at our table with a hand on her hip and said, "Look boys, if you want fun, you come to me. If you want girls, then go down the street!" And with that, she nodded and walked away.

"They're transvestites!" said Hank. "Can't you tell?"

No, I couldn't! I was red with embarrassment, and quickly finished my beer and ordered another one. I guess I had been in the land of camels and concealed women too long!

Han got a job with his oil rig, and I was placed on an agricultural commune near the Sea of Galilee. In Israel, an agricultural commune is known as a kibbutz, and they were first started in 1918 just near Tiberius. My kibbutz, Kibbutz Bet Zera, was the second oldest kibbutz in Israel, started in 1920. I worked for awhile in a plastics factory, and then was fortunately transferred to the avocado planting and date harvesting portion of the kibbutz.

We volunteers, young travelers from Europe, North America, Australia and other countries, all lived in one compound on the edge of the kibbutz. We all shared a room with two or three other volunteers, worked during the day, and partied during the night. It was a fun life; one of friends, relaxation, playing volleyball, swimming in the pool, and meeting at the communal dinning hall. The dusty crossroads of Jordan and Syria, Turkey, Afghanistan and Iran seemed far behind me now. Women in halter-tops and shorts were a common sight, and I eventually became more comfortable around them.

It wasn't long before I became involved with a kindergarten teacher from Sweden. Her name was Catharina, and we fell in love. It wasn't easy being in love on the kibbutz, though, as there was very little privacy. Wild parties often ended in scandals, and the kibbutzniks took a dim view of some of the shenanigans that went on in the volunteer's quarters.

One day, while harvesting dates, I accidentally dropped a tractor-trailer hitch on my foot, and crushed the top of the foot. My entire foot swelled up, and I was reluctantly taken to the hospital by the kibbutz for an X-ray. Miraculously, nothing was broken, but I would be unable to walk for several weeks.

I rested, but unfortunately, I had an infected mosquito bite on my foot , which then spread all over my swollen foot. The kibbutz nurse thought little of it, giving me an ointment and admonishing me not to be such slackard. The kibbutz couldn't afford to have unproductive members, and I needed to get back to work as soon as I could. After a week, my foot hadn't gotten better, and I insisted on seeing the doctor. It turned out that he only came on Thursdays, so I would have to wait until the next week.

When I finally did see him, a Jew from Argentina with a cigarette with an inch of ash dangling from his mouth, he gasped in surprise! "You must go to the hospital right away!" he informed me. "This is a very bad infection. If you want to save that foot, the kibbutz must take you to the hospital tomorrow!"

Reluctantly, the kibbutz dropped me off at the hospital in Tiberius the next day. After several hours of waiting, I was able to see a doctor. With a tall, blonde nurse standing next to me, the doctor examined my swollen, infected foot. "We'll have to operate today," he announced. "Your foot is going gangrene."

"Gangrene!" I cried, and nearly fainted. "Oh, no, you have to save my foot!"

"We'll do what we can," he said, and ordered me into an operating room, where the surgery would take place.

While I waited in the side room, I met an elderly Jewish woman from New York City who had been hit by a car on the streets of Tiberius earlier that day. She was all right, but might have some internal injuries. The doctor had also told her to wait for him in this small operating room. We talked for hours, and I wondered what was going to happen to me. When we had been there for such a long time that I seriously began to wonder if the doctor was coming to see us, I hopped out on one foot into the main office.

Looking around, I saw no one. Even the lights were off. I hopped out into the hallways of the hospital. They were deserted, with not a person in sight! A sinking feeling started to settle deep down in my stomach. It was a Friday afternoon, and the sabbath begins at dusk, lasting until Saturday at dusk. Suddenly I saw the tall, blonde nurse heading toward the outside doors of the hospital.

"Hey!" I shouted. "Hey! Have you forgotten about me? What about my operation?"

She turned and stared at me in horror. "Oh, no!" she cried. "We did forget about you! The doctor has gone home already! It is too late!"

"Too late to save my foot?" I asked. Standing on my good foot, I leaned against the wall for support. I wanted to die. I was a walking person. I walked everywhere. I needed that foot! I couldn't go on without it!

"You'll have to come back on Sunday, that's all there is to it," she said firmly. I told her of the New York woman still back in the room. Later, back at the kibbutz, I drank sabbath wine until my thoughts of the amputation of gangrenous foot became a drunken blur. Suddenly the fun of the kibbutz had turned into my personal nightmare.

On Sunday, the kibbutz took me back to the hospital for surgery. Catharina came with me for moral support. I was given anesthetic, and went unconscious while held my hand. The last words that I muttered, Catharina told me later, were "you must save my foot, I still haven't found the Ark of the Covenant yet..."

I was jerked up out of the anesthesia by the pain of the scalpel entering the top of my foot. The firm hand of the nurse forced me back down. I felt Catharina squeeze my hand. I passed out again. When I came to an hour or two later, my foot was bandaged.

"They saved your foot," Catharina told me. "The kibbutz will be here in a few minutes." And so began my convalescence over my crushed foot. I had not lost my foot, and over time, they told me, it would be as good as new.

One day, while I was still on crutches, an older kibbutznik named George came to the volunteer's quarters on a Saturday. George was a widower who originally came to Israel after World War II from Czechoslovakia. Now in his late sixties, George like to come by the volunteer's quarters to see who would like to take a nature walk with him to some of the wild areas to the west of the kibbutz in the Jordan Valley. Being much better, and almost able to walk again, I joined him and a few French girls on his walk.

George was obviously a very well-educated and interesting person. His patience in explaining things, and his obvious delight and wonder of the marvels around us were a pleasure to share. We walked down a ravine and George stopped to show us some interesting stones and artifacts that could be seen in the wall of the ravine. "This is the bottom of an ancient road," George informed us.

"Really?" asked one of the French girls. "How can that be? This wall is vertical!"

"It was flat, many thousands of years ago," answered George. "But, some time in the distant past, a great cataclysm destroyed this area. At that time, there were elephants in this area, and it was much greener than it is today. Elephant bones have been found along this road. Probably elephants were used for transport, just as they are in India and Southeast Asia today."

"But that is impossible," said another French girl. "We know the history of this area. Historians have never said what you are saying."

"Nevertheless, it is true," insisted George. Long before Abraham or Moses, this was a green land, with cities, roads and many animals. Then a cataclysmic change destroyed it. The Dead Sea was created, as this is the Great Rift Valley that goes all the way to East Africa. There are sunken cities in the Dead Sea. Even the story of Sodom and Gomorrah is related to such events. Many ancient civilizations have come and gone throughout history. Even ours may be destroyed in a like manner at any time."

"That is not a pleasant thought," said one of the French girls. George nodded, and we continued our nature walk. I looked carefully at the wall. It was a stone paving, with cement between the stones, not unlike many cobblestone streets. I was grateful for George reminding me of how impermanent all things are, and of how genuinely lucky I really was.

§§§

By the time I could walk again, I was ready to leave the kibbutz. They were glad to have me go, as I had ceased to be a productive member more than a month before. I packed my things and went to see Catharina. She was leaving to go back to Sweden in a few days herself. What romance we might have had before had pretty much been lost by my accident. I thanked her for coming to the hospital with me. It had meant a lot to have someone there who cared about me. Certainly none of the kibbutzniks had really given a damn.

After the lingering pain of the my foot, I was pretty much numb to the emotional pain of parting with someone whom I loved. Yet, as a tear came to her eye and she said goodbye, I found my smooth exterior a bit ruffled. I turned and shouldered my pack. With a slight limp, I headed out of the kibbutz gate, and walked down the dirt road that led to the main road into Tiberius.

I just bummed around northern Israel for awhile, stopping in at Sefad, a Jewish stronghold in the war against the Romans, later a crusader castle, and the center for Jewish kabbalism in the 16th century. I continued onto Acre (Akko) where the remains of an ancient Phoenician port are found. Today, the main tourist attraction is the huge crusader fort. Acre was the capital of the Crusader Kingdom when Jerusalem fell to the Moslems.

I stayed with friends on another kibbutz, and then continued to Haifa, the main port of Israel, and one of its most beautiful cities. I spent one day visiting the Bahai Temple, which is the world center for the Bahai faith. Bahai means 'glory' in Persian, and the religion was banned in that country. Today, it is extremely dangerous to be a Bahai in Iran, meaning almost instant death at the hands of fanatical Muslims. The Bahai faith was founded only in the early 1800's by the Bahai prophet, Baha-Ullah, who spent 24 years in a Turkish prison in Acre, a few miles north, and died there. Bahais believe that all religions preach brotherhood, love and charity, and that the purpose of religion is to bring mankind together, not raise barriers between them. It is little-known to most people (even Bahais, incredibly!) but Baha-Ullah taught the doctrine of reincarnation. The gold-domed roof of their main temple, surrounded by beautiful gardens, can easily be seen from the city.

From Haifa, I stopped in Nazareth, and then went on to Megiddo, the famous fortress that is the source of our word for the last battle on earth: Armageddon. Megiddo was an ancient fortress that commanded the important passage between North Africa and Asia Minor. Many ancient battles were fought at Megiddo, for whoever held the fortress had control of the caravan trade that passed along the eastern Mediterranean. While Israel/Palestine was the crossroads of the known world, Megiddo was the fortress that commanded those crossroads.

Remains of the huge stables can still be seen, large enough to house hundreds of horses. An incredible tunnel was built from a spring on the plain to secretly bring water into the city in times of siege. A huge grain silo was also built within the city for the same purpose. The oldest remains are of a Canaanite Sun Temple.

Pharaoh Thutmose III ordered his victories over Megiddo in 1478 B.C. to be carved in detail on stone. Solomon kept an important garrison at the fortress, and Josiah, a king of Israel, was killed at Megiddo in 610 B.C. trying to stop the advance of King Neco of Egypt, who had nevertheless told him: "My quarrel is not with thee." (Goes to show what can happen when you interfere in other people's business).

Later that evening, as I was back sitting outside the Wailing Wall in Old Jerusalem, I thought about that final battle of Armageddon. According to Revelations, after the Jews return to Israel and the third temple is rebuilt, then the final battle on earth will happen before the Millennium of Peace or new Golden Age. This battle is called in the Bible "Armageddon". The Jews have already returned to the land, yet, the third temple is yet to be rebuilt. Not because the Jews wouldn't like to, but the Dome of the Rock, the oldest mosque in all of Islam stands on the spot. Without the holy mosque on the temple mount, nothing could stop the Israelis from building the third temple. In fact, it is rumored that the Israelis have *already* built the temple, but it has not been assembled. It is in pieces and lies scattered throughout the world. When the time comes to rebuild the temple, the Israelis will gather the parts from around the world and assemble the temple within a matter of days! This quick assembly is important, so that by the time world is aware and can protest the construction, it will be too late.

Assembling the temple from stones cut and prepared in the rest of Israel or even overseas is not so strange, as this is exactly the way in which Solomon constructed the first temple himself! In I Kings 6:7 we read: "And the (Lord's) house . . . was built of stone made ready before it was brought thither: so that there was neither hammer nor axe nor any tool of iron heard in the house...."

What needs to happen before this awesome feat of construction is that the Dome of the Rock Mosque must be destroyed. This could happen during an earthquake, which would not be so unusual. Earthquakes have rocked Jerusalem in the past, and the Great Rift Valley just to the east of Jerusalem is an area of intense seismic activity. If the Dome of the Rock were destroyed in an earthquake, something which would naturally affect the entire city of Jerusalem, the Israelis would not hesitate to pour every single dollar and effort that they have into rebuilding the temple.

Just how important the rebuilding of the temple is to most Jews can be witnessed by sitting in front of the Wailing Wall, I thought, as I looked at the unusual sight in front of me. The "Wailing Wall" is the western wall of King Solomon's Temple, and the only part of the original temple that is still left. While the Dome of the Rock Mosque lies on top of the temple mount, Jews are forbidden to visit it. Instead, they may stand at the Wailing Wall below the temple mount and mosque and pray.

It was a Friday afternoon that I sat by the wall, and at dusk, the start of the sabbath, Orthodox Jews by the hundreds streamed down from the Jewish Quarter and the New City to the wall to pray. To these people, nothing is more important than their faith in God and the rebuilding of Solomon's Temple.

Every once in awhile, one reads in the *Jerusalem Post* about a radical Israeli who tries to blow up the mosque with a couple of cases of dynamite. Israeli authorities are very much on the lookout for such Jewish radicals, but the possibility of some underground Jewish group blowing up the Dome of the Rock Mosque is quite a reality. Either way, the mosque must come down before the temple is rebuilt. For the Jews, it is just a matter of time.

I discovered that some things had changed since I had last been in Jerusalem; such as the new construction in the Jewish Quarter and in fact that St. Mary's Bath House, where I had stayed before, no longer existed. Where the old Arab with his bottle of arak had gone, I'll never know.

At the small private Youth Hostel in the Armenian Quarter I met Simon, a young British law student who had been on a kibbutz and was now taking a few days to see what he could of Israel before he had to return to Britain. We decided to travel down to the Dead Sea and go to the ancient Jewish Fort of Massada on our way to Eilat on the Red Sea. So after breakfast one day, we took off out of Jerusalem, heading for the Great Rift Valley.

It was late in the day when we arrived at the desert fortress of Masada. With our packs we hiked to the top and found the fortress deserted. Knowing that a guard would sweep the ancient city, we hid until sunset, and then watched the sun set in the Negev Desert to the west.

"Why should there be such a fortress out here in the middle of this desolate desert?" asked Simon. "Surely there is no one to fight out here?"

Surveying the orange horizon to the west, I had to agree with Simon. Who indeed would need a fortress here? It seemed impossible that some major battle should be fought over this land. Yet, Massada was living proof that someone would build it, and a three-year siege once took place here. Yet, they are incidents that are not connected.

Massada was begun by Judas Maccabeus in the second century B.C. and Herod later turned it into his own personal palace-fortress. It was here that he planned to retreat if the Jewish patriots should ever revolt. Since he was not the legitimate heir to the Judean throne, he was always in such danger, and indeed, it was because of this fear that he had all the male children slain in his kingdom when he learned that the true heir had just been born. His fears were unrealized however, and he died in his own bed in Jerusalem of his various diseases.

It was left for those Jewish patriots that he sought to defend himself against to make their last stand at Massada: the famous Jewish Revolt and last stand of the Zealots in A.D. 70. For three years they held out against the overwhelming odds of the Roman Legions camped outside the fortress. Finally, the Romans built a huge ramp up the side and then set the city on fire. The Zealots knew that their time had come, and chose ten men to execute everyone, rather than be taken into slavery by the Romans. After the ten men had executed all the Zealots, they themselves drew lots, and one man killed the other nine. He then fell upon his own sword. Two women and five children had hidden in a cistern, however, and recounted the story to the Romans who swarmed over the walls. They were so moved by the story, tha their lives were spared.

Today, Massada is a symbol of Jewish/Israeli fortitude and strength. Outside the youth hostel at the base of the fortress is the sign, "Massada shall not fall again".

Simon and I spent most of the next day walking around Massada, and I was amazed at the huge cistern near Herod's hanging palace. We had lunch down at the youth hostel, and then starting walking down the main highway toward Eilat. The sun was hot, beating down on us relentlessly with not a cloud in the sky. I covered my head with a scarf and Simon held his guide book over his head to shield his face.

"Hey, it says here that there is a pillar of salt around here that is called *Lot's Wife*." said Simon. " Can you see it?"

I scanned the hills around us, but the late afternoon sun was in my eyes. I could see nothing. We got tired of walking and sat down on a bench that was a bus stop. Simon read to us about Sodom and Gomorrah while checked the bus schedule for the next bus.

Simon suddenly quoted from the Bible, "And the Lord said, Because the cry of Sodom and Gomorrah is great, and because their sin is very grievous....Then the Lord rained upon Sodom and upon Gomorrah brimstone and fire from the Lord out of heaven; And he overthrew those cities, and all the plain, and all the inhabitants of the cities, and that which grew upon the ground. But his (Lot's) wife looked back from behind him, and she became a pillar of salt....and, lo, the smoke of the country went up as the smoke of a furnace." (Genesis 18:20; 19:24-26, 28)

"Well, this is where it happened," said Simon, looking around around. "Do you think that it is true?"

"Sure," I said, "I believe that it is considered historic fact."

"Fact!" he said. "Fact! How can it be fact? It's impossible!"

"Hardly impossible," I replied. And with that, while we waited for a ride or a bus, we discussed the destruction of Sodom and Gomorrah. This Biblical passage has come to epitomize the destruction of God's wrath on those places which sin. The Bible is very specific about the site of Sodom and Gomorrah, plus several other towns, they were in the Vale of Siddim which was located at the southern end of the salt sea. Other towns in the area, according to the Bible, were Zoar, Admah and Zeboiim (Gen. 14:2). In the middle ages, a town that was also called Zoar existed in the area.

The Dead Sea is 1280 feet below and an incredible 1200 feet deep. Arab tradition has it that so many poisonous gases came out of the lake that no bird could fly across it, dying before it reached the other side. The southern end of the lake is a tongue that is only fifty or sixty feet deep. It is possible to see entire forests of trees encrusted with salt beneath the water in this southern "tongue" of the lake. It is theorized that the cities of the Vale of Siddim were destroyed when a plate movement caused the Great Rift Valley, which the Dead Sea is a part, to shift, and the area at the southern end of the Dead Sea subsided. In the great earthquake, there were probably explosions, natural gases issued forth, and brimstone fell like rain. This is likely to have happened about 2,000 B.C., the time of Abraham and Lot.[47]

Pilots flying over the Dead Sea have claimed to have seen cities beneath the water, and we know that there are forests in the southern end, forests where, perhaps, Lot's cattle once grazed. In the world of the cataclysmic geologist, the inundation of a small area at the end of the Dead Sea, and the loss of the few cities is pretty easily explained, and piddly-shit on a geological level. Tectonic plate movement on a much larger scale is relatively common, so the cataclysmist believes. What is a bit more mysterious is personalities involved, such as an Angel warning Lot to leave before the destruction and his wife turning into a pillar of salt when she looks back.

Perhaps this is all merely allegory, but it remains intriguing, none the less. It has been suggested that the cities were going to be "nuked" by extraterrestrials. They warned Lot to get his family out, but his wife looked back, and was blinded by the atomic flash. Perhaps Sodom was on the wrong side at the tail end of one of the devastating wars between Atlantis and Rama. Lot and his family were mere visitors, and warned to leave before the first strike on the city, but his wife was not so lucky, wanting to fetch something that she had left behind...

"That's ridiculous!" exclaimed Simon. "I don't believe that for a minute!"

"Well, I'll admit that it does seem sort of farfetched," I replied. "Perhaps it would be more likely that Lot had been warned, psychically perhaps, of the impending earthquake. After all, animals seem to sense an earthquake before it happens. The Chinese now use animals to help them predict earthquakes. So he gathered his cattle and family together and split just before the destruction. We'll just have to assume that his wife being turned into salt is an allegory for having hesitated or turned back and was then caught in the destruction, perhaps covered with hot asphalt and brine."

"Perhaps it's just a warning for wives to obey their husbands!" laughed Simon. Suddenly, he jumped up and waved at a bus coming our way. It stopped and we boarded it with our packs.

With that, we were off south, just as the sun was setting again in the west over the desert. Just as it got dark, I saw something on our left that made me look twice.

"What is that!" exclaimed Simon. "Some sort of space station?"

Indeed, it was like something from another planet. It was the closest thing that I had ever seen to an alien base. Strange towers shot up out of the desert. Bizarre buildings with domes and spires were covered with multi-colored lights. I expected to see a flying saucer land at any moment. It was the Dead Sea Chemical Works. During the day, it looks more normal, like an oil refinery or something, but at night, the lights that are strung about the facility make it seem otherworldly.

It was late that night when we reached Eilat. It was too late to find a youth hostel or a hotel, so we did what we had done at Massada, camped illegally at the beach. In the morning, we forced to get up at sunrise, for the first of the tourists and joggers were hitting the beach.

We found our way to the Youth Hostel, and checked in. This would be my last stop in Israel before crossing the frontier to Egypt. Simon would return to Jerusalem and then fly back to London. That afternoon while lying on the beach, I thought of how Eilat was once a great sea port; so great that more than twenty tons of gold was brought to Israel through it by King Solomon! Where, many historians have asked, did this awesome quantity of gold come from? The answer to that is, the lost city of Ophir.

In 1943, the famous Israeli archaeologist discovered near Eilat at Tell-el

Kheleifeh copper smelting furnaces and metal forges that dated back to the tenth century B.C. These furnaces were part of King Solomon's port city of Ezion-Geber, from whence King Solomon launched his Phoenician ships for the mysterious land of Ophir.[7,48] In the Bible, we are told:

> King Solomon made a fleet of ships in Ezion-Geber, which is beside Eilat on the shore of the Red Sea in the land of Edom. And Hiram sent in the navy of his servants, shipmen that had knowledge of the sea, with the servants of Solomon. And they came to Ophir, and fetched from thence gold, four hundred and twenty talents, and brought it to King Solomon.... Once in three years the fleet came in bringing gold, silver, ivory, apes, peacocks...a very great amount of red sandalwood and precious stones.
>
> *I Kings 9:26-28, 10:22, 11.*

Solomon had borrowed heavily from his friend and father-in-law the Phoenician King Hiram of Tyre to build the Temple that was to house the Ark of the Covenant. He offered Hiram twenty cities in the Galilee area as payment, but, as the Bible records, King Hiram visited the cities "and liked them not." Therefore, Solomon built the refineries and port on the Red Sea, and with Phoenician naval expertise, made the dangerous voyages to Ophir.

The Phoenicians, called "ships of Tarshish" after the Phoenician port in Spain called Tartessos, were well suited to the venture, having sailed throughout the Atlantic, and were the remnant of the legendary "Atlantean League", a great civilization of seafarers descended from the times of Osiris, Atlantis and Rama.

What these great seafarers brought back to Ezion-Geber was a literal fortune in gold. Four hundred and fifty talents is almost twenty tons, or 40,000 pounds of gold! At $500 an ounce, that would be about 320 million dollars! The source of that much gold would lure many an adventurer, even today, to cross an ocean or two. But they also brought back silver, ivory, apes, peacocks and more. Yet, the land of Ophir, where all this wealth came from has eluded historians for centuries.

When the large ruins of Zimbabwe were discovered in the last century, it was at first thought that these were the ruins of Ophir. To most historians though, Zimbabwe has been proved to be of more recent date, not 3,000 years old, as King Solomon's Ophir would have to be (more on Zimbabwe in chapter 15). Zimbabwe was about as far as most historians could stretch their imaginations, and when it turned out that Zimbabwe couldn't be Ophir, they looked closer to home, rather than farther.

It was suggested that Tartessos in Spain itself was Ophir, after all, they were "ships of Tarshish". Others felt that Ophir was in Somalia, or maybe Sofala, on the Mozambique coast near Zimbabwe. Later, it was believed that a certain ancient gold mine found along the Red Sea in Saudi Arabia was the source of Solomon's Gold. At this mine, there was still gold left, and the Saudi Arabian government began reworking it!

The myopic attitude of Biblical scholars and historians may be summed up by this statement from Manfred Barthel, the German scholar who wrote *What the Bible Really Says*: "Zimbabwe was too recent, India too far away, the Urals too cold...It does seem likely that Ophir was somewhere on the shores of the Red Sea."[43]

Similarly, the more openminded German scholars Hermann and Georg Schreiber say in their book *Vanished Cities*: "At one time the idea arose that the Ophir of the Bible may have been in what is now called Peru. But that is out of the question; no merchant fleet sailed so far in the tenth century B.C. The Solomon Islands, north of Australia, have also been suggested. But that is pure nonsense; all these have in common with King Solomon is the name, by which they were first called in 1568. Moreover, there is no gold in the Solomon Islands. It is also impossible to equate Ophir with the Spanish port of Tartessus, as a modern church lexicon does; no one who wanted to sail from Palestine to Spain would build his ships on the Red Sea!

"Flavius Josephus, the great Jewish historian of the first century A.D., guessed that Ophir was located in Farther India. But India was more interested in importing than in exporting gold.

"The search for the famous land of gold has therefore been restricted fundamentally to southern Arabia and the African coast. Arabia, however, is also out of the question. If Ophir had been situated there, as some scholars still maintain, Solomon would not have needed the assistance of the King of Tyre; he would simply have used the ancient caravan routes of the Arabian peninsula."[7]

The Schreibers have made many good points these last three paragraphs. And at the same time they fall into the strange logic of the isolationist: those who believe that ancient man never ventured far from land or the known. These are the "Columbus discovered America" types. Manfred Barthel is typical of the isolationist at his worst. For Barthel, even India is too far away. Even though it took three years (that's right, three years!) to go to Ophir and back, India was too far. Seafarers three thousand years ago couldn't sail from the Red Sea, around Arabia to India! That would be impossible!

Never mind that stone-age Indonesians migrated across the Indian Ocean to Madagascar or Polynesians colonized vast tracks of the Pacific Ocean, including remote places like Easter Island or Hawaii; huge, sophisticated navies cannot go from Eilat to India! This is the kind of anthropological myopia that dominates mainstream academia. Needless to say, seafaring traffic between India, Africa, Egypt and Red Sea ports is an established fact.

The Schreibers make a very good point about India, however, and that is that India was a very sophisticated country in itself, and it would be an importer of gold, not an exporter, just like Israel or Egypt at the time. Another clue here are the other items brought back from Ophir: apes, ivory, peacocks and sandlewood. While the apes and ivory have typically led researchers to suspect ports in Africa, both are found in India and Southeast Asia, and more importantly, peacocks especially, and sandlewood incidentally, come from India and Southeast Asia.

The Schreibers almost make the good point that Arabia, the most popular location for Ophir cannot be the site because it is far too close to Ezion-Geber, and Solomon would hardly have needed a fleet to reach such gold mines. Nor would ivory and apes, and especially peacocks or sandlewood have come from there. Furthermore, the thing that none of them have touched upon, is the fact, clearly stated in the Bible, that it was a three year journey to Ophir and back!

To go to a port in Arabia might take three weeks, but not three years! Even to go to Somalia or Zimbabwe for that matter, would only take a few months, monsoon winds not withstanding. Indeed, it is the ancient historian Flavius Josephus who comes closest in his guess as to the location of Ophir when he said *Farther India*. Probably he meant Indonesia by *Farther India*.

Orville Hope makes a very good point in his book, *6000 Years of Seafaring*,[45] that the reason it took three years to make the journey to Ophir was that it took one year to make the journey. It took another year to grow a crop for the journey home, and it took a final year to make the journey homeward. During the middle year, the mining and refining of the ore would be done, while the crops were also being raised.

Barry Fell, in his books *America B.C.*[20] and *Saga America*,[21] maintains that the Egyptians were mining gold on Sumatra. Phoenicians would have taken these mines over, and mined for King Solomon. Indeed, in 1875 an article in the

Anthropological Institute Journal was on the subject of Phoenician characters in Sumatra.[33] Were these Phoenician characters left by the sailors of Solomon's fleet?

Likewise, Thor Heyerdahl shows clearly in his book, *The Maldive Mystery*,[49] how ancient seafarers like the Phoenicians would have had to pass through the equatorial channel in the Maldives, stop at the Sun Temple there, and then sail on to Sumatra, and of course, further on to Australia, New Zealand and the Pacific. That the Egyptians and Phoenicians, as well as the ancient Indians and Chinese, were exploring the vast Pacific many thousands of years ago is discussed in my book *Lost Cities of Ancient Lemuria & the Pacific.*[35]

Suddenly, we see how the idea of Solomon's ships even reaching Peru is not so farfetched at all, and is in fact, quite plainly within the achievement of the ships. Phoenicians were brave and excellent sailors, and their ships were large and well equipped, much better than the ships of Columbus. The land of Ophir had to be a land of gold, and Peru certainly was such a place. Even the Solomon Islands may well have reached by Solomon's ships, yet, without gold mines, they would have been of little interest to the ancient seafarers. Perhaps the islands had only 20 tons of gold in them...it would be interesting to know just how it is these islands did get their name. I am personally not aware of the story, though I would concur that it probably has nothing to do with Solomon himself.

Orville Hope, in his book *6000 Years of Seafaring*[45] asserts that King Solomon's Ophir was actually in New Mexico, where Hebrew inscriptions, refineries, and fortifications have been found near Albuquerque. While these ancient Hebrew inscriptions may be authentic, it is unlikely that Solomon's ships would have sailed around Africa and up to the Gulf of Mexico on their way to such mines. A similar argument would made for Solomon's mine in Brazil; it is just impractical to sail from the Red Sea to North and South America, especially if one has a port city, as Solomon had, on the Mediterranean. It may well be that the Hebrews, sailing with their buddies, the Phoenicians, did sail up and down the Americas as well, but does not seem to be the location of Ophir.

There is one other location for Ophir that has been overlooked by everyone. It is Australia! Australia is one of the most mineral-rich countries in the world, and it would be reached by sailing via India and Sumatra. There have been discoveries of ancient mines in both Western and Northern Australia. At the town of Gympie in Queensland, an alleged pyramid was found, now destroyed, along with a three-foot-high statue of the Egyptian god Thoth in the form of a baboon, along with numerous Egyptian and Phoenician relics (see *Lost Cities of Ancient Lemuria & the Pacific* [35]). What is more, the same place became known as "The Town that Saved Queensland" because of a gold rush there in the late 1800's.

Personally, I think that Solomon's fleet could have reached Peru, but I doubt that it was actually Ophir. More likely, the fleet left Ezion-Geber, sailed through the Equatorial Channel of the Maldives and then on to the old Egyptian gold mines of Sumatra. Probably, they also visited ancient mining sites along the coast and rivers of Australia. On the return trip, after growing food for the return voyage, they would stop at ports in Indonesia and India and gather what peacocks, apes, ivory and sandlewood they had not already gotten nearby the mines that they worked. And so the El Dorado of the Bible came to finance the building of the temple.

THE WOODEN, GOLD-ENCRUSTED ARK OF THE COVENANT CARRIED THE STONE TABLETS OF THE TEN COMMANDMENTS. HELD SACRED BY THE HEBREWS, THE ARK WAS ONCE CAPTURED BY THE PHILISTINES. WHEN THEY REGAINED IT, THE HEBREWS SUPPOSEDLY PLACED IT IN KING SOLOMON'S TEMPLE. THE COPTIC RELIGION CLAIMS IT'S IN THEIR CATHEDRAL AT AKSUM, ETHIOPIA. MANY THEOLOGIANS BELIEVE THIS COULD ONLY BE A COPY, THAT THE ORIGINAL ARK WAS LOST CENTURIES AGO. BUT NO ONE IS SURE.

Miscellaneous Detail of King Solomon's Temple. Showing the Ark of the Covenant, the Table
Shewbread, the Golden Candlestick, the Altar of Incense, the Pillar of Brass, the Louver, the Molte
Sea, the Altar of Burnt Offering, the Cheribein (Cher-U-Bim), the Tree of Life, and the Key Stone.

The Tabernacle in the Wilderness as depicted by
the Bible Temple Publishing company. Copyright 1975.

Moses faces the Holy of Holies on top of the Ark of the Covenant
while inside the Tabernacle in the Wilderness. From an old print.

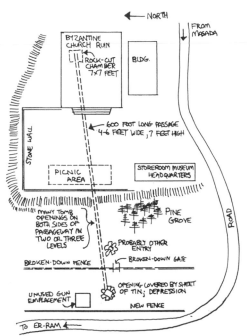

Map of the Byzantine Church and Monastery on Mount Pisgah and the tunnel Croster claims to have used to enter the secret cave inside the mountain where he claims he saw the Ark of the Covenant.

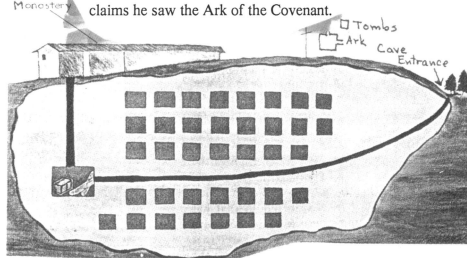

Croster's diagram of the tunnel he used to access the underground cave below the Catholic Monastery on Mount Pisgah, Jordan. Jordan cancelled all archaeological exploration in their country following the publicity of Croster's expedition.

Croster's "exclusive" photo of the Ark in a cave on Mount Pisgah in Jordan. It is almost certainly a replica.

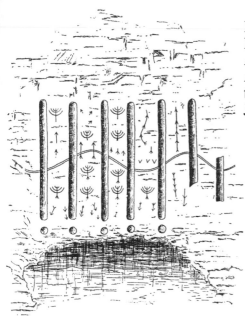

Left: Petroglyphs found at the Finke River in Australia in 1873. From *Picture Writing of the American Indian, Vol. 1* (Smithsonian Institution, 1893). They are identical to Jewish candlebra. Ophir?

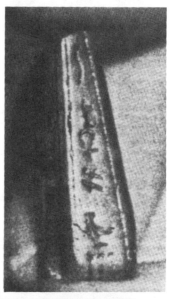

Right: An amber-glass obelisk-shaped pin found in a field at Kyogle, Australia in 1983. It is said to be 5,000 years old.

Two statues dug up near the Hawkesbury River, NSW. The man on the left is bearded. Both seem to be of ancient Middle Eastern manufacture, and are said to be of Phoenician origin.

General plan of the city of Megiddo (American Schools of Oriental Research, *The Biblical Archaeologist*, 33, 1970, p. 70)

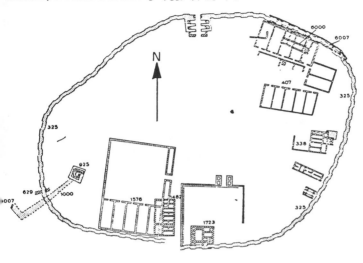

Reconstruction drawing of the Palace of Megiddo, Building 1723 (Oriental Institute of University of Chicago)

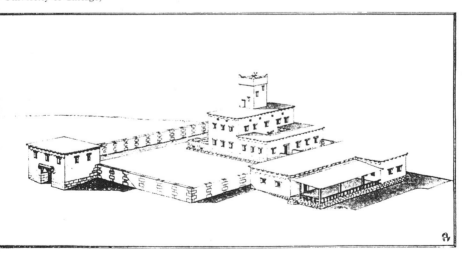

Lt. Lynch, USN, first circumnavigator of the Dead Sea, discovers 'Lot's Wife', the pillar of salt at Usdum, near the drowned cities of Sodom and Gomorrah.

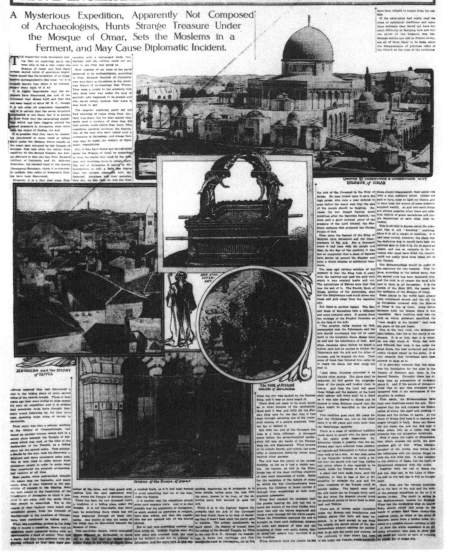

Magazine Section
Part Five

The New York Times.

SUNDAY, MAY 7, 1911.

Magazine Section
Part Five

HAVE ENGLISHMEN FOUND THE ARK OF THE COVENANT?

A Mysterious Expedition, Apparently Not Composed of Archaeologists, Hunts Strange Treasure Under the Mosque of Omar, Sets the Moslems in a Ferment, and May Cause Diplomatic Incident.

New York Times report on Captain Parker's expedition and his alleged discovery of the Ark of the Covenant, May 7, 1911.

The mysterious Sphinx, older than the pyramids.

Chapter 4

NORTHERN EGYPT:
RIDDLE OF THE SPHINX

If a man make a pilgrimage round Alexandria in the morning,
God will make for him a golden crown, set with pearls,
perfumed with musk and camphor,
and shining from the East to the West.
                                                    –Ibn Dukmak

Going from Israel to Egypt wasn't like crossing from one country to another.
It was like crossing from one continent to another, like crossing from one age to
another – from the modern Europe to Ancient Arabia! Israel is not really a
Middle Eastern country, it a European country transplanted into the Middle East
by European refugees after World War II. The British made every effort to keep
Jewish refugees from returning to Palestine. Among the many areas where the
British tried to transplant refugees were Vietnam, the Island of Mauritius, and
uninhabited parts of Africa.

Egypt is a land that is ancient, and remains that way, despite the incongruous
spectacles such as crowded rusty buses, air-conditioned Mercedes sedans and
camel drivers in sunglasses. Egypt is a land whose culture and history fade back
into the dim mists of antiquity. Egypt was ancient even to the ancients, who
viewed Egypt in much the same way as we view ancient Greece or Rome today.
Recorded history goes back more than five thousand years in Egypt, while the
great library in Alexandria, burned down by the invading Muslims more than a
thousand years ago, held records of an even more ancient Egypt and of
civilizations before it. It was at that library in Alexandria that Plato read about
Atlantis, and related to his fellow historians and philosophers Egypt's records of
our dim past.

In incredible splendor Egypt basked in the light of its ancient god, Aton, the
sun, for at least twenty-seven hundred years. Modern history has its beginnings in
Egypt. With the United States barely two hundred years old and already on the
brink of destroying itself and the rest of the world, I could hardly help but think
that there might be something to learn from the ancient Egyptians.

The glory of ancient Egypt was still before me as I headed for Alexandria by
bus. Simon was probably still tanning on the beach on Eilat while I watched the
desert outside the window of the bus. It is many hours across the desert to the Nile
Delta. We passed the Suez Canal, and headed toward the Mediterranean Coast. It is
interesting to note that the Egyptians had built a canal connecting the Mediterranean via the Nile with the Red Sea in 610 B.C., stretching from the Nile Delta
town of Bubastis near present day Zagazig to the Bitter Lakes. The Pharaoh
Necho made this his pet project even though there was a prophecy that the canal
would be of more use to invading barbarians than the Egyptians. After more than
100,000 workers died during the project, he abandoned the idea.

The project was taken up again by one Egypt's Persian rulers, Darius, a
century later (one of those foreign "barbarians"?). Cleopatra and her navy, after

their defeat by the Romans, attempted to escape to the Red Sea by way of this canal, but were unable to do so because of the low flood level of the Nile at the time.[50] What is even more amazing is that before this canal was constructed, ships were apparently moved across the isthmus to the Red Sea on a ship railway constructed of polished granite blocks as rails.[33] Todays innovations, are ancient history to the Egyptians!

After we stopped for lunch for half an hour at the canal, everyone piled back into the bus in the late afternoon sun for the last leg of the journey to Alexandria. Out the window the cotton fields, rice paddies and wheat fields of the lush Nile delta passed by in a green blur; donkeys and shirtless men were tilling the fields with water drawn off the Nile in a labyrinth of tiny canals. We passed Tanis, the city in *Raiders of the Lost Ark* where the Ark of the Covenant is discovered. Tanis is believed to be the Biblical city where the Hebrews were persecuted by the Egyptians before fleeing though the Red Sea (or Reed Sea, as the case may be) on their way to the Promised Land. For many centuries it was one of the most important cities on the Nile Delta, and some scholars think that it may also be ancient Zoan, where Mary and the baby Jesus fled to take refuge in Egypt with the Essene Brotherhood when King Herod was having all the male children slaughtered in Israel. Another location for Zoan, however, is the suburb of Cairo now known as Heliopolis (City of the Sun).

Tanis was probably used as the location of the Ark of the Covenant in *Raiders of the Lost Ark* because according to tradition, a sandstorm covered the city for 40 days, completely burying it, until it was lost to the world. This made it a perfect place for the Ark of the Covenant, at least for the movie.

It was just after evening when the bus pulled into the main square at Alexandria, the Midan Ramli. I swung my pack over my shoulder and looked around. I hardly had time to notice that one of the famous colonial hotels of Alexandria was right in front of me before I was attacked by a gaggle of taxi drivers.

"Where do you want to go?" asked one.

"Ten pounds," said another one, without even asking where I was going. I had paid less than that to come all day on the bus from the Sinai. I felt a bit nervous with all these Egyptians in their robes with cigarettes dangling from their mouths and their hands waving and grabbing me by the arm and pulling me here and pulling me there...

"Wait!" I suddenly shouted, pulling myself away from two or three of them. "I don't want a taxi! I'm staying here at the Cecil Hotel!" And with that, I marched across the street into the lobby, where a thin, uniformed man held the door open for me. I really wasn't going to stay at the Cecil, it was far too expensive for me. But I did want to use the phone, as I had the number and address of the sister of a friend, who lived in Alexandria with her Egyptian husband.

At the desk, they seemed somewhat relieved that I didn't want to check in, but only wanted to make a phone call. After phoning the surprised sister of my friend, I was told that I could stay at their home in an eastern suburb of the city for a few days. I was grateful to stay with Mary Anne, her husband Diaa and their two children. Mary Anne greeted me at gate of their large house, and I was immediately whisked inside to the dinning room where I had left over lentils and rice. The two children, aged nine and seven ran amok through the house until Diaa came home from the shipyards where he had a tanker of wheat coming from Toronto.

"The kids get restless in the house," said Mary Anne. "They don't have a large yard to play in like in Toronto."

Their father Diaa played with them for awhile, and then put them to bed. We talked for awhile, and then I made my way to their living room couch. Unlike the kids, I didn't need to told when I was tired and needed to go to sleep.

§§§

There is an island in the surging sea,
which they call Pharos, lying off Egypt.
It has a harbor with good anchorage,
and hence they put out to sea after drawing water.
–*Homer, The Odyssey, Book IV*

The next day I walked the streets of Alexandria, a stranger in a very strange land. Out on the street, I was faced with a sensory bombardment. The street was alive with smells, sounds, sights I had never seen before. Men walked along in long striped robes, which practically swept the street as they walked. The shops were small, with little bare light bulbs in the back, casting long shadows in the dusty, litter-strewn streets. I took a tram downtown to the harbor and stopped in a tea shop where old men were playing backgammon and smoking honey and tobacco through three foot tall water pipes that sat on the floor. I had a cup of tea and watched in utter fascination the people walking by in the street.

Alexandria is not really an Egyptian city, it is Greek. As one might easily guess, Alexandria is named after Alexander the Great, the Macedonian king who first conquered the city states of Greece and then set out to conquer the rest of the world, starting with Persia. Persia was also Egypt's traditional enemy, and so Egypt fell willingly into Alexander's hands. He went to Memphis near modern-day Cairo and then descended the Nile to the small Egyptian town of Rhakotis. Here he ordered his architects to build a great port city, what was to be Alexandria. Alexander then went to the temple of Ammon in the Siwa Oasis where he was hailed as the reincarnation of a god, which is to say, some great figure from ancient Osiris or Atlantis. Which god, we do not know. He hurried on to conquer the rest of Persia and then India. Eight years after leaving the sight of Alexandria, he returned to it; dead. He never saw the city, though his bones rest there to this day.

The first thing that I wanted to check out in Alexandria was the famous light house. One of the Seven Wonders of the World, the Pharos Light House, built on the island overlooking the harbor at Alexandria, is truly a wonder of technology and architecture. It was designed by an Asiatic Greek named Sostratus and probably dedicated in 279 B.C. It stood in a colonaded court and had four stories. The square bottom story was pierced with many windows and contained the rooms, estimated at 300, where the mechanics and attendants were housed. There was a spiral ascent-probably a double spiral-and in the center there was hydraulic machinery for raising fuel to the top.[51]

The second story was octagonal and entirely filled by the spiral ascent. Above that was the circular third story, and above that the lantern. There was a mysterious mirror at the top that was apparently a gigantic telescope lens which could be used to view ships from a distance, and as a magnifying glass to direct intensified sunlight that could actually light ships on fire. There is a legend that the Byzantine emperor wanted to attack Alexandria after the Arab conquest in 641 A.D., but was afraid because of the "magic mirror" that would detect and burn his ships. Therefore, he sent an agent who gained the Caliph's confidence and told him that beneath the Pharos the treasure of Alexander the Great lay buried. The Caliph commenced demolition, and before the inhabitants of the Alexandria, who knew better, could intervene, the two upper stories, including the magic mirror, had fallen into the sea. Henceforth the lighthouse was only a stump of its former self with a bonfire at the top.[51] Even less remains today.

As I walked along the harbor, I smiled, letting the warm sun hit my face. Fathers passed with their young sons along the retaining wall of the harbor. Egyptian sailors walked hand in hand down the walk way; a peculiarly Egyptian sight. Alexandrian women, their heads covered with scarves that blew in the wind, ignored the steady gazes of young Egyptian men loitering on the waterfront.

I walked past the aquarium to the Arab fort that is all that remains of the base of the lighthouse, and marveled at what it must have been like in those ancient days. I paid one Egyptian pound to go inside, and spent an hour walking around the outer walls of the fort and inside what little is left of the base. Probably everything that can be seen today at the lighthouse is of Arab construction, though E. M. Forster believes that parts of the outer base may be from the original lighthouse.

Of all the mysteries of Alexandria, however, none are more intriguing than the prehistoric harbor which lies to the west of the Pharos lighthouse near the promontory of Ras El Tin. Discovered at the turn of the century by the French archaeologist M. Jondet and discussed in his paper *Les Ports submerges de l'ancienne Isle de Pharos*, the prehistoric port is an ancient, massive line of stones that today is completely submerged. Except for the brief mention of an ancient harbor by Homer in the *Odyssey*, there is no mention of it at all in the historical records that have survived.

The Theosophical Society, upon learning of the submerged harbor of megaliths, quickly ascribed it to Atlantis. M. Jondet theorized that it might be of Minoan origin, part of a port for the Cretan ships. It may also be of ancient Egyptian origin, Forster theorizes in his excellent guide to Alexandria, built by Ramses II circa 1300 B.C. Most of it lies in 4 to 25 feet of water and stretches for 70 yards from east to west, curving slightly to the south. Near it was the legendary Temple of Poseidon.[51]

Probably the origin of the massive, submerged harbor, which was definitely at least partially above water at one time, is a blend of M. Jondet's theory of Minoan builders and the Theosophical Society's belief that it is from Atlantis. Once more we are back to the little-talked-about civilization of Osiris that existed in the Mediterranean when it was a fertile valley before the sinking of Atlantis. Both Crete and Egypt, as well as Malta and other areas of the Mediterranean with prehistoric megaliths, are theoretically part of this civilization. Yet, there would not be a port where the prehistoric harbor stands now, if the Mediterranean was a fertile valley.

Therefore, in that theory, it must be post-cataclysmic. With the Mediterranean slowly filling up with water, it must have stabilized after a few hundred years, and then the remnants of the Osirians, using a technology and science inherited from Atlantis, built what structures and ports they could. Later, in another tectonic shift, the port area, probably used by what we would call "pre-dynastic Egyptians" like those who built Abydos and the Temple of the Sphinx, was submerged, and was then generally useless. It is interesting to note, in accord with this theory, that a Temple to Poseidon was located at the tip of Ras El Tin. Atlantis was known to the ancients as Poseid, and Poseidonis or Poseidon, was a legendary king of Atlantis. Similarly, Poseidonis and Osiris are thought to be the same person. The main temple at Rhakotis, the Egyptian town which Alexander found at the ancient harbor, was naturally dedicated to Osiris.

In the book, *The Alexandria Project*,[52] the author Stephan Schwartz describes several scuba dives about the eastern harbor. In the last dive, they discover underwater huge cut-stone blocks, including what they believed to be "The Crown of Osiris", part of a statue of Osiris that was huge and about 20 feet underwater. They also discovered huge pedestals, presumably for statues, cut out of massive granite blocks with sockets in them. Says Schwartz in his book, which is largely about experimental psychic archaeology, "Like everything else in the Pharos area, this pedestal is on a very large scale."[52]

I walked down to the Ras El Tin area from the lighthouse, to see if I could walk out to where the prehistoric harbor was. Following an Egyptian with a wheelbarrow full of bricks, I headed past a construction site towards the tip of the promontory. Suddenly a bayonet was pointed at me, and a uniformed soldier told me in Arabic to halt.

With a bayonet directed at my gut, this seemed like a good time to practice my Arabic, which I had been slowly picking up ever since I arrived in Syria. "Can I go any further?" I questioned in halting, incorrect Arabic.

"No. This is off limits. Military Zone," he stated. Later, Mary Anne told me that there is a Military Hospital and a palace at the tip of Ras El Tin, and foreigners (or Egyptians without passes for that matter) are not allowed into the area. Well, that about sums it up for explorers wanting to check out the submerged prehistoric harbor at Alexandria: it is impossible. This is now a military zone, and not an area for maverick archaeologists.

I had dinner that evening with Mary Anne, Diaa, and their kids, telling them about my adventures that day. The next day, I visited the few actual Egyptian remains to be seen at Alexandria: "Pompey's Pillar" at the citadel of Rhakotis. It was here that one of the great libraries of Alexandria was kept. It seems that there were two; one was at the Serapeum (Osiris Temple) at the citadel of Rhakotis, and the other was in the Greek part of town, supposedly where the modernday library stands today.

From almost the founding of the city to 391 A.D., Alexandria was the most learned spot on earth. Then the split in the Christian church occurred. The original Christian church maintained that Christ was the Archangel Melchizedek, and as an archangelic being, had merely "borrowed" the body of Jesus. This original church, headquartered at Antioch and then Constantinople in present-day Turkey, taught reincarnation, karma, that Jesus was the son of Joseph and Mary, and that Jesus and Christ were two separate personalities. That Jesus and Christ were two separate personalities, plus the doctrine of reincarnation and virgin birth, were eventually debated hotly in Alexandria, especially by the Alexandrian Patriarch Theophilus who opposed the view that Christ and Jesus were two separate personalities. This view of oneness between Christ and Jesus became known as Monophysitism, and during the councils of Nicea (325 A.D.) and Constantinople (381 A.D.) it won the day, resulting in what is known today as the Nicene Creed which proclaimed the doctrine that is carried on today by the Roman Catholic Church. Nestorus, then Patriarch of the original church, was forced into exile in Baghdad, where his doctrines today are known as Nestorian Christianity. Gnostic Christianity and Coptic Christianity also stem from the original church, upheld by Nestorus.

At any rate, during one of the heated debates that raged in Alexandria, even after the Nicene Creed, the Patriarch Theophilus led an angry mob against the library (or both libraries, shall we say) and destroyed the books, in the name of the "one and only true church". What may have been left of the many thousands of books and scrolls in the libraries were burned by the Moslems a few hundred years later, though there was not likely much of importance left. Coptic Christianity, as opposed to Roman Catholic Orthodoxy centered in Constantinople, then became an important off-shoot from the Church, and is still the main Christian faith in Egypt today, much to the chagrin of the Patriarch Theophilus, who was unable to accomplish in Egypt what he did for the northern part of the Mediterranean. Coptic Christians believe in reincarnation, karma, and Christ as an Archangelic being, just as Nestorus taught.

I paid fifty piastres, and visited what is left of the citadel of Rhakotis, now known as "Pompey's Pillar". The tall pillar, standing 84 feet high on the hill, and made of Aswan granite, has nothing to do with Pompey. Forster believes it was given this name by the Crusaders, but was from some other building, long since destroyed. It was erected about 297 A.D. by the Roman Emperor Diocletian.

It is impressive to look at, and after I took a few obligatory photos, a guard opened an iron gate for me, and I walked down a flight of stairs into a series of long subterranean galleries, excavated in the rock and lined with limestone. Some believe that these underground galleries were part of the library, but their actual purpose is unknown.

My final day in Alexandria, I visited the museum, which is largely of Greek artifacts. There was an interesting room with relics from the crocodile cult of Sobek, an Egyptian deity depicted as a crocodile. There were mummified crocodiles and statues of men with crocodile heads, not the sort of gods most westerners could get into worshiping.

Also interesting were some mummies from the Christian era. There were several of Christians wrapped in cloth with a crosses on their necks. A sign on one said that "...Embalming was disapproved of by Christians, but there were some exceptions." Why did early Christians disapprove of embalming? The reason, to be explored in more depth in the next chapter, is because of the Coptic Christian belief in reincarnation and the belief that mummification prohibits an ego from reincarnating, so long as the previous body is still preserved.

In this light, the ancient Egyptian practice of mummification was wrong, and by some accounts, diabolically evil. It was this controversy in ancient Egypt that led to the so-called heresy of Atonism by Pharaoh Akhenaton and his Queen Nefertiti, though most Egyptologists, in their infinite wisdom, do not understand this. It is why no mummies have ever been found of any true Atonist. Early Christians (at least in Egypt), like Atonists, believed in cremation, like Hindus or Buddhists, and embalming was merely just another form, though more primitive and less expensive, of mummification.

Following this line of thought, mummification continues to this very day in the form of embalming the dead. No discussion of the mysteries of Egypt is complete without coming to grips with the cult of mummification as seen in the light of the philosophy and religion of the times. It is a fascinating subject, and it plays a greater part in the history of Egypt than usually admitted. Further discussion must wait for a discussion of Atonism itself in the next chapter.

§§§

There are many wonders in this world;
but none are greater than man!
–*Socrates*

After three days of wandering about Alexandria, I was anxious to get to ancient Egypt, the part of Egypt that was ancient to the Greeks. After thanking Mary Anne and Diaa for their wonderful hospitality, I caught a bus up to Cairo, a four-hour ride in a fairly new Mercedes bus packed end to end with striped-robed Egyptians and businessmen with brief cases going to Cairo to make a deal. I sat next to an old man who gazed dreamily out the window, spitting occasionally on the floor and smiling at me from time to time. We took the desert highway, heading south on the edge of the sand dunes of the western desert, with the Nile Delta to our east. We passed through the suburb of Giza on our way, and the pyramids, Egypt's main tourist attraction, loomed as gigantic man-made mountains along the side of the road. I gazed in awe from the window, but in moments we were speeding into downtown Cairo.

In Cairo the bus let us all off at Tahir Square, which can be considered the center of Cairo, Africa's largest metropolis by far, at nearly ten million. Cairo can be rather overwhelming, I found out, as I stepped off the bus into a huge crowd of people and was swept away for nearly a block with the surge of people along the street. It turned out to be rush hour. Totally confused and frustrated, I made an effort to grab a taxi. This proved to be impossible, as every cab, and there were quite a few, was packed full, with several people in each one.

Fortunately, it turned out that I didn't need a cab. The cheap hotels for travelers were just down the street: Tahir Street, in fact, an easy walking distance from the square. I walked several blocks down to Al Azhar Square where I took a room at The Golden Hotel for two dollars and fifty cents a night and got a tiny single on the third story. It had its own bathroom and shower, stuffed into what should have been a closet. It was an old run-down hotel whose better days were ancient history. I checked the view from my room; I had a great view of a brick alley about ten feet away, but at that price and with a bath, I couldn't complain. The mattress was stained and lumpy, but I didn't care. I washed the dust of the Libyan desert off in the shower, which didn't have a spray nozzle on it anymore. The lukewarm water just came out of a pipe, in a steady stream at least. The toilet, a hole in the floor, was flushed by a bucket of water that sat beneath a faucet. In an Egyptian toilet, you had to bring your own toilet paper, or use what the Egyptians, and other Africans and Arabians use: your left hand and water.

In Africa and Arabia, as well as most of Asia, people cleanse themselves with their left hand after the toilet, and eat and pass food with their right. It is important to remember while traveling in the Muslim areas of Africa and Arabia not to pass food or eat with your left hand, as this is extremely impolite, and could be taken as a dire insult by someone who is not aware of western ignorance of this basic courtesy.

Mohammed once said, "The Europeans are savages; do they not wipe their ass and eat with the same hand?" He had a point back then, when Europeans were savage compared to the relatively enlightened Islamic Kingdoms. Europe was going through the Dark Ages, and science, literature and the arts were being preserved in the Middle East and North Africa.

I was soon off wandering the markets and side streets of Cairo, gazing into little shops filled with all kinds of brassware or leather stools, eating in small pastry shops with flaky honey-gooey sweets covered with sesame seeds, and then drinking tea afterward. I visited the Mohammed Ali Mosque at the Citadel, in central Cairo. Built like many of the mosques in Cairo, almost entirely out of the limestone casing blocks from the Great Pyramid, it was the scene of one of the most foul ambushes of history.

Mohammed Ali, the Greek-born adventurer who governed Egypt for the Turks, had originally subdued the country with the help of the Mameluke Beys, converted Christian slaves of the Turks who were trained to police, tax and control Egypt under the nominal suzerainty of the Ottoman Sultan. When the Mamelukes themselves posed a threat to the power of Mohammed Ali, he concocted a fiendish plan for their total demise.

In 1811, the Mameluke Beys and their horsemen were invited to a great feast at the citadel, and when the entire troop was crowded into the narrow streets the Albanian mercenaries of Mohammed Ali opened fire to wipe out the Mameluke Beys once and for all. The streets of the citadel flowed with blood, and the only man to survive was the leader of Mameluke Beys was the Amir-Bey who whose horse was reputed to have jumped from the battlements, falling several hundred feet, and then carrying its master to safety in Syria.

The next day I visited the Egyptian Antiquities Museum and walked about aimlessly for six hours gazing at myriads of artifacts, gold-encased coffins, statues, papyrus texts and wooden palladiums. There is so much stuff in the museum that the back half isn't even cataloged. I found myself wandering about in the back rooms among piles of statues; Horus-headed falcons staring moodily at the ceiling, glassy-eyed Pharaohs looking down upon the tourists. I was lost for some time among these relics of times gone by. What an incredibly grandiose civilization! They left a lot for posterity, and here it was, five thousand years later, not just junk, but fantastic, beautiful stuff. In five thousand years, if modern civilization was suddenly wiped out, what would we have to show? Not a whole lot. It is inconceivable that any buildings would remain standing; they're routinely condemned every hundred years or so, not made to last for thousands of years as Egyptian buildings were. Even the Golden Gate Bridge would corrode and collapse within a thousand years, if not sooner. Could it have been that the ancient Egyptians were of a civilization more advanced than ours? They were certainly more advanced than we were a mere three hundred years ago. Like the mysterious "Atlantean League" that built huge megalithic cities, pyramids and ports around the world, the Egyptians were what is generally known as "Builders"; that is, they belonged to an ancient group who made buildings of gigantic size, and they built them with the idea of these structures lasting many thousands of years, even through devastating earth changes.

With my keen eye, I looked through the exhibit of Tutankhamen, the most popular in the museum. Tut's solid gold mask, and the exquisite jewelry are on display and worth seeing again and again. What caught my eye, that no one else seemed to notice, was an unusual display of boomerangs taken from his tomb. There on the wall were two large glass cases of finely carved and inlaid boomerangs. Not just three or four, but at least thirty of them! There was no mistaking them as boomerangs either. There was even an Australian aboriginal boomerang between the two cases for comparison. The young pharaoh Tutankhamen apparently has more than a passing interest in boomerangs, he was a boomerang nut!

It is a little-known fact that the Egyptians enjoyed hunting with boomerangs. They are an ingenious device: what other weapon comes back after being thrown at an animal or enemy? Again, I am suddenly reminded of Egyptian gold mines in Australia and the lost land of Ophir—are they one and the same? Did boomerangs go from Egypt to Australia, or from Australia to Egypt? Or did they develop independently?

It is also interesting to note that a district in Egypt was called a *Nome*, and the governor of a *Nome* was called a *Nome-march*. Interestingly, at the remote and mysterious Pacific island of Pohnpei where the megalithic crystal city of Nan Modal can be found, the chief of the different districts of the island are called *Nanmarches*. Was it the Egyptians who originally settled the island? Kahuna tradition in Hawaii also claims that the Hawaiians are originally from Egypt (for more information on the Egyptians in the Pacific, see *Lost Cities of Ancient Lemuria and the Pacific*[35]).

Further evidence of Egyptian sojourns around the world can be seen in the mummy room. While it has been illegal to show mummies in Egypt for some years, because the late Egyptian president Anwar Sadat felt that it was putting their ancestors on display (a few Greco-Egyptian mummies can be seen in Alexandria, for some reason), there are some mummies of animals in the mummy room. There is a small display in this room of certain grains and dehydrated plants that have been found inside tombs and pyramids. One of the grains is corn (maize) which was only found in North and South America (at that time). The card for the display notes that corn (maize) has been found inside tombs, though, curiously, it points out that claims that the seeds have been germinated successfully are unfounded. While this may be an attempted denial of "pyramid power", the point of where this corn came from in the first place is totally missed. Obviously, as Barry Fell,[20,21] Orville Hope[45] and Thor Heyerdahl[19,49] claim, the Egyptians made frequent trips to the Americas, probably using Libyan sailors who were allied with the Phoenicians & Carthaginians.

§§§

With the help of a caretaker in a faded gray uniform and dusty, thick spectacles, I found my way back into the sunlight and the twentieth century again on Tahir Square. After a 10-piastre sandwich of fried chickpea (*falafel*) with some salad in it, the fast food staple of Egypt, I grabbed a bus out to Giza and the Pyramids.

Buses in Egypt are quite incredible. Old and run-down, they are packed from end to end with Egyptians and the occasional tourist. Sitting next to me with his head in my armpit was another traveler, a young white male with curly brown hair. It turned out he was from Scotland.

"Cheers, mate," he said, looking up from his cramped position on the edge of a seat. His nose was right in my armpit, my arm reaching up to grab the luggage rack above his head.

"Howdy," I smiled. "I just had a shower this morning, in case you couldn't tell."

"No problem," he said. "I've been in Egypt for a week now. I'm used to this."

"The buses are pretty crowded."

"You can say that again. The bus driver doesn't even stop at the bus stations because he can't pick up any more passengers."

"That can be a problem if you're trying to get off the bus, I suppose," I said.

"You can say that again," he said. "I've just come here from India, and I can't believe how much Egypt reminds me of the sub-continent; crowded, poor, dirty, old, but incredibly interesting!"

We ended up at the last stop: the Pyramids at Giza, a western suburb of Cairo, a forty-five minute bus ride from downtown. The Great Pyramid, as tall as a forty-story building, towered above us out of the sand and into the sky like some alien structure. Behind it was the pyramid of Chephren. Although smaller than the Great Pyramid and built afterwards, it looks taller than the Great Pyramid in many photos because it stands on a small plateau or rise. To the southeast is the Sphinx, small by comparison, staring silently out toward the Nile in the east.

The Scottish fellow's name was Derek. Together we walked past a couple of camels and up to the Great Pyramid. I stood there for a moment in awe. It was huge, a literal man-made mountain. "A lot of effort just for some dumb tomb," commented Derek.

Actually, though it is usually said that the pyramids were tombs for the pharaohs, the evidence is generally against this theory. As the normally traditional Archaeologist Kurt Mendelssohn says in his book *The Riddle of the Pyramids*:[58] "While the funerary function of the pyramids cannot be doubted, it is rather more difficult to prove that the pharaohs were ever buried inside them...

"Leaving out Zoser's Step Pyramid, with its unique burial chambers, the nine remaining pyramids contain no more than three authentic sarcophagi. These are distributed over no fewer than fourteen tomb chambers. Petrie has shown that the lidless sarcophagus in the Khufu (Cheops) Pyramid had been put into the King's Chamber before the latter was roofed over since it is too large to pass through the entrance passage.... One would like to know what has happened to the missing sarcophagi. The robbers might have smashed their lids but they would hardly have taken the trouble to steal a smashed sarcophagus. In spite of careful search no chips of broken sarcophagi have been found in any of the pyramid passages or chambers. Moreover, it has to be remembered that from the Meidum pyramid onward the entrance was well above ground level. At the Bent Pyramid even the

97

lower corridor is located 12 meters above the base and bringing a heavy sarcophagus in or out would have necessitated the use of a substantial ramp....

"The fact that sarcophagi in the Khufu and Khafre (Chephren) pyramids were found empty is easily explained as the work of intruders, but the empty sarcophagi of Sekhemket, Queen Hetepheres and a third one in shaft under the Step Pyramid, pose a different problem. They were all left undisturbed since early antiquity. As these were burials without a corpse, we are almost driven to the conclusion that something other than a human body may have been ritually entombed.

"We have already referred to the fact that Snefru seems to have had two, or even three, large pyramids, and he can hardly have been buried in all of them....

"While very few people will dispute that the pyramids had some connection with the afterlife of the pharaoh, the general statement that the pharaohs were buried in them is by no means undisputable...Quite possibly each pyramid once housed the body of pharaoh, but there also exists...an unpleasantly large number of factors that speak against it. It is on the basis of these complexities and contradictions that Egyptologists had to try and find a solution to the most difficult problem of all: why were these immense pyramids built in the first place?"[58,53]

"Well, if the pyramids weren't tombs," asked Derek, "What were they?"

I ran down the possibilities with Derek. "Well, there is the theory that they were astronomical observatories. Another idea is that the pyramids, especially the Great Pyramid, were geodetic markers and time capsules in the sense that higher knowledge, such as sophisticated geometry and mathematics were incorporated into the structures. Others have claimed that the pyramids were centers of initiation. It has also been suggested that a secret library from Atlantis or something is hidden somewhere inside, underneath or nearby the pyramids of the Giza. This library is generally called the Ancient Hall of Records. Then, there is belief by some that the Ark of the Covenant was contained inside the Great Pyramid for some time, and then taken by Moses when the Israelites split for the Promised Land. There is even the belief that the pyramids were an energy device for tapping the energy of the Van Allen Belt, where the bulk of the pyramid was a protective baffle, like insulation around an electric cable. Then, of course, there is always pyramid energy."

"Jeeze," said Derek. "Which theory do you believe?"

"Well, I don't really know," I confessed. "But it seems to me that the pyramids could well be a combination of all these things. They are wonderful embodiments of higher mathematics and learning, plus they could easily have been used many purposes, including mystical initiations and as a repository for ancient records or artifacts."

We each bought a ticket to go inside, for 60 cents, and climbed the stairs that led to the entrance of the building. The Great Pyramid: first wonder of the ancient world and the only one left! For thousands of years the pyramids had been attracting travelers and hitchhikers from all over the world. Some Great Pyramid enthusiasts even brought their armies and took over the country so that they could visit the pyramid. Now that's quite a tourist attraction!

On entering the pyramid, we first descended a narrow passage. Presently we came to a split where there was a narrow passage leading upward. This passage was originally blocked by a huge granite plug. The first Arabs to dig into the pyramid, led by the Caliph Al Mamun, dug around this granite plug to get to the passageways beyond.

We ascended the low passage, steep and dark, until it opened into the Grand Gallery a hundred yards or so up the passage. The Grand Gallery was forty feet high at least, and since we had to crouch in the passageways, at first it seemed even higher. Back into the passage for another few hundred yards, and then, with one last, great step, we emerged into the King's Chamber, located about in the center of the structure. Pyramid, in fact, is a Greek word, and means "fire in the center."

The King's Chamber was entirely empty except for a lidless granite tomb at the northern end of the room, large enough for twenty-five or thirty people to crowd into. The walls were completely bare of hieroglyphics or art of any form, nor was there any ornamentation of any kind on the sarcophagus. In fact, it is rather strange that there was not even a lid to it. The room is just as the Arabs found it when they entered centuries ago.

"Is this it?" said Derek, echoing what the Arabs must have said when they first entered the pyramid. "Let's go."

Derek left, but I stayed for a while, the only person in the room. The tomb was just big enough for a normal-sized person to lie in comfortably. On an impulse, I got into the tomb and lay down. According to esoteric legends, initiates of the Egyptian Mystery Schools would spend the night in the King's Chamber, sleeping in the tomb. They supposedly gained entrance by still-undiscovered secret passages in the pyramid. Looking up at the ceiling, I was filled with awe, not just for the pyramid and its builders, but for life in general. It was a great feeling lying there; I felt very much alive, yet contemplative on the mysteries of life, particularly mine. What was my purpose, I asked myself?

Some people entered the room, and I listened to the odd echoes that their voices made as I lay in the tomb. I got out, sat quietly in a corner for awhile, and then returned back down the passages to the exterior. Derek was waiting for me.

"You were in there a long time!" he said. "I've waited out here for forty-five minutes!"

"Forty-five minutes!" I answered back. "Wow, it hardly seemed like any time at all to me." Time had flown.

"How do you suppose this pyramid was constructed?" asked Derek. "Now I suppose you're going to tell me that they didn't use ramps at all, but levitated the stones!"

"Well, levitation of the stones was hardly necessary," I replied. "Herodotus claims that he was told by priests two thousand years or more after their construction that teams of ten thousand men each labored for ten years to build a ramp for the blocks, then another twenty years to build the pyramid, and finally a further ten years to put the casing stones on the pyramid starting from the top down. Herodotus claimed that Cheops financed the construction by having his own daughter work as a prostitute. An inscription read to Herodotus at the base of the pyramid by priests told of the number of onions and radishes it took to feed the laborers.

"Yet, it appears that Herodotus was being told a story. No trace of a ramp has actually been discovered. Most scholars believe that the ramp Herodotus refers to is the causeway leading up from the Nile, past the Sphinx. All the pyramids had this causeway, and it had nothing to do with their construction. There are no actual wall drawings of a pyramid being constructed, but there are drawings the portray the transport of gigantic obelisks and giant statues weighing more than a hundred tons by men pulling sleds.[53]

"According to John Anthony West, even though it was possible to muster up enough labor to build the pyramid over time, some sort of lifting device needed to be employed, which so far no one has come up with. Other engineers claim that no lifting device was needed and that the ramp merely had to reach the top of the pyramid. However a Danish engineer named P. Garde-Hanson has calculated that such a ramp would require seventeen and a half million cubic yards of material, seven times more than that used in the building of the pyramid itself! Garde-Hanson believes that a ramp going half way up the pyramid would be better, but it is still necessary to use a lifting device, which brings us back to the old problem.[53]

"The placing of the limestone casing stones, which weigh up to 10 tons or more, is an even bigger problem, as they are cut and fitted with such precision. Cheops did not even sign his own pyramid—the only marks in the pyramid are quarry marks on granite blocks on the inside of the pyramid. These were only discovered with the tearing apart of the pyramid, and were never meant to be seen. A possible other method is the ingenious theory that the pyramid itself was a hydraulic pump, and the blocks were floated into place on rafts.

"Another theory that has a certain attraction to mystics is that the blocks were levitated, using what the Egyptians called Ma-at, a force of mind power akin to the Sanskrit Mana power of the mind."

"I knew you'd get back to the levitation stuff," said Derek. "Do you really believe that these blocks were levitated?"

"Well," I confessed, "I think that the levitation of stones is quite possible, and probably even used in ancient times. However, I think that the construction of the pyramids was done in another fashion which I have not yet mentioned."

Derek sat down on one of the five ton blocks exposed around the base of the pyramid. "So what is that theory?" he said, rolling his eyes.

I then told Derek the curious theory of the pyramids by an authority on ancient construction techniques named Dr. Joseph Davidovits. Davidovits has been saying over the last few years that the Great Pyramid of Egypt, as well as other pyramids in Egypt, was not constructed out of cut stone as has always been assumed. Davidovits believes that the large blocks were actually poured into place, and that they are an advanced and ingenious form of synthetic stone that was cast on the spot like concrete.

Davidovits reported on his research at a meeting of the American Chemical Society. He is the founder and director of the *Institute for Applied Archaeological Sciences* located near Miami. He claims that a new deciphering of an ancient hieroglyphic text has provided some direct information about pyramid construction and that it supports his theory that synthetic stone was the construction material.

The text, called the "Famine Stele," was discovered 100 years ago on an island near Elephantine, Egypt. It consists of 2,600 hieroglyphs, about 650 of which have been interpreted as dealing with stone-fabrication techniques. The text claims that an Egyptian god gave the instructions for making synthetic stone to Pharaoh Zoser, who is said to have built the first pyramid in 2750 B.C.

Included were a list of 29 minerals that could be processed with crushed limestone and other natural aggregates into a synthetic stone for use in the building of temples and pyramids. Like the chemists of the 17th and 18th centuries, the Egyptians named these minerals according to their physical properties. The materials were called "onion ore," "garlic ore" and "horseradish ore" because of their distinctive smells.

Davidovits believes the minerals in the ores contained arsenic. Other ingredients for making synthetic stone—phosphates from bones or dung, Nile silt, limestone and quartz—were also readily available.

According to the theory, the ingredients were mixed along with water and placed into wooden forms similar to those used for concrete. Davidovits said the cement used in the pyramid stone binds the aggregate and other ingredients together chemically in a process similar to that involved in the formation of natural stone.

Portland cement, by contrast, involves mechanical rather than molecular bonding of its ingredients. Thus, pyramid stone is extremely difficult to distinguish from natural stone. He also maintains that this "Egyptian cement" would last for thousands of years, while ordinary cement has an average life span of only 150 years. Organic fibers, having accidentally fallen into the mixture, have been found in the stone blocks of the Great Pyramid.

"Now, I like that theory," said Derek. "Hey, maybe the onions and radishes that Herodotus mentioned were the onion ore concrete and stuff?"

"That's an idea," I replied. "Here look at these stones." We were sitting on the pyramid itself, and the stones we were speaking about were all around us. On careful examination, they were as Davidovits said: conglomerates of shells, sand, coral, and all kinds of stuff, particularly shells.

"Wow," said Derek! "It's just like that guy Davidovits said! I'll bet that this stuff is cast like concrete!"

"And who was the "god" that gave the formula to the Pharaoh Zoser? Perhaps some Atlantean refugee or even an emissary from one of the secret brotherhoods, landing in a vimana for lunch and a chat with the Pharaoh? One thing to keep in mind is that the Egyptian civilization lasted for many thousands of years, and it waxed and waned as a culture during that time, just as western culture has for the last two thousand years or so." Derek and I laughed and headed down towards the Sphinx.

The Sphinx is one of the three most controversial structures in Egypt, along with the Great Pyramid and the Osirion at Abydos. Carved out of solid rock, the Sphinx seems to typify the mystery of Egypt as it gazes silently out away from the pyramids. The age of the Sphinx is a matter of great debate. The body is severely eroded, though the Egyptian government is now reconstructing it.

What could have caused this severe eroding? The controversial German Egyptologist Schwaller de Lubicz observed that the the dramatically severe erosion on the body could not be the result of wind and sand, as universally assumed, but was rather the result of water. Geologists agree that in the not so distant past Egypt was subjected to severe flooding. This period is usually held to coincide with the melting of the ice from the last Ice Age, circa 15,000 to 10,000 B.C.![53]

This would indicate that the Sphinx was first constructed about this time, and would make it easily the oldest structure in Egypt, long before the recognized Egyptian civilization began. Suddenly, we are back to the tales of the ancient Osirian Empire, Atlantis, and the cataclysmic pole shifts that have rocked our planet every ten thousand years or so.

The Sphinx is often said to be the likeness of the Pharaoh Chephren, of whom several statues, one in the form of a sphinx, were found up-side-down in the temple next to the sphinx. The sphinx of Chephren seemed to have a similar face as the Great Sphinx, and so the construction was attributed to Chephren, even though a stele found on Giza, called the Inventory Stele, says that Cheops built the Great Pyramid next to the Sphinx, which was before Chephren's time. Egyptologists who favor the Chephren theory, brand this ancient Egyptian stele a forgery.

The Valley Temple of Chephren next to the Sphinx is also an unusual structure. It is constructed of huge granite and limestone blocks weighing up to a 100 tons a piece. No inscriptions of any kind are to be found in the temple, and the blocks are perfectly fitted together in a curious jigsaw pattern that interlocks the blocks. This is a trademark of "The Builders", a type of megalithic construction that is not only extremely difficult to make, it is also difficult to tear down. Blocks that are interlocked in such a fashion are earthquake proof, and shift as a mass. Because they are interlocked, they cannot sheer like bricks or square blocks of masonry.

It is especially interesting to compare the construction technique found at the Valley Temple of Chephren and that found at the massive cities in the high Andes around Cuzco, Peru. Here at Cuzco, Sacsayhuaman, Ollantaytambo and even

Machu Picchu, the same type of construction is used. Typically, this kind of building is called Atlantean Construction.

There are also said to be secret passages beneath the Giza plateau. These passages go to the pyramids, allegedly starting from the Sphinx, and are part of the ancient Mystery Schools of Egypt. A strange shaft in the Giza plateau between the Pyramid of Chephren and the Sphinx is known as Campbell's Tomb or Campbell's Well. This shaft is now blocked by a grate, but one can still look down it. The shaft is about 15 feet square on each side and about a 100 feet deep. On each side of the walls, one can see numerous tunnels, passages and doors cut into the solid rock. These passageways are part of the tunnel system that goes beneath the Giza plateau. It is purposely dangerous to attempt to reach the pyramids or the secret underground rooms that can be found in the tunnels. Their existence, and what lies in them, is a matter of legend and prophesy.

What could lie in these Halls of Record? According to some, the Halls of Record are libraries of knowledge from Atlantis preserved in the form of encoded quartz crystals, much like a hologram might be encoded by a laser. Also in these secret chambers, sealed from the rest of mankind during the dark age of Egyptian History when the evil priesthoods sought to control mankind, are supposedly machines and devices from that forgotten age.

According to the rare book *Migdar – The Secret of the Sphinx*,[56] there is an underground city beneath the Giza plateau, with an entrance somewhere near the Sphinx. Tunnels do indeed exist beneath the Giza plateau; Campbell's Well is evidence of that. Where they go, and how the secret chambers may be opened is a matter for time to tell. Yet, at this very moment, teams of explorers are searching for these hidden Halls of Records.

> Man fears time.
> But time fears the Sphinx.
> *–Arab proverb*

§§§

Derek had to get back to the Youth Hostel, so he left. It was getting late anyway; it would be sunset soon. I decided to hang around for awhile, and maybe catch the Sound and Light Show that was later in the evening. Killing time, I walked back around the Great Pyramid, and while on the far western side, decided to climb to the top. Most of the tourists were gone, and I didn't see anyone, so I tried it even though it is against the rules to climb on the structures.

Nearly an hour later, after plenty of huffing and puffing and dragging myself up over countless three- and four-feet-high blocks, I was sitting on the flat top of the Great Pyramid. To the east, I had a great view of Cairo and the Nile. To the west, the sun was just setting in the desolate western desert, an area of sand and a few oases.

I watched the sun silently while sitting cross-legged on the granite stones, when I suddenly heard some heavy breathing to my left. I figured it was some tourist policeman, come to arrest me, and started to think of a good story to tell him. With a great gasp of breath, followed by much panting, a figure hauled himself up into the top with both hands, and a knee over the edge. Crawling on his hands and knees, he came over to where I was sitting crosslegged on the western edge.

He wasn't an Egyptian. A trim man in his fifties, I guessed, he had short brown hair and brown plastic-rimmed spectacles. Wearing a white shirt with the sleeves rolled up, and a pair of black wool pants, he didn't seem dressed for pyramid climbing.

"Quite a climb," I said. "Yes, indeed," he panted. The two of us sat silently on top of the pyramid for several moments and looked off to the west. Beyond the sand and other ancient structures, a giant orange sun was slowly settling down into the horizon.

"Ah, this is a great pyramid," I sighed.

"Yes," the man laughed, "yes, it is."

"Have you been inside yet?" I asked.

"Earlier today," he replied.

"It's interesting," I said, "that the pyramid was empty when first opened by the Arabs a thousand years ago. And that the King's and Queen's Chambers were totally devoid of hieroglyphics or ornamentation."

"That, you might say," said the man, turning his head back to look at me, "is the Riddle of the Sphinx." He reached into his back pocket and brought out his billfold. "Who, why and how the Great Pyramid was built has always been a mystery. The Great Pyramid is very different form the other pyramids. Evidence suggests that is was built for some other reason than to serve as a tomb."

"I've heard that. Well, the Great Pyramid has always fascinated me." I said, my eyes on the orange glow to the west.

The stranger pulled a bill out of his wallet. It was an American one-dollar note. "The Great Pyramid can be found on every one-dollar bill," he said. "This is the Great Seal of the United States." I had never noticed it before, but there it was, the Great Pyramid! On top of it was a sort of all-seeing eye, and in the dim light I could read a Latin inscription below it. "Novus Ordo Seclorum." I translated this via my high school Latin as "New Order of the Ages."

"That's pretty interesting," I said.

"The Great Pyramid is very interesting," he said. "Among other things, it encompasses a number of incredible mathematical concepts. Aside from the Golden Mean and The Pythagorean Theorem, the pyramid rests on a square that is exactly 36,524.2 pyramid inches in diameter..."

"Pyramid inches....?"

"Indeed. One hundredth of that is the precise number of days in a year. The pyramid inch, by the way, is one five-hundred-millionth of the diameter of the earth through its polar axis."

"...one five-hundred-millionth..."

"Right! The smallest one weights nearly three tons and there are over 2.3 million blocks of stone in the pyramid! Yet, these blocks of stone were cast into place like concrete, not cut like the granite or limestone blocks. Not bad for a culture that was supposedly one step from the stone age!" He suddenly turned to me. His face was scorched by the last light of the sunset and alive with a terrific grin. There was no doubt about it, it felt great to be up there. "The builders of the Great Pyramid were superior scientists and architects. Their expertise, precision, and mathematical knowledge was unmatched until only a few years ago. For nearly six thousand years, the builders of the Great Pyramid were unequaled in excellence!"

"Wow, and we're sitting on top of it right now." I said. Glancing back to my left, I could see the Sphinx still visible in the dim glow of the sunset. The cool winter night was coming alive with stars, dancing in the desert to the music of the spheres. "What is all this new order stuff, then?" I asked.

He thought carefully for a minute and then said, "Well, the pyramid wasn't just built as a scientific marvel. Some ancient records say that it was built to hold the Biblical Ark of the Covenant. That is what the empty tomb in the King's Chamber is for. And it was also built to formalize a prophetic message, that mankind will once again live in a new golden age, a new order, after thousands of years of war and chaos."

"Can't wait," I said. "When will all this take place, do you suppose?"

"Oh," said the stranger, standing up and stretching, "pretty soon, I would think. We can't take too much more of this chaos and war."

It was already dark when we started down. The last rays of the sunset had already died. At our backs, the lights of Cairo were illuminating the sky. We carefully climbed down the pyramid, lowering ourselves from block to block, the man shining the flashlight for me as I went ahead. At the bottom, we shook hands.

"Thanks for the conversation," I said.

"My pleasure," he said, turning toward the road. "Good night."

I walked to the sphinx just as the sound and light show was starting, and sat on a sand dune to watch the show. Afterwards I caught a crowded bus back into Cairo. Just before the bus came I stood and looked skyward at the African stars in the cloudless desert sky. Practically across the street was the Great Pyramid, looking like a space ship about to take off. On its far side sat the sphinx, still waiting silently for an answer to its riddle.

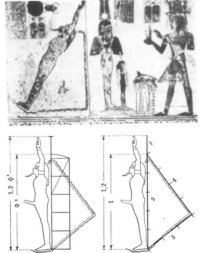

In his book *Serpent in the Sky: The High Wisdom of Ancient Egypt*, John Anthony West correctly points out that this relief from the tomb of Rameses VI shows the direct evidence of Egyptian knowledge and use of the 3, 4, 5 Pythagorean triangle. West writes that Egypt waned and ultimately fell through the misuse of knowledge. He states:

This is but one valid reason for keeping certain types of mathematical knowledge secret. There are many others pertaining to the course of development and initiation of the individual: the man found incapable of keeping a simple secret cannot be entrusted with a more complex, more dangerous secret.

In every field of Egyptian knowledge, the underlying principles were kept secret, but made manifest in works. If this knowledge were ever written in books — and there is mention of sacred libraries whose contents have never been found — then these books were intended only for those who had earned the right to consult them.

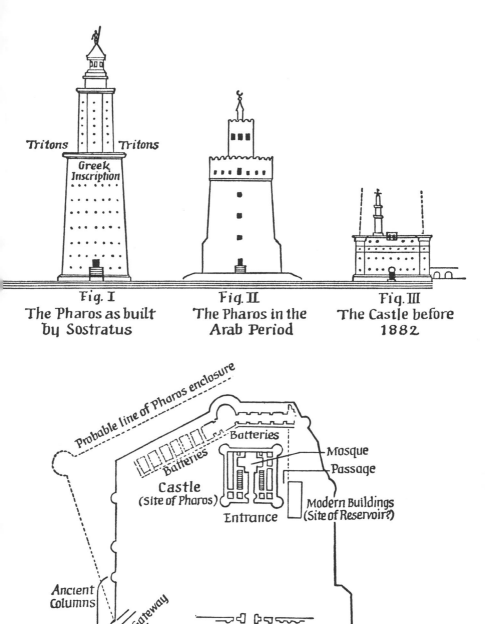

Fig. I
The Pharos as built
by Sostratus

Fig. II
The Pharos in the
Arab Period

Fig. III
The Castle before
1882

Tritons   Tritons
Greek
Inscription

Probable line of Pharos enclosure

Batteries
Batteries

Castle
(Site of Pharos)

Mosque
Passage

Modern Buildings
(Site of Reservoir?)

Entrance

Ancient
Columns

Gateway

Causeway

Entrance

The Pharos lighthouse as it is today, an Islamic fort,
and the probable outline before it destruction.

The crown of Osiris found in the Pharos area. The area is littered with the ruins of both a temple and the lighthouse. Only extensive research will be able to sort these materials out. Note the raised edge midway down the shape. This is a pharaonic form of the crown, showing the combining of upper and lower Egypt.

One of the enormous pedestals found in the Pharos area. This massive block of granite was found lying on its side. At one time the "socketlike" indentation would have had a statue "plugged" into it. But either the seawater dissolved the marble or, more probably the statute was carried off after the lighthouse and temple were destroyed by earthquakes. The smaller holes, inside the bigger indentation, were used to move this block of stone. Like everything else in the Pharos area, this pedestal is on a very large scale.

From *The Alexandria Project* by Stephan Schwartz.

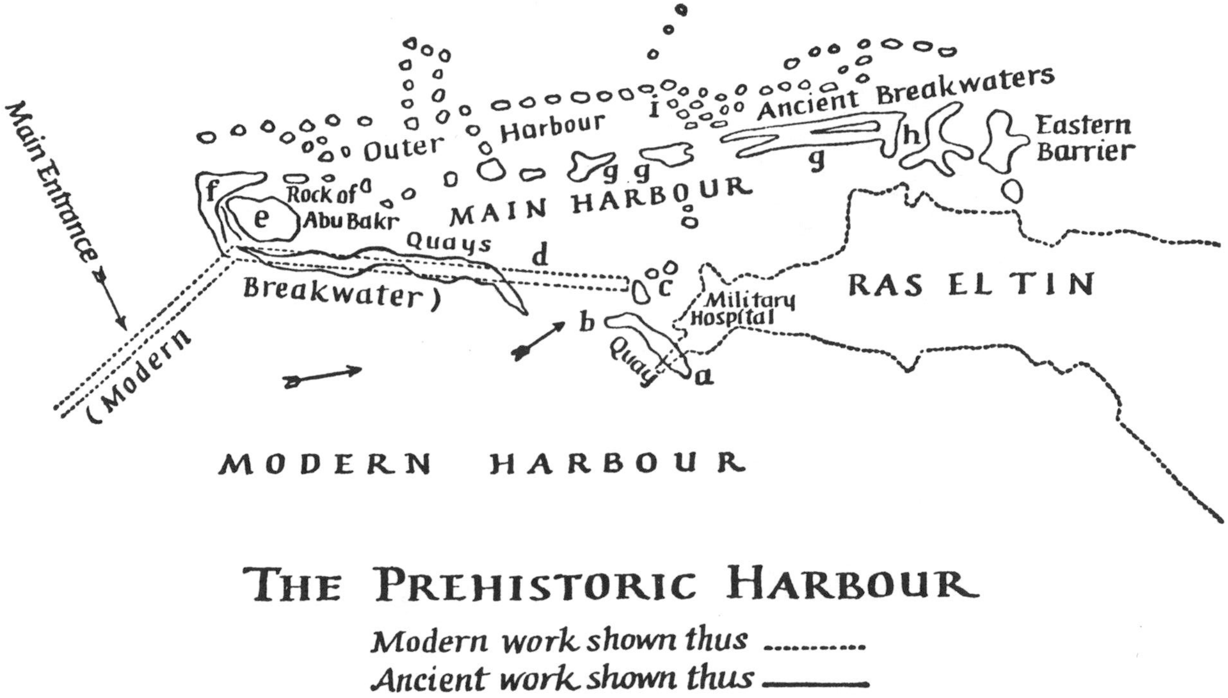
Main Entrance
Outer
Harbour
Ancient Breakwaters
Eastern
Barrier
Rock of
Abu Bakr
MAIN HARBOUR
Quays
Breakwater)
(Modern
Military
Hospital
Quay
RAS EL TIN
MODERN HARBOUR
a
b
c
d
e
f
g
g
g
h
i
THE PREHISTORIC HARBOUR
Modern work shown thus
Ancient work shown thus

Rare map of the megalithic, submerged harbour at Ras El Tin in Alexandria. This area is now off limits.

An old print from Napoleon's conquest of Eqypt in 1798. His scientists found only the head of the Sphinx above the sand. The nose had been shot off by

The Giza pyramid complex photographed from 4,000 feet just before sunset. the south-east slope of the Great Pyramid reveals a V-Shaped depression, which occurs on each side. It is the only known pyramid in Egypt with this odd construction feature.

The hollowing-in feature of the Great Pyramid was also noticed by Napoleon's artists. A century later, the structural engineer David Davidson related this phenomenon to the three lengths of the year—solar, sideral, and anomalistic.

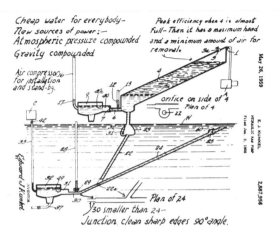

Kunkel's *Hydraulic Ram Pump.* This drawing shows the ram pump based upon the design of the Great Pyramid, exactly as filed with the United States Patent Office in 1955.

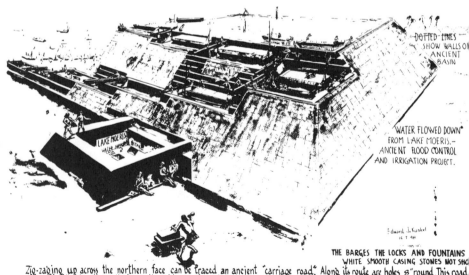

An illustration from the book, *Pharaoh's Pump* by Edward Kunkel. He theorized that the Great Pyramid was constructed by using a system of water locks to float the blocks in place.

The Complex of Pyramids at Giza. This diagram shows the north-south meridian through the center of the Great Pyramid, and the alignment of other buildings in the complex.

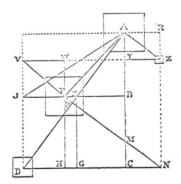

Pythagorean Triangles Formed by Perimeters and Centers of the Three Major Pyramids. This drawing has been supplied courtesy of Alexander Abramov, the Soviet space engineer, and shows some of the relationships among these pyramids.

THE VIGIL IN THE KING'S CHAMBER. *The soul in the form of a man-headed hawk departing for Amenti.*

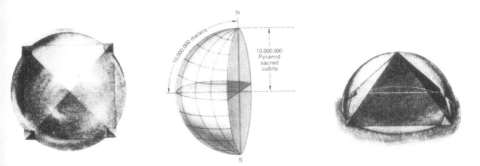

The Great Pyramid's Measurements. These measurements incorporate a value for pi (the constant by which the diameter of a circle may be multipled to give its true circumference); the fundamental proportions of the "golden section" [AB/AC = AC/CB = 1.618]; and the "sacred" 3-4-5 and 2-root 5-3 triangles ($A^2 + B^2 = C^2$) found in its main chamber. These facts are not surprising when one considers that all Masonic Lodges to this day are erected on this geometry.

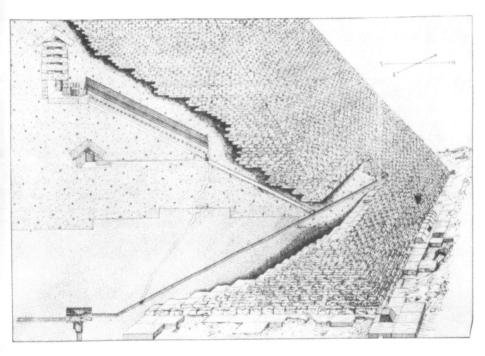

Interior section of the Great Pyramid, as illustrated in *Destiny* magazine.

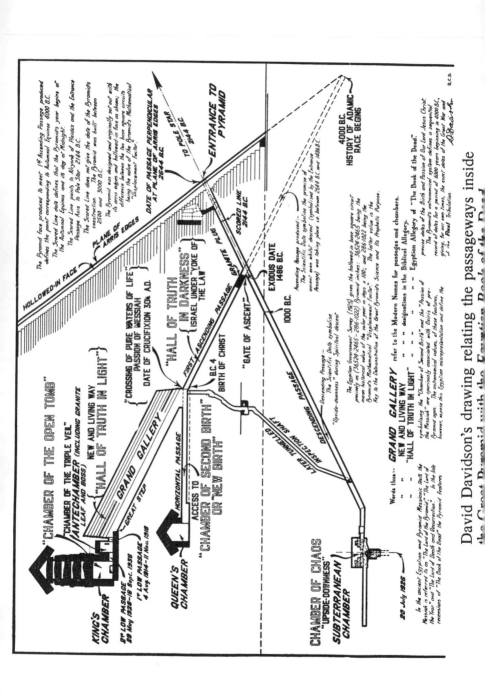

David Davidson's drawing relating the passageways inside
the Great Pyramid with the *Egyptian Book of the Dead*

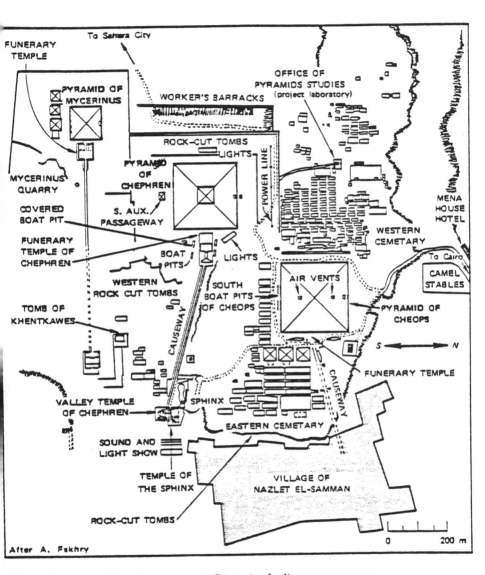

Plan of some of the structures at Giza, including causeways.

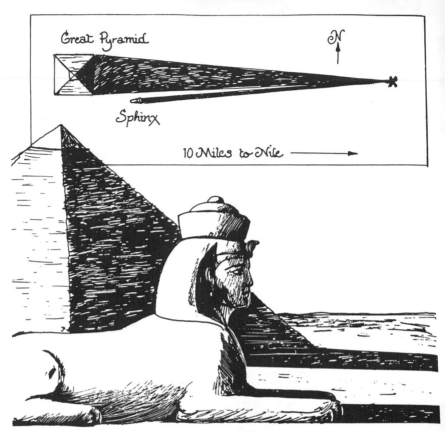

Rene Noorbergen's sketch of a spot where the shadows of the Great Pyramid and the Sphinx meet on a day in late October as the sun sets. Is there a secret chamber to be found here?

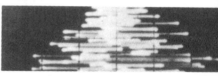

Location of Possible New Chamber in Gre *Pyramid.* (a) Acoustic echoes, looking throu the floor of the King's Chamber. (b) Sket showing location of possible new chamber relation to other structures within the Pyrami

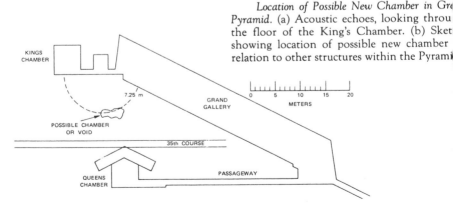

Secret passages, chambers and tunnels are said to exist beneath the Giza plateau, as well inside the Great Pyramid. Esoteric information claims that a special "Hall of Records", an Atlantean library, exists somewhere near the Sphinx.

Diorite statue of Chephren in the Cairo Museum. He built the second pyramid of Giza, the Chephren Pyramid. He also put the "Dream Stele" in front of the Sphinx. Some Egyptologists claim that Chephren's likeness is that of the Sphinx, though the Sphinx was built many thousands of years earlier.

Perfectly cut megalithic construction in granite, an extremely hard crystaline rock, at the Valley Temple of Chephren, next to the Sphinx. Attributed to Chephren, the temple is probably much older, contains no Egyptian hieroglyphs of any kind and, like the Sphinx and the Osirion at Abydos, predates Egyptian civilization. It is probably from the Osirian Civilization circa 10,000 B.C.

The stone masonry technique employed in a corner of the Valley Temple adjacent to the second largest pyramid at Giza, Egypt (allegedly the Egyptian Kephren's), is obviously similar to the stone masonry technique employed in the corner of the temple of the three windows on the other side of the Atlantic Ocean at Machu Picchu, Peru. One major difference in construction techniques between the two is the use of polygonal blocks and the absence of cement between the stones in Peru. Another difference between the two is that the stone masonry at Machu Picchu, Tiahuanaco, and other South American sites employed a keyway or slot cut into one stone in the wall opposite a keyway cut in the stone next to it. Metal clamps fitted into the slots joined the stones together. Thus, each stone employed in the building of the temple was joined together into a stone masonry system admirably suited for stability during a period of earthquakes.

STILL·MORE·SECRETS·FROM·THE·PYRAMIDS

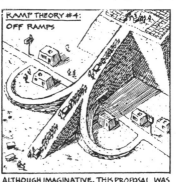

RAMP THEORY #4: OFF RAMPS

ALTHOUGH IMAGINATIVE, THIS PROPOSAL WAS ABANDONED SINCE NO RESTAURANT WOULD TAKE A PARTY OF 25 AND A 2½ TON BLOCK OF LIMESTONE WITHOUT A RESERVATION.

SLED BUMPER STICKER
GIZA c. 2500 BC

LIMESTONE W 160 cm. ht. 181 cm.

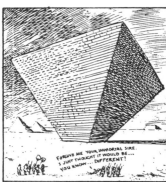

FORGIVE ME YOUR IMMORTAL SIRE. I JUST THOUGHT IT WOULD BE... YOU KNOW... DIFFERENT!

THERE WAS VERY LITTLE OPPORTUNITY FOR SELF-EXPRESSION IN THE DESIGN OF PYRAMIDS

WELL IT BELONGS SOMEWHERE! YOU'LL JUST HAVE TO TAKE IT DOWN AND TRY AGAIN!

· THE PYRAMIDS ARE FAMOUS FOR THEIR SIZE AND PRECISION ·

Chapter 5

SOUTHERN EGYPT:
ATLANTIS & THE CURSE OF KING TUT

Do not be proud of your knowledge.
Listen to the ignorant and the wise.
Art has no limit, and no artist's skills are perfect.
Truth may lie as hidden in the earth as copper,
or it may be found at play upon the lips
of maidens bent above their grindstones.
—*Ptah-hotep, Egyptian teacher, circa 2540 B.C.*

I hung around a few more days in Cairo; seeing the Mohammed Ali Mosque, built largely out of the limestone casing stones from the Great Pyramid; the bazaars; Heliopolis, thought to be Zoan, the Essene retreat in Egypt where Mary and the infant Jesus took refuge; Sakara, where the gigantic coffins of the Bulls of Apis are stored underground in the Serapeum. These coffins weigh more a hundred thousand pounds, and are astonishing to look at. One is actually stuck in the underground passages, so huge, that it could not be moved afterwards. As I stood looking at the gigantic basalt coffin, I could only conceive of levitation moving it. It was so heavy that the number of men needed to move it by pulling a sled by rope could not have stood within the tunnel's confines. I tipped the guard a few piastres baksheesh, and fought my way back through the camel-ride hawkers to where I could get the local bus back to Cairo.

Cairo was starting to get on my nerves after a few days, and back at my hotel, I realized that it was time to leave and visit the lost city of Akhetaton further south.

I took a share taxi from near the Giza railway station, and was shortly zipping along the west bank of the Nile, dodging camels, donkey carts and women with long robes, scarves and typically some large jar on their heads. I was sandwiched in the back seat of the Mercedes with several other Egyptians, including a tall, swarthy man in robes and a turban. Behind sat his wife and daughter, or maybe his wife and mother-in-law; I couldn't be too sure.

We arrived at at the agricultural Nile city of El Minya around noon. I looked around for a hotel, and found a cheap single at the Ibn Khassib Hotel near the train station. That night I toured El Minya by horse drawn carriage, and then left for Tel-El-Amarna the next morning.

It was a forty-minute taxi ride to the ferry station on the Nile where one crossed the river to the small Arab village on the other side of Tell El Amarna. I bought a ticket to see the tombs, and jumped on board a wagon with benches that was drawn by a tractor.

"Pretty fancy set up for the tourists?" said a large, dark haired woman. Her name was Francis, and she and her friend, Della, were visiting Egypt from Chicago.

I laughed at her comment. We were the only people in the trailer. "It doesn't seem like Akhetaton gets very many tourists, does it?" I said. This was indeed the case. Yet, why would any person come to Akhetaton? There were no great buildings or pyramids here.

The answer is because of its history, and an unusual history it is. The New Kingdom of Egypt came about after the rule of the mysterious Hyksos, the "Shepherd Kings" who were in control of Egypt for several hundred years (1783-1550 B.C.). History does not know who these foreign rulers, who infiltrated Egypt, peacefully took over the government, and finally left, either voluntarily or by force, were. They have been given credit by some authors for building the Great Pyramid, and for carrying the pyramid inch around the world to such places as Britain and the Middle East.

With the rise of the New Kingdom came a new prosperity for Egypt. Egypt became an empire with far-flung foreign influence. There is a great deal of evidence that the Egyptians were off exploring the whole world at this time. From the ports on the Red Sea, ships went to Punt, India, and apparently even to Australia and the Pacific. The colonization of the Pacific by Egyptian-Libyan fleets most probably occurred at this time. There was a renaissance of building and art, and Egypt controlled large areas of North Africa and the Middle East. Egyptian culture had not been so advanced since the early days of the Old Kingdom.

Something was going to happen, though, during this period that was to shock Egypt greatly, and lead historians and philosophers to debate the subject for thousands of years afterward. Even though the Egyptians themselves went to great lengths and extremes to wipe this period of Egyptian history from the records, it is probably the best known period in all of Egypt's many thousands of years of history.

In the year 1353 B.C., the Pharaoh Amenhotep III (also called Amenophis III) died. His oldest son Amenhotep IV then became the Pharaoh by actually marrying his own mother, Tiy. Because of this action, Immanuel Velikovsky claimed that the Oedipus myth of the Greeks was based on the Akhenaton story.

However, Amenhotep IV's father had many wives. One of his wives was a young princess from Mittani, which was a small and little known kingdom in present day Syria. Mittani was settled by Aryans who had come out of Central Asia, and had strong ties with the ancient India of the Rama Empire, as well as the Aryan Vedic invaders of 3,000 B.C. This Mittanian princess was the famous Nefertiti. Amenhotep IV also married Nefertiti, who was legally his step-mother, and she began having children by him.

They had a stream of daughters, and by the year 1360 B.C., Amenhotep IV decided to change his name to Akhenaton, *Glory to Aton,* and he and Nefertiti would build a new city on a plain where no city had ever been before. They chose the area at Tel El Amarna to build their city Akhetaton, *City of the Horizon of Aten,* for several reasons. Firstly, it was in the geographical center of Egypt. Secondly, it was a fertile plain that was crescent-shaped, with cliffs coming close to the Nile on each end, which would make it easy to defend during a civil war in Egypt. Just such a civil war was brewing, and did in fact occur, though it was largely conducted in secret.

Akhenaton became known as the heretic king because he suppressed the Amun priesthood that had long been the true rulers of Egypt. Akhenaton proclaimed that the many gods of Egypt were false and that there was only one god, and that God was Aton, the Sun! Like monastic utopianists, Akhenaton and Nefertiti liberated the people of Egypt (whether they knew it or not) from the thousand-year yoke of the Amun priesthood and the diabolical practices that they followed.

To this day, most Egyptologists do not understand what the battle between the Atonists and the Amun priests was really all about. Probably they will never understand it, because they do not understand the principles underlying the battle. The Dead Sea Scrolls taught that there is a cosmic battle between the "Sons of

Light" and the "Sons of Darkness", and such was the battle being fought on Egyptian soil during the end of the New Kingdom.

For a thousand years the Amun priests had ruled Egypt by their manipulation of the people through the worship of various gods and goddesses, most of them personified people, such as Osiris and Isis. Even the god of evil, Set, from which we get our word Satan, was one of the many gods of Egypt. The religion of Set, like many religions today, was a religion of fear, where, if one gave generously to the temple, then nothing harmful would happen to the person (who was true and faithful).

The other insidious device that was fostered on the Egyptian people was the cult of mummification, so well identified with ancient Egypt. Mummification was an extremely expensive and time-consuming process. It took 40 nights or more, and generally would cost the deceased's family their entire fortune. And of course, the richer the person (such as a Pharaoh) the more costly the mummification and accompanying ceremonies.

The Atonists abhorred mummification for several good reasons. They believed in reincarnation, as did many Egyptians, and certainly Mittanians like Nefertiti, and the cult of mummification was in direct opposition to their religious beliefs. Atonists believed that the body of a person must completely deteriorate (ashes to ashes, dust to dust) before a person can reincarnate again. This is why Hindus, Buddhists, Jains, Zoroastrians, Coptic Christians, etc. prefer to be cremated or dismembered upon death, so as not to leave any of the former physical body behind.

However, with mummification, the physical body did not deteriorate after death, and in fact, mummies that are thousands of years old have been shown to still have living tissue in them! In theory, it would be possible to clone or take the genetic codes out of the cells of mummies in Egypt. This, then, at least the Atonists believed, was the most diabolical part of the Amun priest's nefarious activities: the rendering of their victims so that would be unable to reincarnate again, and be trapped on the astral plane "for eternity".

With the suppression of the Amun priests, Akhenaton and Nefertiti basked in the peace and quiet of their new capital city in the desert. Things were going well for awhile, until Nefertiti's mother-in-law began interfering with her marriage to Akhenaton. Tiy pointed out that Nefertiti had so far borne six daughters, but no males. Where was the heir to the throne?

Meanwhile, the Amun priests were not idle, and they took care to plot the destruction of Akhetaton, and the demise of those who had suppressed them. According to Philipp Vandenberg in his book, *Nefertiti*,[59] it was actually she who was the power behind the Atonist movement, and it was well known in Egypt that this was true. Nefertiti, as a Mittanian, would have been well taught the ancient solar religion of Atlantis and the Rama Empire: a religion of reincarnation, karma, and of different planes of existence. Angels and Archangels live on higher planes above us, they believe. Archangels are said to live on the sun. Similarly, the sun is the sustainer of all life, as evidenced in the Hymn to Aton.

Akhenaton became estranged from Nefertiti, who was given her own palace at Akhetaton to live in. This was divorce Egyptian-style, and Akhenaton married another woman, whose name has not come down to us. According to legend in Egypt, the Amun priests made a poison ring with which to kill Nefertiti. A messenger was told to take it to the Queen at Akhetaton. The messenger, however, became confused, and the poison ring was given to Akhenaton's new wife, rather than to Nefertiti, for whom it was meant. Akhenaton's new wife then died from the poison.

Whether this poisoned queen was the mother of Tutankhamun is debated by historians. At about this time, Smenkhare, Akhenaton's son or brother (no one is quite sure which) became co-regent with Akhenaton and married the oldest daughter of Akhenaton and Nefertiti, Meritaten. It was common in Egypt for brothers and sisters to marry and even for father and daughter or mother and son, as was the case with Akhenaton.

About the year 1348 B.C., Vandenberg says that Akhenaton married his daughter Ankhesenpaten, who was only 12 years old at the time. This was probably to further establish the Akhenaton/Nefertiti royal line, because, at about the same time, Smenkhare died at about the age of twenty, probably poisoned by Amun priests, and Akhenaton died at about the same time, probably also of poisoning. At some point here, possibly just before Akhenaten's death, probably upon the death of Smenkhare, Meketaten, second oldest daughter of Nefertiti and Akhenaton, apparently attacked the Karnak Temple in Thebes (Luxor), which was the stronghold of the Amun priests. She and her army were defeated and killed.

Ankhesenpaten then married the young Tutankhaten, who was later to change his name to Tutankhamun, which symbolized the defeat of Atonism, and the triumph of Amunism once again. Nefertiti lived alone in her palace, and Tutankhamun moved back to Thebes, where he too was poisoned at a young age. Legend has it that he was given a poisoned flower, which scratched his cheek. The general Horemheb then became Pharaoh, and asked Nefertiti to marry him. She refused. What happened to Nefertiti after that is not known, though it is presumed that she lived a lonely life in a deserted city. Her dream of a glorious New Age city had lasted but a few short years. With her death, the Amun priests had the city completely destroyed, and monuments in Thebes and elsewhere with any reference to Akhenaton, Nefertiti or the Aton, were defaced, destroyed, or buried.

The works of Akhenaton reach right through the doors of time to us today. In the early 1940's, a Chinese representative of the Rosicrucian Brotherhood visited the Rosicrucian Fraternity in San Jose, California. He brought with him a manuscript which had been kept in a secret Asian archive for thousands of years. It was said to have been authored by the Egyptian Pharaoh Akhenaton, historically the founder of monotheism. The Rosicrucians translated this book and published it under the title *Unto Thee I Grant*. That a book by Akhenaton would be in a secret library in Tibet is evidence of the activities of certain secret societies (good and bad) that continue to operate today.

§§§

The cattle are content in their pasture, the trees and plants are green,
the birds fly from their nest. Their wings are raised in praise of your soul.
The goats leap on their feet. All flying and fluttering things live
when you shine for them. Likewise the boats race up and down the river,
and every way is open, because you have appeared. The fish in the
river leap before your face. Your rays go to the depth of the sea.
–Hymn to Aton

He causeth the grass to grow for the cattle, and the herb for the
service of man... where the birds make their nests: as for the stork,
the fir trees are her house. The high hills are a refuge for the wild goats...
So is this great and wide sea, wherein are things creeping innumerable,
both great and small beasts. There go the ships.
–Psalm 104

"Over here are the tombs," said our guide, a young Egyptian law student named Mohammed.

"Is everyone in Egypt named Mohammed?" asked Della, the stockbroker from Chicago.

"Only the men," replied Mohammed, and we all laughed. We had been talking about Akhenaton and Nefertiti on the slow ride out through the desolate plain of the ancient city. There was very little to see, merely brown dust everywhere. To the east was a crescent range of cliffs with tombs cut in them. The tractor and trailer stopped at the foot of the cliffs, and we got out.

We climbed up a path to a large walkway along the cliffs. Here watchmen with their large keys opened a few of the tombs for us. We went through the various tombs, and I was interested to find a picture of a breadfruit in one, or something that looked like a breadfruit. Breadfruit grows only in Tahiti, and it was the collection of breadfruit trees in Tahiti by the *HMS Bounty* to led to the famous mutiny.

None of the tombs had a coffin. There were interesting paintings on the walls, though much had been scratched out, by the Amun priests possibly, or if not, by pious Moslems later. "Were any mummies ever found in any of these tombs?" asked Della.

"No," replied Mohammed. "They were probably stolen."

"No way," said Francis. "There were never mummies in these tombs! The Atonists didn't believe in mummification. Probably there bodies were just laid inside a coffin, or maybe they were cremated."

"Where is the tomb for Akhenaton?" I asked.

"It is over there, up a ravine. It is by itself, very far away from here."

"Can we go there to the tomb?" I asked.

"No one is allowed to go to that tomb," said Mohammed. "We do not even have a key. You must get special permission to see that tomb from the authorities in El Minya or Cairo. There was a woman who came here about a half a year ago. She lived in the village of Tel El Amarna for three weeks. She told me that Akhenaton was a prophet, and that she was the reincarnation of one of his followers. She got permission to go to his tomb from El Minya. She went inside and lit some incense, and said some special prayers."

"Was there anything inside the tomb?" asked Della.

"No, nothing. There is nothing inside any of the tombs. Maybe they were always empty," replied Mohammed.

"What about Nefertiti," I asked, "where is her tomb?"

"There is no tomb for Nefertiti," Mohammed replied.

"Of course not," replied Francis. "That's because she believed in cremation!"

We walked back down and got back in the tractor trailer to run around the remains of the ancient city. We stopped by the palace of Nefertiti, where the remains of a swimming pool could be seen, and then to the main palace and Temple of Aten near the south end of the crescent plain. From there we visited a stele that marked the boundary of Akhetaton. It was carved into solid rock and had Akhenaton and Nefertiti basking in the rays of the sun, a sun which had little hands, reaching down to touch life on the earth. When Akhetaton was first set up, it was very much like the capital of the United States, the District of Columbia. Akhetaton was a separate Nome within Egypt, and the boundaries of city were marked by six steles placed alone the boundaries.

"I find the stuff about mummification and reincarnation very interesting," I said as we moved along the rocky plain.

"It makes sense to me," said Della.

"Sure," said Francis. "The Coptic religion of Egypt is the heir to the Atonists. They too believe in reincarnation and cremation. They also believe that Christ was the Archangel Melchizedek, and that he, and other Archangels, live on the sun."

I then read to the ladies and Mohammed part of the book *Behind the Mask of*

Tutankhamun by Barry Wynne.[60] Wynne was a close friend of Lord Carnarvon's youngest daughter, Lady Evelyn, and did extensive research on the book. He relates the fascinating tale of Count Louis Hamon, known as the most famous palm reader in the world under his pseudonym of Cheiro, a word derived from the Greek word *Cheir*, meaning hand, and pronounce Ki-ro, like the capital of Egypt.

Cheiro worked at the Valley of Kings near Luxor (ancient Thebes) in the early 1890's. He had a strong friendship with an Egyptian guide, who had once saved his life, and the night before Cheiro was to leave Luxor and return to Europe via Cairo, the guide presented him with a very strange gift. To the somewhat startled surprise of Cheiro, the Egyptian guide produced the right hand of a mummy in an excellent state of preservation. It was delicate, obviously that of a woman, the finger nails perfect in shape, covered with gold leaf. On the first finger a gold ring shone, upon which he could see minute hieroglyphics. The hand itself appeared solid, nut-brown, as if carved from a solid block of wood. The bones, where it had been severed at the wrist, were white and gleaming. The Egyptian then recounted the bizarre story behind the hand, the story and the hand, having been handed down from generation to generation for centuries.

This was the hand, said the Egyptian to Cheiro, of Meketaten, the daughter of Akhenaton and Nefertiti. She had raised an army and attacked the Karnak Temple of Luxor, attempting to break the power of the Amun priests once and for all. She was killed, and her hand was hacked off at the wrist and mummified. It was then placed in Karnak Temple for all to witness, where it remained until the decline of Egypt and the Egyptian's family kept it for hundreds of generations.

"Within this hand is the Ka of Princess Meketaten," the Egyptian told Louis Hamon/Cheiro, "and I hand it to you for safe keeping."[60] Hamon then returned to England where he became quite famous as a psychic. He was friends with many famous people of the time, including Mark Twain, various English royalty, and Lord Carnarvon himself, the philanthropist who financed the excavation of Tutankhamen's tomb.

In 1922, Hamon and his wife were living in Dublin, but Ireland was racked by political strife. They decided to return to England. Throughout the previous thirty years the hand had been kept on a purple velvet cushion and was always on display in Cheiro's salon. While packing their things for the move to London, Cheiro noticed that the hand was becoming soft, and that small droplets of blood had oozed out of the knuckles and from beneath the finger nails. He took it to a chemist, who declared that the blood was indeed human, and that perhaps a coating of shellac and pitch would restore the hand to its former rigidity. This worked for a short period of time, but soon, the bleeding started again. On the 31st of October, 1922, the Hamon's were packed up and ready to leave for London the next day. The only thing that remained unpacked was the strange mummified hand of Meketaten.

On an impulse, Cheiro took the object, and after his wife quoted a few passages from the Book of the Dead, he thrust hand, cushion, stand and all into the fireplace. Immediately flames of great brilliancy shot up around it and there was a distinct aroma of spices in the air. They watched for quite some time until all was consumed. Then, their task completed, they decided to retire to bed.

It was a clear, calm night, but at the very moment they had turned to mount the stairs, a sudden wind brought them to a halt. The wind seemed to gather tremendous force and battered at the heavy oak doors. As the Hamon's watched, terrified, a great pressure seemed to build up against the doors, until, with a crash, they burst wide open! However, there was nothing to be seen but the moonlit garden.

Then, beyond the door, something began to materialize. An undefinable shape it started to move through the porch and into the hall where gradually its outline and figure began to develop into that of a woman. However, the form stopped short at the hips. It was very beautiful; a remarkable face finally looked towards them. Nobility, grandeur and pride seemed to mark every line.

She wore what appeared to be a headdress, formed by the wings of beetles, fashioned in beaten gold, the ends of which rested gracefully on her shoulders. In the center of her forehead was a golden asp, the emblem of Egyptian royalty. The figure moved into the house and gazed directly at Hamon and his wife. Both hands were clasped together, as if in a moment of ecstasy. She appeared to be trying to speak and her lips began to move, but no sound emanated. Suddenly she threw back her head, lifted her hands slowly in the form of an arch and bowed towards them. Then, as quickly as she had come, she began to drift away, still remaining in the same posture, but retreating through the door until she was lost in the night.[60]

It had been a very frightening experience for Hamon and his wife, and they dared not go to bed that night. In the cold rays of the dawn, Hamon raked out the dead ashes of the fire and found the individual bones of the hand, white and calcined. He collected the ash and found the gold ring, untarnished. The Ka of Meketaten had been released from the occult prison created by the Amun priests of Karnak. Yet, this strange adventure was only the beginning in an incredible saga of reincarnation, fabulous gold treasure and the strange curse that claimed the lives of a number of people who sought to free the Ka of a boy king who was held prisoner in the Stygian darkness of the underworld.

§§§

"That's a pretty strange story," said Della as we traveled by bus to Abydos further south along the Nile towards Luxor.

"It makes sense to me," said Francis peeling an orange and looking out the window at the sugarcane fields. We were passing the central Nile town of Asyut, a place where riots between Moslems and Coptic Christians have occurred frequently in the last few years. The rise of fundamentalist Islam has caused fervent Moslems to demonstrate against local Christians, as well as foreigners in the area. The Egyptian police and army have been called out to stop the uprising, and it is typical to see armed police or Army officers guarding hotels, banks, and government offices.

The police themselves have been a problem at times. The Giza Holiday Inn hotels were looted and burned by rioting police officers a few years when a false rumor about extended service and lower pay swept through the barracks in Giza. A drunken mob took to the streets and set fire to every hotel they saw.

I told Francis and Della about the police riot and Della said, "That's great, rioting police are going to protect us from rioting Islamic fundamentalists! Is there a safe place in this country?" I couldn't answer that question.

At Abydos, we visited the Osirion, a structure which, like the Sphinx and Valley Temple next to it, are believed to be associated with the ancient Osirian Empire of the Mediterranean. In fact, it is from the Osirion that we get the name for this lost civilization from the time of Atlantis.

Abydos was, for a long time, the most important temple in all of Egypt, even more important than the vast complex of Karnak in Luxor. The cult of Osiris was centered here, and funerary boats were usually placed in tombs so that the deceased could journey to Abydos and back after death, a pilgrimage to the Tomb of Osiris. Two of Tutankhamen's funerary boats can be seen at the Luxor Museum. They were built circa 1336 B.C. and a card there states that they were to take Tut to Abydos and back.

127

What made Abydos such important temple and pilgrimage point was the presence of the Tomb of Osiris, which was only uncovered in the last century. When the shifting sands had been removed from the south side of the main temple, a massive complex of red granite, was discovered, most of it in a swamp thirty feet below the surface of the ground. The *Illustrated London News* carried this article by Edouard Naville on the Osirion (or *Strabo's Well*, as it was also called then) on May 30, 1914: "It consists of a rectangle, the inside of which is about a hundred feet long and sixty feet wide. The two long sides are north and south; east is the side of the temple of Seti; west the doorway with the lintel, fifteen feet long, which had been discovered in 1912. The enclosure wall is twenty feet thick. It consists of two casings: the outer one is limestone rather roughly worked; the inner one is in beautiful masonry of red quartzite sandstone. The joints are very fine; there is only a very thin stratum of mortar, which is hardly perceptible. *Here and there the thick knob has been left which was used for moving the stones. The blocks are very large – a length of fifteen feet is by no means rare; and the whole structure has decidedly the character of the primitive constructions which in Greece are called cyclopean, and an Egyptian example of the which is a Ghizeh, the so-called temple of the Sphinx.* [Italics mine]

"...As for the pool, it is probably one of the most ancient constructions which have been preserved in Egypt. It is exactly in the style of the so-called temple of the Sphinx...and one of the characteristic features of which is the total absence of any inscription or ornament. But the pool is even more colossal. In the temple of the Sphinx the pillars are four feet square; here they are eight and a half feet. It is impossible, in spite of the havoc made, not to be struck by the majestic simplicity of the structure, chiefly in the corner where the ceiling has remained."

The structure was called the Tomb of Osiris (Osirion) for no exact reason, except that it was next to the Temple of Osiris built by Seti I, and seemed much older. Being underground, it was likened to a tomb. The name stuck, but complete excavation proved to be impossible. The foundations are cut many feet below the current level of the water table, which has risen some twenty feet since New Kingdom times. Water flowed into the excavated temple faster than the pumps could pump it out, and excavations on the underground side of the structure abutting Seti's temple of Osiris also had to be abandoned when sand kept pouring in from above and it was feared this would undermine the foundations of the main temple.[53]

The discovery of the massively simplistic Osirion created a controversy among Egyptologists, until it was believed that Seti I of the XIXth Dynasty, who had constructed the Temple of Osiris, had also built the Osirion. This was reasoned because a massive block near to Seti's temple was decorated with astronomical scenes by Seti. Egyptologists then decided that Seti had built the entire complex.[53]

What is more logical, especially since its architecture is completely different from anything Egyptian, before or since, is that Seti himself discovered the structure while building the Temple of Osiris. It is even more likely that Seti's Temple of Osiris was built because of the older ruins found in the swamp. Seti then carved some astronomical symbols on the ancient stonework, a practice that was very common in ancient Egypt. Many pharaohs would routinely have the cartouche of another Pharaoh removed from some obelisk or monument, and substitute their own, claiming that they had raised the structure. Seti, however, does not claim to have built the Osirion at all, he merely did a little astrological doodling on one of the blocks.

The story of Osiris himself, as related by the Greek historian Plutarch, is revealing in our search for lost cities and vanished civilizations. Osiris was born of the Earth and Sky, was the king of Egypt and the instrument of its civilization.

He weaned the inhabitants from their barbarous ways, taught agriculture, formulated laws and taught the worship of the gods. Having accomplished this, he set off to impart his knowledge to the rest of the world.

During his absence, his wife Isis ruled, but Osiris' brother and her brother-in-law, Typhon (Set, known to us as Satan) was always ready to disrupt her work. When Osiris returned from civilizing the world (or attempting to, at least), Typhon decided he would kill Osiris and take Isis with whom he was madly in love, for himself. He admitted seventy-two conspirators to his plot and had a beautiful chest made to the exact measurements of Osiris. He threw a banquet and declared that he would give the chest to whomever could lie comfortably within in the chest. When Osiris got in the box, the conspirators rushed to the chest and fastened the lid with nails. They then poured lead over the box and dumped it into the river where it was carried out to sea. When Isis heard of Osiris' demise, she immediately set out to find her beloved.

The box with Osiris in it then came aground at Byblos in present day Lebanon, not too far from the massive slabs at Ba'albek. A tree grew where the box landed, and the king of Byblos had it cut down and used it as pillar in his palace, Osiris still being inside the chest, which was now inside the tree. Isis eventually located Osiris, and brought him back to Egypt, where Typhon (Set/Satan) broke into the box, chopped Osiris into fourteen different pieces, and scattered him about the countryside.

The loving Isis went looking for the pieces of her husband and each time she found a piece she buried it — which is why there are temples to Osiris all over Egypt. In another version, she only pretends to bury the pieces, in an attempt to fool Typhon, and puts Osiris back together, bringing him back to life. Eventually she found all the pieces, except the phallus, and Osiris, one way or another, returned from the underworld, and encouraged his son Horus (the familiar hawk headed god) to avenge his death. Scenes in Egyptian temples depict the hawk headed Horus spearing a great serpent, Typhon or Set, in a scene that is identical to that of St. George and the dragon, though depicted thousands of years earlier. In the happy ending, Isis and Osiris get back together, and have another child, Harpocrates, though he is born prematurely and is lame in the lower legs as a result.[53,72]

There are many important analogies in the legend of Osiris, including resurrection, the vanquishing of evil by good, and perhaps a key to the ancient Osirian civilization. I have already theorized the idea that the Mediterranean was once a fertile valley with many cities, farms and temples. It was noted by Naville in the *London Illustrated News* article that here and there was a "thick knob...which was used for moving the stones. The blocks are very large – a length of fifteen feet is by no means rare; and the whole structure has decidedly the character of the primitive construction which in Greece are called cyclopean, and an Egyptian example of which is at Ghizeh, the so-called temple of the Sphinx."

Naville is directly relating the Osirion with the gigantic, and prehistoric construction in Greece and at the temple of the Sphinx. Other such sites around the Mediterranean are on the island of Malta, (in fact, virtually every Mediterranean island has prehistoric megaliths on it) and those found at Ba'albek and other areas of eastern Mediterranean. Furthermore, the knobs, which may or may not be for moving the stones, are the same sort of knobs that occur on the gigantic stones that are used on the massive walls to be found in the vicinity of Cuzco, Peru.

The lack of inscriptions indicates that the Osirion, like the valley temple of the Sphinx, were built before the use of hieroglyphics in Egypt! The Osirion is evidently a relic from the civilization of Osiris itself. Who was Osiris? According to esoteric legend, he was a king of Atlantis who was tricked and murdered by a group of infiltrators (seventy-two of them?) who sought to use Atlantean higher knowledge for their own nefarious purposes. Osiris attempted to stop them, but it was too late. It is interesting to note how Osiris is washed up at Byblos, so near to Ba'albek, and that he was cut up into fourteen pieces. Perhaps an analogy indicating that the Osirian Civilization broke up into fourteen different parts?

We continued south down the Nile, this time by taxi, to the Temple of Hathor at Dendera. Like the Temple of Osiris, it is a beautiful and massive edifice with huge columns that tower over one's head like redwoods. The temple is of quite recent origin, built in the 1st century B.C., but it encloses earlier temples. An inscription in one of the subterranean vaults says that the temple was built "according to a plan written in ancient writing upon a goatskin scroll from the time of the Companions of Horus." This is a curious inscription, essentially stating that the Ptolemaic (Greek) architects of the 1st century B.C. were claiming that the actual plan of the temple dated to the legendary prehistoric era when the "companions of Horus" ruled Egypt. This long era extended for many thousands of years, and in a sense takes us back, once again, to the legendary civilization of Osiris.

Probably most interesting to me was an incised petroglyph in the room designated No. XVII that depicts a strikingly unusual scene with what appear to be electrical objects. The famous British scientist Ivan T. Sanderson discusses ancient Egyptian electricity in his book *Investigating the Unexplained*.[61] In the petroglyph, attendants are holding two "electric lamps" supported by *Djed* pillars and connected via cables to a box. The Djed columns are explained as insulators (though they are probably electrical generating devices themselves). Djed columns are interesting, as they are usually associated with Osiris. They are said to represent the column in which he was found at Byblos in Lebanon by Isis. The odd "condenser" design at the top of the columns makes them seem strikingly like electrical devices.

An electrical engineer named Alfred Bielek explained the petroglyph to Sanderson as depicting some sort of projector with the cables being a bundle of many multi-purpose conductors, rather than a single high voltage cable. Another depiction from a papyrus scroll showing a Djed column with an ankh with hands holding up an orb was thought to be a static electricity generator commonly known as a Van de Graff machine, named after the inventor (or re-inventor, as the case may be). In such a device, static electricity builds up in the orb, and, says electrical engineer Michael Freedman, "...what better 'toy' for an Egyptian priest of ancient times? ...such an instrument could be used to control both the Pharaoh and fellahin (peasant), simply by illustrating, most graphically, the powers of the gods; of which, of course, only the priests knew the real secrets. Merely by placing a metal rod or metal-coated stave in the general vicinity of the sphere, said priest could produce a most wondrous display, with electric arcs and loud crashes. Even with nothing more elaborate than a ring on his finger, a priest could point to the 'life-symbol,' be struck by a great bolt of lightning, but remain alive and no worse for wear, thus illustrating the omnipotent powers of the gods–not to mention himself–in preserving life for the faithful."[61]

Sanderson goes on to mention that the Egyptians had to be familiar with electricity, as the Nile is full of an electric fish known as *Gymnarchos niloticus*. He also mentions how electroplating was used by the Egyptians. Part of the evidence for ancient Egyptian electrics is the mystery of why tombs and underground passages are highly painted and decorated, yet there is no smoke residue or evidence of torches on the ceilings! It is usually assumed that the artists and workers would have to work by torch light, just as early Egyptologists did in the 1800's. However, no smoke is found on the tombs. Did the Egyptians use electric lighting? As fascinating theory, but another ingenious theory was that the tombs were lit by series of mirrors, bringing sunlight from the entrance. Many tombs, it is said, are far too elaborate with deep and twisting turns, for this to work. Even more interesting is the suggestion by some people that the ankh was also a device which was similar to an electrified tuning fork which was used to "acoustically levitate" stones!

That the Egyptians and Babylonians electroplated objects is generally considered scientific fact.[80] However, it is not generally known that an aluminum belt fastener, believed to be at least 1,700, years old was discovered in China. Aluminum was discovered by the west in 1803 and not refined until fifty years later, a process which requires electricity! Ancient Hebrew legends tell of a glowing jewel that Noah hung up in the Ark to provide a constant source of illumination and of a similar object in the palace of King Solomon about 1000 B.C.[80]

And, speaking of ancient electrical devices, behind the Temple of Hathor is a Temple of Isis. According to the *Lemurian Fellowship*[36] and *The Ultimate Frontier,*[37] the Ark of the Covenant was kept in the Temple of Isis before it was moved into the Great Pyramid. Was this the temple where the Ark was kept? There are many Isis Temples in Egypt, and it could have any one, though such a temple would have to be the oldest Isis Temple in Egypt.

§§§

Francis, Della and I arrived late the same night in Luxor, the ancient city of Thebes, an area that is famous for its temples and tombs. We found a hotel down by the Nile, and sacked out from a long and exhausting day of traveling. Basically a small, sleepy tourist town, Luxor is popular because of Karnak Temple (a huge temple complex which was successively added to by all the pharaohs, most of which is still standing) and the Valley of Kings, five miles across the Nile to the west.

The gals from Chicago and I decided to rent bicycles and ride out to the Valley of Kings, a necropolis of sixty-four pharaohs' tombs including the tomb of Tutankhamun, the only one not looted by grave robbers. We paid twenty-five piastres for a ferry across the Nile and then rode out to the tombs, about six miles away. We visited all the tombs that tourists are allowed to see, including Tutankhamun's tomb, where his golden mummy lies encased in his original three nested gold coffins. His tomb alone remain unlooted for three thousand years, mainly because the rubble from the excavation of another tomb had covered up the entrance to his.

As I was deep inside one of the tombs in the valley, I suddenly felt sick to my stomach. Though I was several hundred feet below ground in the catacombs of the tomb, I suddenly bolted to the surface to keep from vomiting. As I sat on the terrace of the restaurant in the valley, I was reminded of what a dangerous profession it was to rob tombs, as the ancient Egyptians placed poisonous gases as well as other boobytraps and death pits in them to discourage grave robbers, many of whom did meet a rather grisly end while attempting to rob a pharaoh's possessions. It was typical to fly a parrot or toss a lit torch into the tomb to check for poisonous gases. If the torch suddenly went out, or the parrot flew about the tomb for a bit and suddenly fell dead, then the presence of deadly gases were confirmed. This, however, was not the most dangerous way of "cursing" a tomb.

According to Barry Wynne,[60] Amun priests would curse a tomb by taking a slave or vagabond, and having him help with the long mummification process that would take several months. When the mummy was ready and tomb about to be sealed, they would torture the poor fellow to near death. Then, he would be hypnotized and told that it was his duty to stay with the tomb and guard it for eternity. In this way, an "earth-bound" spirit would be created, who, from the astral plane, would guard the tomb.

On the 26th of November, 1922, the British Egyptologist Howard Carter, with his sponsor, Lord Carnarvon, and Carnarvon's daughter, Lady Evelyn, broke open part of the plaster wall that concealed the inner chamber of the young Tutankhamun's tomb. Barry Wynne interviewed Lady Evelyn in the early 1970's when she was seventy years old, she told him of her thoughts, which were of Tutankhamun's wife and of the flowers that she laid on the coffin as the last person to leave the tomb. As she gazed through the small opening created by Carter, she saw those flowers on the tomb, just as Ankhesenpaten, daughter of Akhenaton and Nefertiti, had laid them more than three thousand years earlier. Flowers were indeed laid on the gold coffin, and it is believed by Egyptologists that they were laid there by his young wife just before the tomb was sealed.[65]

Lord Carnarvon was a serious believer in the occult, though he was not a strong willed person, and constantly sought the advice of spiritualists, and sometimes held seances in his castle. Carter attended a few of these. At one seance, Carter, a Coptic Christian, identified the language of a "spirit entity" as ancient Coptic. At one point during the long and tedious opening of the tomb, Count Louis Hammon/Cheiro received a message from the "Ka" of Meketaten, "Lord Carnarvon not to enter tomb. Disobey at peril. If ignored would suffer sickness; not recover; death would claim him in Egypt." Unfortunately for Carnarvon, who took such advice very seriously, it was too late. He had already secretly entered the tomb with Carter and Lady Evelyn, though this fact was not disclosed until after his death.

Later, after the fabulous tomb had been officially opened, Lord Carnarvon was leaving the tomb, and he was bitten on the cheek by a mosquito. The next morning while shaving, he nicked the bite with his razor. A strange form of blood poisoning set in, and Carnarvon's health began to deteriorate at a very rapid rate. He went to Cairo, and his health continued to wax and wane. Worried that he would die, his son was called from India to be at his side. Shortly after his son arrived in Cairo, late one night, there was a complete electrical blackout in all of Cairo. The lights were out for five to eight minutes all over the city. During the black-out Lord Carnarvon's nurse came rushing to his bed. He said, "It is finished. I have heard the call and am prepared." With that he died.[60]

The head of the Cairo Electricity Board was a Colonel Cornwall, who was awakened by his servant at approximately 2:00 A.M. and told that there had been a power failure. Colonel Cornwall immediately drove straight to the power station, but the lights came on before he reached it. He later stated to Carnarvon's heirs, "When I arrived it was to find that all my officials were completely baffled. There seemed no reason for the failure whatsoever. The breakers were in position, the fuses intact....I am afraid there is no technical explanation whatsoever."[60]

This was all big news for the Cairo newspapers the next day which proclaimed that King Tut's Curse had killed Carnarvon. Howard Carter was quick to denounce any tales of a curse. What makes the story all the more fascinating was that after Tutankhamen's mummy was X-rayed, it was discovered that the boy king had a fresh scab on his cheek in approximately the same position where Lord Carnarvon had been bitten by the mosquito which ultimately led to his death.

Carnarvon was not the only person to supposedly succumb to the curse. More than thirty persons died "untimely deaths" that were linked to the opening of the tomb. Many of these people had never even been to the tomb, yet they themselves at least *believed* that they were victims of the curse. One "victim" was Lord Westbury, whose son, Richard Bethel, secretary to Howard Carter, had died a year before Lord Westbury hurled himself out of a seventh story window and killed himself. Just before he did so, he was quoted as saying, "I cannot stand the horror any longer. I am going to make my exit!"[60,62]

Philipp Vandenberg in his book, *The Curse of the Pharaohs* [62] said that there were three principle causes for the deaths of those who dared to ignore the curse: fever with delusions, strokes accompanied by circulatory collapse, and sudden cancers that were quickly terminal. Vandenberg believes that the Egyptians used a form of prussic acid, the same used in gas chambers today, derived from peach pits, to poison the tombs. He also theorizes the use of minute hallucinogens such as rye mold and even forms of radioactivity that would accelerate cancer in tomb robbers. However, his book, though rather fantastic, stays within natural science without venturing too far into the occult.

"It is obvious that this is a tale of reincarnation and karma," stated Francis, sipping a coke.

"Really," said Della. "How so?"

"Well," said Francis, "It is obvious that the persons dealing with Tutankhamun's tomb are the reincarnations of certain people of the time."

I took a sip of a Stella lager beer, and looked out in the blazing sun of the Valley of the Kings. What she said made some sense. "Well," I volunteered, "Barry Wynne does hint that Lady Evelyn Herbert, Carnarvon's daughter, was the reincarnation of Ankhesenpaten, Tutankhamun's wife. Apparently she thought so, though he only hints at it, rather then saying it outright. After all, she was still alive when the book came out."

"She couldn't be the reincarnation of Tut's wife if there was a mummy of her," said Della.

"There is no mummy of Ankhesenpaten, as far as I know," I replied.

"It would also explain why she did not succumb to the curse," said Francis. "Since she was the beloved wife of Tutankhamun, she was not an intruder into the tomb."

"Well, who then was Lord Carnarvon the reincarnation of?" I had to ask, since he of all people was apparently the main object of the so-called curse.

"Why, it is obvious," cried Francis. "Carnarvon was the Amun priest who had cursed the tomb in the first place. What better way to work off your karma than to die of your own curse. It was he who had placed an earthbound spirit at the tomb, and it was he who first entered it, along with Carter and Lady Evelyn, Tut's old wife."

"OK, Francis, that was a clever answer," said Della, "But then, who was Howard Carter in your tale of reincarnation and karma?"

"That one is not so easy," she confessed. "What do you think, David?"

"Hmm," I replied. "Well, in your theory, Francis, I suppose that Carter was the reincarnation of some Atonist, I suppose. He of all people should have succumbed to any curse, but he did not, living to a ripe old age. Well, he was a Coptic Christian, which is supposedly the successor religion to Atonism. He believed in reincarnation, and he was well-versed in Egyptian magic. Barry Wynne says that he spoke to Tut's Ka doubles in front of the inner chamber every night before he opened it. He must have been aware of the way the Amun priests attempted to curse the tombs, yet, he was very strong-willed, unlike Carnarvon, and was less likely to succumb to psychic attack."

"Could it be," asked Della suddenly, "that Carter, Carnarvon and Lady Evelyn were there to try and release Tutankhamun from the prison he had been placed in by the Amun priests? Maybe that was their goal, though they may not have known it, to release Tutankhamun, just like Meketaten had been released by Count Louis Hammon?"

We looked out at the tomb where King Tut's *Ka* still lay in suspended animation, hardly a hundred yards away from where we sat. If that had been their goal, they had not succeeded.

Back at Luxor, we went for a walk along the Nile and up to the Temple of Karnak, the 3,400-year-old temple of Amenophis III. It is crowded with the monumental architecture of Egypt's Age of Empire. Within the 60-acre complex are temples, chapels, giant stone pharaohs, commemorative obelisks, and the largest columnar structure ever raised, the Temple of Amun. We caught the nice sound and light show that evening. On our way back to the hotel, there was a sudden black out all over Luxor. We stopped in the pitch blackness to let our eyes get accustomed to the dark.

There was a nervous tension for a moment, and then I said, "Do you suppose someone else has just succumbed to the curse of the pharaohs?" And with that we had a much-needed laugh.

§§§

A few days later we said good-bye. The ladies were off by train to Cairo, and I was taking the train to Aswan, only a few hours south along the Nile. I went second class air-conditioned, in comfortable reserved seats that were much like airline seats, reclinable and spacious. We were constantly amused on the train. Even a magician came to our car and did magic tricks, and then passed a hat. At other times I was engaged by some Egyptian wishing to practice his English and learn about America.

Aswan is the site of the astonishing granite quarries that supplied much of the hard, red stone for Egyptian monuments. At the northern end of the quarries is the largest obelisk ever cut in Egypt. It was left unfinished, probably because it split before it was completed, and was therefore of no use. If completed it is estimated that it would have weighed between 1,168 to 1,650 tons. The labor involved seems unimaginable, and just how the obelisk would have been moved out of the cut rock trench around it is not known. Some believe that the solid rock walls enclosing it would have had to have been cut away before it could have been moved. Some of the mysteries that come into play here are whether some form of levitation, as some people claim, was to be used on the obelisk, whether the obelisk was cut with some advanced cutting beam or by simple wooden wedges which are placed in holes drilled into the rock and then soaked to split the obelisk away from the bed rock. In theory, the obelisk was then polished with dolerite pounding balls. Dolerite is even harder than granite, one of the hardest known stones, and a serious problem is just how Egyptians even created these pounding balls, as dolerite is almost impossible to cut.

Also in Aswan is the beautiful temple of Isis at the island of Philae near the first cataract. I wondered if this was the Temple of Isis that that was alleged to hold the Ark of the Covenant when it was first brought from Atlantis and before it was allegedly placed inside the Great Pyramid. [37]

Aswan is a very pleasant, and, for Egypt, relaxing town. Here locals do less begging of tourists for "baksheesh" (tips), and there are no huge crowds of people fighting to get on buses. It is a peaceful spot, good for walks along the Nile, afternoon sailing among the islands of Aswan in a felucca, or just watching those lovely sailboats from the balcony of a hotel.

I checked into the Aswan Palace and shared a room with a young French traveler, his head shaved, perhaps because of lice, and with one of the smallest backpacks I had ever seen. The room had a great view of the Nile, and we used an old but clean bathroom down the hall. I spent several days in Aswan just relaxing. I wandered in the market, ate Egyptian sweets, looked for 3-piastre eggs, drank tea and sat for hours every day by the Nile watching sailboats, called "feluccas," sail up and down the Nile.

It was very pleasant, but eventually my dwindling supply of money—I had less than $150 left—made me realize that my stay in Egypt had to come to an end. I would soon be taking the ferry to Sudan. I wanted to get as far into Africa as I could before I ran out of cash. I bought some provisions for the trip and went by train, only half an hour's journey, to the New Harbor just past the High Dam, built by the Russians in the early sixties: a monumental task Egyptians like to compare with the building of the pyramids.

There is a twice-weekly steamer on Lake Aswan between Sudan and Egypt. It leaves the docks at the High Dam a few miles north of Aswan and takes two nights and a day to make the trip to Wadi Halfa, the northernmost town in Sudan.

On the evening of the departure I was standing on the docks at sunset, waiting to board the ferry. I got a space on an upper deck sitting with some other travelers, and later in the night the ferry started to move. In the morning I found we had indeed gone somewhere: to a small mooring about a hundred yards away from port! Shortly after breakfast we were really under way, the diesel engines sputtering us along. The shores on either side of the long, skinny man-made lake were absolutely barren and rocky. There was not a plant or village in sight.

The ferry passes by the fantastic, colossal statues of Ramses II and the Great Temple behind it. However, this was at night, so I couldn't see the temples from the boat. There is a hydrofoil that takes tourists from Aswan, but it is often being repaired. It is also possible to fly to Abu Simbel, the name for this site of statues and temples, but that is quite expensive. I just had to pass it up.

On the ferry the travelers chatted and told stories they had heard of Sudan, Kenya, and other countries. We were all excited to be heading into the heart of Africa. On the second day of the journey, in the late afternoon, I leaned against the boat rail. The sun was just setting to the west, turning the sharp barren mountains of the Western Desert a deep orange-red. I thought of the strange experiences I had had in Egypt, it was a land that was much more than temples, tombs and pyramids. It was a journey into the remote and incredible past, a past that was far removed from my own civilization, though it seemed to me no less technologically advanced. I smiled to myself about the many mysteries that I would never be able to solve and watched the orange glow of the sunset. In the morning I would be in Sudan.

The Great Seal of the United States...a prophetic message? A mystical reminder? The Latin phrase at the top translates into "He has looked with favor on our beginning." The bottom phrase translates into "The new order of the ages."

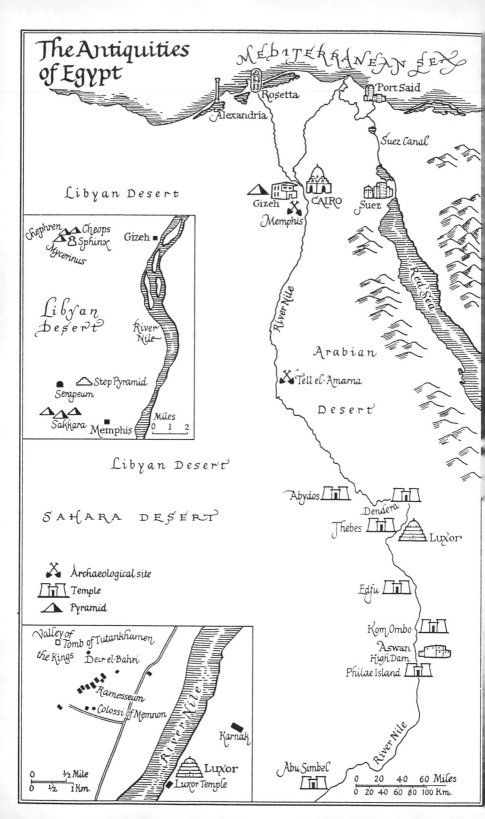

The Antiquities of Egypt

MEDITERRANEAN SEA

Rosetta
Port Said
Alexandria
Suez Canal

Libyan Desert

Chephren · Cheops
Mycerinus · Sphinx
Gizeh
Gizeh
CAIRO
Suez
Memphis

Libyan Desert

River Nile
River Nile

Step Pyramid
Serapeum
Sakkara · Memphis
Miles 0 1 2

Arabian

Red Sea

Tell el-Amarna

Desert

Libyan Desert

SAHARA DESERT

Abydos
Dendera
Thebes
Luxor

✗ Archaeological site
Temple
Pyramid

Edfu

Kom Ombo
Aswan High Dam
Philae Island

Valley of the Kings
Tomb of Tutankhamen
Deir el-Bahri
Ramesseum
Colossi of Memnon
Karnak
Luxor
Luxor Temple

0 ½ Mile
0 ½ 1 Km.

River Nile

River Nile

Abu Simbel

0 20 40 60 Miles
0 20 40 60 80 100 Km.

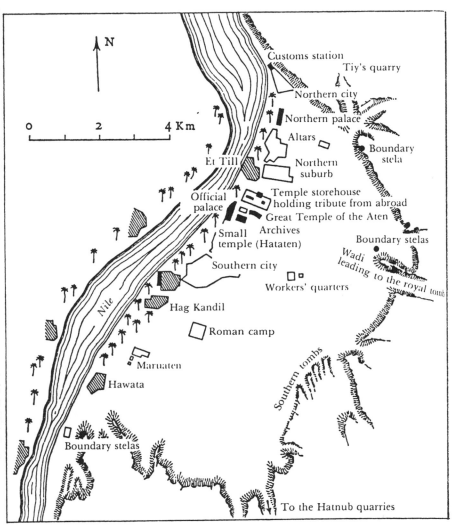

Map of Akhetaton, built on a semi-circular plain in the center of Egypt at a previously uninhabited place.

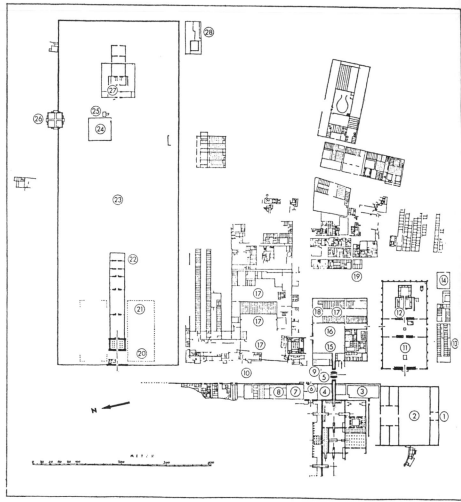

Survey Map of the Inner City of Akhetaten

1. Throne room
2. Great pillared hall
3. Storehouse
4. Southern harem
5. Window of Appearances
6. Courtyard
7. Northern harem
8. Garden
9. Bridge
10. Royal Road

11. Altar
12. Sanctuary
13. Temple storehouse
14. Lake
15. King's house
16. Garden
17. Storehouse
18. Pond
19. Archives
20. Perhai

21. Offering stones
22. Gematen
23. Great temple
24. Slaughterhouse
25. Stela
26. Hall Where
   Tribute Is Received
27. Sanctuary
28. Panehesi's house

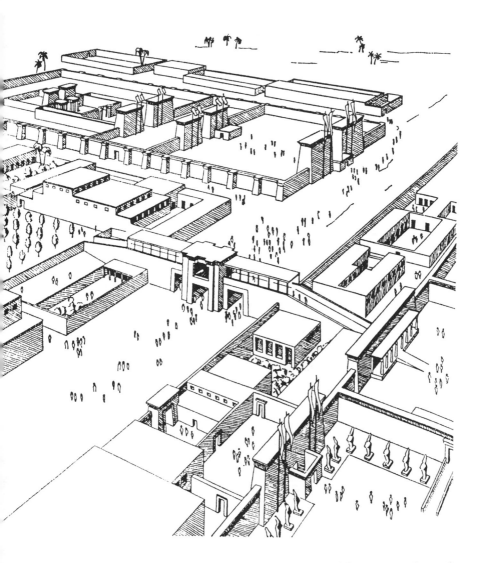

Reconstruction of the town center of Akhetaton by the archaeologist Pendlebury. The administrators of the kingdom were able to pass from the Governmental Palace on the lower right, to to the Residential Palace on the upper left by way a covered bridge.

Akhenaten and Nefertiti (visible in the double profile of Akhenaten) at the great reception that took place in year 12 of Akhenaten's reign. Their six daughters stand behind them. (Drawing by Norman Davies from a wall relief in the tomb of Meriere in Amarna.)

Nefertiti embraces Akhenaten, who steers the chariot of state with a sure hand, while Meritaten leans across her father's quiver in order to spur on the horses. (Norman Davies's restoration of a damaged ink drawing from the tomb of Ahmose in Amarna.)

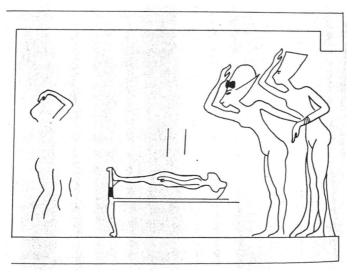

A reconstructed drawing from a tomb at Akhetaton. Here, presumably Akhenaton and Nefertiti view Meritaten, who was killed by Amun priests after she attacked Karnak Temple in Luxor. Her right hand was cut off and mummified.

A reconstruction showing the great hall, the pool which is
Strabo's well and the "tomb of Osiris"

A view of the Osirion, or "Tomb of Osiris" (also known as
Strabo's Well as reconstructed. This structure, like the
Sphinx and the Valley Temple, predate Egypt, have no
hieroglyphs, and are presumed to be remnants of the ancient
Osirian Civilization. From the *Illustrated London News.*

Napoleon's Artist's Impression of the Temple of Denderah in Operation.

According to John Anthony West in his scholarly work *Serpent In the Sky: The High Wisdom of ncient Egypt*:

The aim of all initiatic religions is the same the world over and has always been: to guide man om his natural state of consciousness (which is called 'illusion' or 'sleep') to a higher state. This igher state, (called 'illumination' or 'the Kingdom'), his destiny and birthright is, within the ontext of the natural world, 'unnatural'.

It is impossible to construct a rational argument to compel the unwilling and the immune into cknowledging either the existence or the importance of this higher state. It is impossible to rove' to the skeptic that this has any direct bearing upon hmself or his life. Initiatic writings are erefore comprehensible only to initiates, to those who have at least put one foot down upon e long road. And the further along the road, the greater and deeper the understanding.

شكل «١» منظر يمثل الآلهة
ايزيس على عرش وهى ترضع ابنها
حورس .

The goddess Isis upon her
throne, nursing her son Horus.

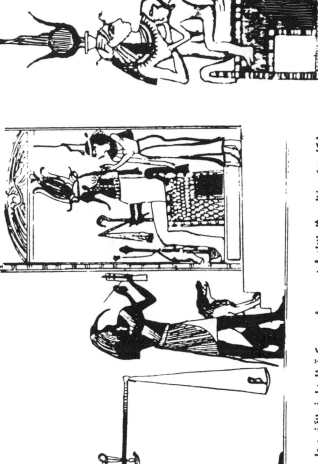

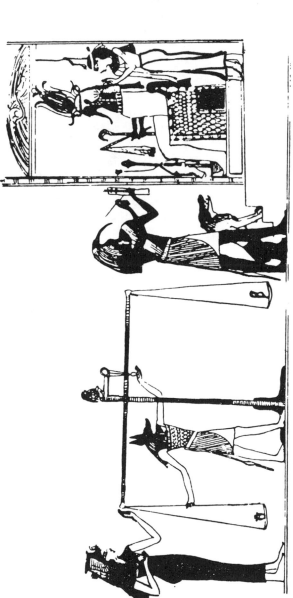

شكل «٤» منظر يمثل الإله اوزيريس رئيس محكمة العدل فى الآخرة على عرش وامامه الميزان وهو يراقب وزن قلب المتوفى لفحص أعماله فى الدنيا . الإله انوبيس

The god Osiris upon his throne before the balance presiding over the weighing of the heart by Anubis, at the
Last Judgment.

شكل «٥» صورة تمثل الشهيد سركيس وهو يدوس بجواده التنين او الشيطان وهو شبيه بإله الشر «ست» «المصري القديم».

Picture of the martyr Sarkis crushing the dragon or the demon identical with Set, the ancient god of evil.

شكل «٧» منظر فرعوني فريد منقوش على جدران معهد هيبس بالواحة الخارجة توضح الإله حورس وهو يطعن الحية العظمى التي ترمز إله الشر «ست».

Iconographic pharaonic scene in the temple of Hibis at kharga Oasis depicting the god Horus killing the great serpent, which symbolises Set god of evil.

In this detail from the Ani Papyrus, Isis, wife of Osiris and Nephthys, are worshipping a Djed Column on top of which is an ankh holding an orb. Is this an electric generator with a lamp on top?

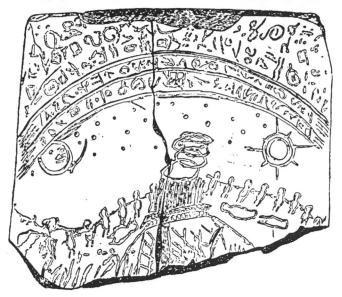

Djed Column on the Davenport Stele of Iowa, as translated by Barry Fell.

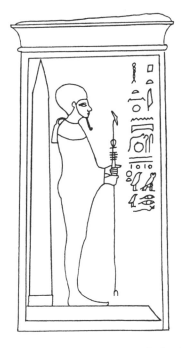

Ptah, Master Architect of the World. He holds a Djed pillar combined with an ankh and stands next to an obelisk.

Djed Columns

Coptic Christian icon of the Archangel Michael holding the balance in his left hand and the Djed cross in his right. Coptic Christianity, and Christianity in general, takes much of its symbolism from the ancient Egyptian Mystery Schools.

Relief from the underground chamber in the Temple of Hathor at Dendera. It shows priests carrying devices attached by a braided cable to an altar. Also supporting the devices (projectors or lamps?) are Djed columns, the theoretical direct current electricity generators. The snake inside the "lamps" may be the "filament".

A Djed pillar being raised by Amunhotep III with ropes. On the right the King is making an offering to another personified Djed pillar, probably Osiris.

Howard Carter and an Egyptian assistant remove the sticky unguent material which covered the space between the second and third coffins. The third coffin, made entirely of gold, contained the mummy.

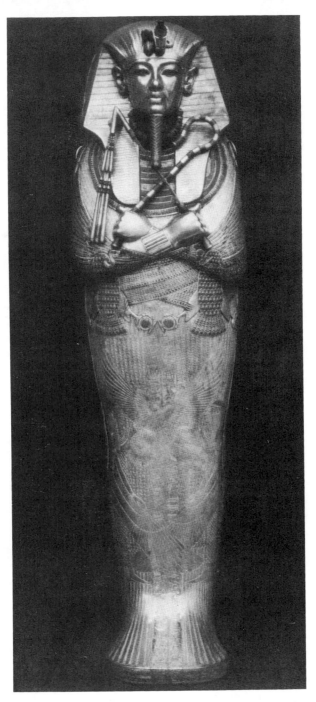

Lid of the third of the three nested coffins from the tomb of Tutankhamun, after cleaning. Solid gold, the king holds the crook and flail, symbols of his authority.

Colonnade to the temple of Amon-Ra. Drawn by Napoleon's savants in 1798. Aimed like a telescope, the colonnade and the temple's axis oriented to the summer solstice.

Chapter 6

SUDAN:
THE LOST WORLD OF KUSH

*The joy of travel is not nearly so much
in getting to where one wants to go
as in the unsought surprises which occur on the journey.*
—Alan Watts, The Way of Zen

The sun was bright and hot at mid-morning when the steamer docked at the Sudanese port of Wadi Halfa. I had no sooner stepped off the boat when the dark, robed, customs officer asked me if I had anything to declare. He squinted at me in the bright morning sun, a faded blue-grey hat covered his dark brown, balding scalp; and short locks of thick, black, curly hair popped out around his ears. Although his skin was quite dark, his nose was long and narrow, though his lips were fuller than an Egyptian's. On the corners of his eyebrows were four tribal scars. He wore a long white robe which was thoroughly in need of washing. If it hadn't been for his official cap, he'd have been just another body in the crowd of white robes, which is what most men in the Sudan wear.

I thought for a moment, and then told him, No, I didn't have anything to declare. He smiled broadly and genuinely. "Welcome to the Democratic Republic of the Sudan. Enjoy your stay."

"Shukran, (Thank you)!" I answered in Arabic. I felt as if I had just been welcomed personally by the president himself. On that positive note, I officially stepped into the largest country in Africa.

Northern Sudan, like southern Egypt, is as barren as desert can be—no sign of vegetation anywhere. The flat landscape is broken by jagged volcanic mountains which rise several hundred feet out of the thick brown dust that covers everything.

In the middle of all this dust and scorching heat squats Wadi Halfa, not one's idea of an idyllic desert oasis. A small town of about a hundred mudbrick houses, the main reason for its existence is as a border post and the northern terminus of the extensive Sudanese railway system. There's also a satellite telecommunications station here, built by the Americans, and of course the small port immigration station on Lake Aswan for the twice-weekly ferry to Egypt. It is really not a spot that encourages tourists to linger.

Just near Wadi Halfa, however, is the ancient fortress of Buhen. A massive structure built out of mudbrick, the same construction material used today in Sudan, Buhen was built at the site of the second cataract by Egyptian Pharaohs of the Middle Kingdom (2040-1783 B.C.) as a fortress to guard the Egyptian Empire's southern borders. To the south of the Egyptian Empire were the mysterious empires of Kush and Abyssinia. The fortress of Buhen was within Nubia, a client state of Egypt itself.

Buhen was a huge fortress, in fact a group of fortresses, with towers, bastions, and redan walls, all very modern in conception. The huge gateways were particularly impressive to the British archaeologist W.B. Emery, who excavated

153

the fortress and drew a reconstruction of it in 1959. Unfortunately, Buhen, like many magnificent monuments along the Nubian Nile, was flooded with the creation of Lake Nasser in the 1960's.[74]

Also near Wadi Halfa, at the fortress of Mirgissa at the second cataract, was a slipway, a sort of primitive canal that allowed Egyptian vessels to skirt the rapids of the Nile by means of a cut through the desert across a loop of the river. A clay channel some two kilometers (1.21 miles) long was kept moist and slippery, and the cargo and ships would be hauled along it separately.[74]

As one who enjoys just sitting in a sidewalk cafe and watching the world go by, I immediately noticed a small teastall off to the left of the docks by the lake. Walking up to it, I asked the young, dark Sudanese kid, in the Arabic I was starting to pick up, the price of a cup of tea. Since I was now in a new country, I wanted to get a feel for prices and a cup of tea was a good place to start.

"Three piasters," said the kid, who was probably about ten years old. His bright white teeth contrasted strongly with his charcoal face.

"Fine," I said in Arabic, "I have to change some money." He smiled and said in simple Arabic, "That's OK, you don't need money. Here's a cup of free tea."

I was truly stunned. A camel could just as easily have come up and kicked me in the head. We certainly weren't in Egypt anymore. There you would be charged double for a cup of tea!

I sat and had a cup of pinkish-red karkadee tea with the young boy. Karkadee tea is made from the red petals of the hibiscus flower. It is served with sugar (lots of it), hot or cold. This was the first time I had ever had any, and it was delicious.

My new friend, the tea seller, was pretty tall for his age, I thought. Lanky and slender, he wore a long, dusty, white robe, like almost everyone else, and a white knit skullcap on his head, indicating he was a Muslim. His hair was short, tight and curly, and he had beautiful dark brown skin, with those customary Sudanese tribal scars on his eyebrows.

Casually sipping my hot tea, I gazed out at the architecture of Wadi Halfa. Low one-story houses made of brown mudbricks, they were square and typically had four or five rooms. Surrounding each home was a nine-foot wall of the same bricks, making each house a sort of small brown dirt fortress. The houses of the more affluent had broken shards of glass all along the top of the wall to keep out thieves. Looking out past the scattered brown buildings was the train station at the far end of town; beyond it was a barren sea of dust.

"Better get a place on that train," said my kind tea server, "it fills up fast." And, in fact, as people got through immigration they were making a mad dash for the train. I'd heard that Sudanese trains can be pretty crowded, so I grabbed my pack and walked across town to get a third-class seat on the train.

There was an English couple already on the train who said they would save me a seat next to them, so I left my luggage there and went back into town to change some money and buy a train ticket. Then I had some more tea and small cookies at my favorite tea stall with that lanky kid and passed away the rest of the hot, dusty day. The train filled up by late afternoon and we started on the trip to the south. It is theoretically a twice-weekly, twenty-four hour trip to Khartoum from Wadi Halfa. I say theoretically because, like all things in Sudan, trains don't really run the way they are supposed to. The Sudanese are used to this, of course, and it is the generally accepted way of life that "if anything can go wrong, it will."

A train is not really late until it is three or four days behind schedule. Boats that run "regular" schedules on the Nile are known to take an extra two weeks because of variable conditions. The first words learned and those most often used in Sudan are "tomorrow" (bukra), "it doesn't matter" (mahlesh), and "if God wills it" (inshallah). It is this characteristic of Sudan that makes its people probably the most amiable, easy-going and patient of any in the world. You will rarely see a Sudanese in a hurry or angry because something hasn't been done. After all, it will happen tomorrow, if God wills it, and it really doesn't matter anyway.

After a fitful first night on a crowded seat in the third-class car, I spent the entire second day on the roof of the train. Having noticed a few Sudanese up there at some of the stops, I climbed up and sat on top of the car. At first I was a bit afraid of falling off the side, but the train's motion was entirely forward and I was assured by the other travelers that there was no danger. It was cool as well, despite some danger of sunstroke. I kept my head covered to protect it from the unrelenting sun. The train generated a huge cloud of fine dust and it would get so bad inside the train, with the choking dust everywhere, that most of the passengers would breathe through a damp handkerchief so as not to inhale a lungfull of Sudan. That evening as the sun was going down, a steady warm wind blew across on the train roof. The small towns along the tracks began to light up, and I began really thinking about Sudan. Egypt was touristy and Mediterranean, but this was the real Africa! My whole body shivered with the joy of being there on that train, heading south into Africa, into the heart of the unknown, into something I knew nothing about; not what to expect, not anything really. With my mind empty of expectations, I was ready to experience Sudan.

Sudan is the largest country in Africa, about the size of the United States west of Colorado, from the Canadian to the Mexican border. Its population of fifteen million is concentrated in the fertile area around the two Niles—the Blue Nile coming from the mountains of Ethiopia to the southeast and the White Nile beginning at Lake Victoria in Uganda and flowing north to meet at the capital of Sudan, Khartoum.

I lay back on the roof and gazed up at the myriads of stars that prickled through the black sky like a trillion tiny holes in a tin lantern. I could not remember ever having seen so many stars. I'd read that Sudan was one of the most underdeveloped countries in Africa. Before 1964, when Sudan gained independence from Britain, it was known as the Anglo-Egyptian Sudan and was administered jointly by Britain and a semi-independent Egypt. After independence, Sudan kept close political and economic ties with Egypt and declared itself, like Egypt, a Socialist State. Like Egypt, it sought aid from the Soviet Union while at the same time fighting a long and economically frustrating Civil War between the Muslim Arabs of the North and the non-Muslim blacks of the South. In the 1970's, Sudan followed Egypt in breaking ties with the Soviets, and developing stronger ties with the west, though remaining "socialist."

What little we know about the mysterious kingdoms of ancient Sudan come from the enigmatic ruins and Egyptian Records. Almost nothing known about southern Sudan; it is the northern kingdoms of Kush and Nubia that are known to us through history. Since at least 2600 B.C., the land of Kush was a source of gold, ivory, incense, ebony and slaves (although the ancient Egyptians reportedly did not use slaves). The Egyptians made many forays into the nearby land south of them, and certainly at times must have controlled a large portion of the Sudan. They may possibly even have conquered Ethiopia and Somalia, known in Egyptian records at the land of Punt. Just how much they held of the Sudan, and indeed, the entire Sahel area of Africa (that area of arid scrub brush just below the Sahara) is open to conjecture. There is evidence that vast areas as far as Lake Chad and Timbuktu were once controlled by ancient Egypt. This is, of course, the same ancient Egypt that mined gold in Sumatra, Australia and Brazil and colonized much of the Pacific, including Hawaii. When modern Egyptologists cannot even imagine the Egyptians controlling the Sudan, it is almost impossible for them to conceive of the Egyptians controlling such farflung lands as Pacific Islands and mining areas in Australia.

Kush and Nubia weren't always controlled by Egypt. These areas were often independent, and at times even controlled Egypt! In the 8th century B.C., Kush conquered Egypt and set up the 25th dynasty. It didn't last long, however, and in 671 B.C., when the Assyrians invaded Egypt, the kings of Kush had to pull back to their old capital of Napata, far south of Egypt, near the fourth cataract. What kept Kush safe from the advances of foreign armies into the Nile Valley was cataracts in Nubia and the vast empty desert that separated southern Egypt from the fertile areas where the two Niles met. The Romans did manage to sack Napata in 23 B.C., but were never able to conquer the country. Soon after 654 B.C., the Kushite kings established a new capital at Merowe, further south than Napata, probably because they were aware that the more northern cities like Napata would come under siege from Egypt's new rulers, the Greeks and Romans.

It is interesting to note that during the period of domination by the mysterious Hyksos in Egypt, a Theban prince organized a revolt. Curiously, the Hyksos sought help from Kush in the Sudan.[74] The Hyksos lost, but this lends credence to the theory that the Hyksos, the so-called Shepherd Kings of Egypt, who took over the country peacefully, were themselves from Ethiopia or the Sudan.[37]

The new kingdom of Merowe was less Egyptian and more Greek, Roman, Ethiopian and Indian. Traders from India came to the Meriotic ports on the Red Sea, and a cursive script replaced Egyptian hieroglyphics. Because this writing remains undeciphered, much of the history of Merowe remains a mystery. The Meroites were big on building pyramids, either as tombs, or for some other unknown purpose. Certainly, the pyramids now found at the "Royal Necropolis" of sandstone pyramids were used at times as tombs. In 1836, the Italian adventurer Ferlini discovered a large treasure in one pyramid. Apparently the treasure of Queen Amanishakheto, it led to the destruction of other pyramids in search of treasure.[74] As far as can be ascertained, mysterious Merowe seems to have been ruled by women, perhaps by the original *Amazons*.

In keeping with the theory that ancient civilizations were far more advanced than generally thought, there is a very strange and interesting sketch that was discovered at Merowe. It appears in Philipp Vandenberg's book *Curse of the Pharaohs* [62] and is reproduced here. The sketch shows two men operating a device that is said to be identical to a radiation condenser or a laser gun. Others believe it is a rocket of some sort, a telescope or some sophisticated ray gun. Readers of this book may decide for themselves what it looks like. Academic "experts" have nothing to say on the subject, except that it can't be a laser, rocket or radiation gun because they didn't have such devices at the time, despite appearances. This is the same logic that one runs up against time and time again in the discussion of lost civilizations and ancient mysteries.

The remains of Merowe, with its pyramids and radiation condensers, were wiped out by the Axumite king Ezana in the 4th century A.D. Merowe became a Christian kingdom until the 7th century when Arabs began to penetrate Sudan and bring Islam with them, though this affected only the northern part of the country.

§§§

It was late afternoon when we passed by the now desolate city of Merowe. From the roof of the train I watched as we passed the pyramids to the north of the city. They are small, at least in comparison to those at Giza, but nevertheless interesting, silhouetted against the bright blue sky and monotonous brown desert.

The train arrived in Khartoum in the middle of the night. Tired and dirty, the English couple and I managed to walk to the Youth Hostel, which is about eight minutes along the railroad tracks to the north from the main station. After a good sleep in a bed, we explored Khartoum the next day. The houses, as in Wadi Halfa, were of baked mudbricks. There were a lot of them, each with its surrounding wall and garden and terrace inside. Because of the hot dry weather, Sudanese like to sleep outside a lot, dozing on their jute-string beds in their gardens. In the downtown are more modern government buildings and hotels, but they look rather rundown and dull, concrete and square.

There is also the confluence of the Niles and the more authentic-feeling city across the Nile, Omdurman. In Omdurman you'll find a great bazaar and camel market. The tomb of the Mahdi is to be found here as well, and a special treat is the Sufi dancing at a Saint's tomb on the northern outskirts of Omdurman.

Sudan has very few roads to speak of. They are mostly dirt tracks that run off through the desert or forest, each truck following the tracks of the truck that went a few days before him, hoping that driver knew where he was going. In all of Sudan there is about fifty miles of paved road, from Khartoum south to Wad Medani. For the adventurous, an interesting way to travel around Sudan is by camel. Camels are relatively cheap in Sudan; you can pick one up at your local used camel dealer in Omdurman or other towns in the north, and ride him all over Sudan. Camels are a great deal more expensive in Egypt, and it can be quite profitable to take a camel to Egypt and sell it. Generally speaking, the further north you go with your camel, the more valuable it is.

As I stood in front of the Presidential Palace where General Gordon was beheaded by the Mahdi's fervent warriors, I thought about the strange events that can sometimes cause mighty empires to fall.

In Arabic, a war led by a religious leader for some sacred cause, such as ridding your country of foreign rulers, is known as a *jihad* or holy war. In 1830, the Sudan was conquered by the Egyptians and a Turko-Egyptian rule was established under the Ottoman Empire. Then in 1881, Mohammed Ahmed Al-Mahdi (the Mahdi) led a successful revolt, a *jihad*, against the Turko-Egyptian rulers. General Gordon, who was employed as the Governor of the Sudan by the Egyptian Government, failed to halt the revolt of the Mahdi's fanatical *mujaheedin* (holy warriors) and they swept over Khartoum, killing Gordon and all other foreigners.

The Mahdi was then in control of Sudan, but he died soon afterwards. He was followed by Khalifa Abd Allahi who resisted the Egyptians and British for more than fifteen years, but in 1898 the Khalifa was defeated at Omdurman and Anglo-Egyptian rule was reestablished for some time.

Whenever possible, I preferred to travel by train. Sudan has an extensive railway network, but trains are slow and chokingly crowded. At the Youth Hostel in Khartoum I met a Chinese-Canadian biologist named Blythe, from Vancouver, who was interested in traveling to the South of Sudan. After a few days of sight-seeing in Khartoum and Omdurman, we decided to buy tickets on the weekly train to Wau, the southern terminus of the Sudanese railway system. Because of my limited funds, I convinced my potential traveling companion to buy a fourth-class ticket with me, one that was totally unreserved. We were advised to board the train early as it tended to fill up very fast—just how fast and full we found out too late. It turned out that most people actually spend the night on the train—the night before, even though the train doesn't leave until noon the next day. We arrived three hours before departure and the train was already completely stuffed from fourth class through second class. It was not only impossible to get a seat, it was impossible to make our way through the cars, as the aisles were full of people, chickens, beds, suitcases, and other belongings. We had been warned! Although we didn't have first class tickets, we took advantage of the fact that we were obviously tourists, and juggled our way into first class and settled down in the hall, where we spent the night.

The next morning at our first stop, I got off the train and had a quick look around. The sight that greeted my eyes was astonishing. The inside of the train had been completely packed; now the top was just as crowded! Trunks and mattresses, chickens and people were tied or hanging onto the top of the train. There were easily several hundred people up there, enjoying the view and squinting in the blazing African sun.

I climbed up the side to the roof (on the theory that there is always room for one more) and spent the rest of the day relishing the best view I'd ever had on any train. Looking forward I could see the whole length of the train, the diesel engine up at the front dragging us west along the semi-arid grassland of central Sudan. The whole top of the train was in high spirits; we'd all wave at the occasional shepherd with his flock or nomadic family on the move. People would share a mango or some bread. Life took on the spontaneity of the moment; it was good to be alive and we all knew it. A certain exhilaration flowed down the length of the train, through us and into the sunset that was already engulfing us.

The scenery was semi-desert, flat with dry grass and bushes and an occasional baobab tree to break the monotony. Baobab trees have large, thick, short trunks that suddenly flare into leafless branches. During the winter, young trees flower with pink blossoms, still without leaves. I lay back on the roof of the train and dreamed of lost civilizations, lost technology and the mighty empires that once ruled this ancient land.

Someone nudged me and I suddenly sat up where I was lying on top of the train and looked at the gentleman sitting next to me. He was tall with a strong figure. His robes were immaculately white, a nice lizardskin dagger stuck in his belt, and a huge white turban wound around his brown, shaven head. His face had three tribal scars on each cheek and as I stared unconsciously he gave me a wink. Was this the Mahdi?

We crossed the White Nile at Kosti and continued toward the junction of El Obeid and Babanoosa. Toward sunset the next day, I noticed a dark swarm moving along the tops of bushes toward the train. The swarm covered the whole horizon and flew together in a wave motion. Soon we were engulfed in a cloud of locusts, thousands of them blocking out the sky, covering the train and all the bushes that we could see. Everyone was covered with locusts. I was freaking out, brushing them wildly off me, but three would take the place of every one I brushed off. Finally the train plowed through that cloud of locusts as one drives through a rain storm, and it was a great relief for me. The Sudanese on top of the train didn't seem to have minded a whole lot, perhaps some even enjoyed a quick meal out of a handful of locusts. It was dark by now, so at the next stop, I got down from my perch and went into the first-class car where my companion was waiting. Blythe, the Canadian biologist, had preferred sitting inside, wisely keeping our precious spots in the aisle, and there we spent another night in the first-class hallway.

South of us were the Nuba mountains and the remote mountain village of Kadoogli. Here in these mountains lived a strange tribe known as the Nuba. Their customs include painting their bodies, ritual scarring and intense ceremonial dancing leading to trance states. Were the Nuba the last remaining vestige of the ancient civilization of Nubia? An Egyptian Nubia, and independent Nubia, could and probably would have extended as far as the Nuba mountains. The very name suggests it. The Nuba, like many civilizations, must certainly have slid back on the ruler of civilization. The technology that was once part of Nubia, whether imported from Egypt or not, was not apparent in the current lifestyle of the Nuba, except in their metallurgy.

The next morning we awoke to find the train not moving. Stepping outside, we saw we were not in a station, but out in the middle of the desert. During the night we had passed El Obeid and were halfway to Babanoosa, the main station in western Sudan. The engine had broken down before dawn, we were told, so we would have to wait for relief in the form of another engine coming from Babanoosa. We waited and the sun beat down.

At mid-morning all the water reserves on the train ran out and by mid-afternoon our own canteens were empty too. I looked around the train and found the army car which is attached to every train going south and west. The soldiers offered me some water and I gratefully filled my canteen. Still stranded in the desert, we spent another night on the train. Blythe and I made dinner out of halva, a sesame seed candy, and bread. About nine the next morning, a put-put car arrived from Babanoosa. After hours of tinkering, the engine was running and an hour and a half after that we pulled into a small desert station and happily replenished our water supplies at a small well pump, while the engine took on water from a tank. Toward evening, we finally arrived at Babanoosa, where we would change trains for Wau.

There was a mad rush of people trying to get a place on the train to Wau. Hoping to get a seat this time, we hurried over to the waiting line ahead of most of the others who had, it appeared, whole households to move. To our dismay, we found the train already full of "squatters" who had been waiting several days (as the train ran once a week at best). While most of the people fought for a spot on the roof, we once more settled in the first-class hallway.

Later, I walked around the station in the evening by myself. It consisted of only a few small buildings and a ticket office, but many people cooking and sitting around fires or selling food. I talked with a young black private in the army, a tall, very dark Negro from the South. Clean-shaven and handsome, he wore a neatly pressed but old soldier's uniform and he spoke remarkably good English. He had gone to a missionary school in Equatorial Province, he said. I commented on how crowded the train had been coming from Khartoum.

"If you thought that was bad, wait until you see this one to Wau—now that's crowded!" It was difficult to imagine, I admitted. He went on to explain our breakdown in the desert; apparently, the conductors had managed to entice some girls from the south into a first-class compartment, became drunk, and let the engine overheat. They didn't bother fixing the engine, so we sat there for a day waiting for a new one to come from Babanoosa. I was shocked at this, so he further explained that the reason the train went so slowly and stopped for such a long time at each station was that the engineers got paid by the hour, so naturally they took as long as possible to get from Khartoum to Wau. Like everyone else in Sudan, they were in no hurry.

We spent another night in the hallway of the first class and when we awoke the next morning, we were still in the station! Finally, by mid-morning, we left. We had been on the train now for four nights and this was our fifth day. In those five days we had gone about 800 miles. "Right on schedule," the conductor told me.

I was glad to be leaving Babanoosa, though it was a nice enough town. It had a movie theater and a hotel, plus a few guys that hung out on the street and sold bowls of beans. Plenty of nomads with lots of character cruised through town as well, though we were afraid to go too far from the train, lest it suddenly depart without notice, as it was wont to do.

We were now heading due south and the transition of semi-desert brush to savanna was becoming apparent. The Arab influence was disappearing and we were getting into black tribal areas. Bare-breasted women could be seen along the tracks, and occasionally a naked man, sometimes covered with white ash. The men were usually carrying two spears, one to throw and the other to keep in case they were attacked by the wounded animal—or person for that matter.

All day I rode on the roof with hundreds of other passengers. Several times the conductor came crawling along, maneuvering over chickens, beds and bodies, to check tickets. I would obediently dig into my pocket to produce a ticket stub—one of the few who ever did produce a ticket up there on the roof—but no one was ever thrown off.

The frequency of stops was about the same, but food was getting noticeably scarcer. At each stop I climbed down to stretch my legs and look about the station, which at best was a village with a small market. Usually the only things to eat and drink were homemade peanut butter rolled up into balls and sold for a penny apiece, watermelon and beer in a gourd that was a sort of fermented sorghum sludge. Strolling about the station, naked men would come up to me with great grins on their faces and eagerly shake my hand, holding their spears in their other hand. No fear of foreigners for these people; they were genuinely glad to see me and honored that I had come to visit their country. However, I was obviously not an Arab, and the rift between the blacks of the South and the Arabic Sudanese of the North was a wide and violent one, as I would soon find out.

On the morning of the fifth day, the train broke down again, but the engineers fixed it themselves this time, and within a few hours we were going once more. At noon we came upon a small river that hadn't dried up, the first since we crossed the White Nile at Kosti. The train stopped and all the men jumped off, stripped away their clothes and plunged in the water. The bath break took about an hour, and then we were again on our way.

That evening we stopped within seventy miles of Wau at a small town named Aweil. I was invited to dinner at a schoolteacher's house by two of his students, who were on the train. Thinking the train would be there in the station for several hours as usual, I didn't hesitate to accept. Half an hour later we were eating beef stew with pieces of bread and fresh onions, being careful to reach in with only our right hands to scoop up some spicy meat or soak up juice with our bread. Suddenly a young lad came riding up on a bicycle, chattering excitedly. The train had just left—with all my worldly goods on board! Fortunately, my Canadian friend was still with the train.

I was rather upset, but the Sudanese did not seem to care about having missed the train. We finished dinner and looked for a truck to take us to the next station. Not finding one, we all spent the night with the teacher, and the next morning got a ride in a Landrover to Wau. The trip took only three hours, not including the two hours spent vainly hunting an elusive gazelle with the chief of police, who would occasionally let out a shot from his rifle while driving.

When we arrived in Wau, we found that the train still hadn't arrived, having (not surprisingly) broken down some ten miles out of town. I ended up waiting at the station until midnight for the train's arrival. My traveling companion breathed a sigh of relief when I jumped on the train to help unload our packs. He was not so much relieved to see me again as glad that this journey was at last over.

§§§

Wau is the second city of the South, Juba being the capital. There is really not much to say about Wau, which is a town of about two streets, each two blocks long. In the market I saw blacksmiths hammering spearheads out of old car springs, and cobblers making shoes out of camel hide. It was a good place to buy a snakeskin knife, some milk gourds or other nomadic knick knacks.

We needed a good rest after that train ride, so Blythe and I spent three days resting. The first two days, we just slept in the comfy beds of the youth hostel. Later, we ventured out into Wau to do the town, so to speak, all two blocks of it.

"What would life be without its little challenges?" said Robert Maker, a student from the Tonj Institute, a school a day's journey to the east. We were staring into the fire of a blacksmith in the market. He was busy hammering out a knife from some odd hunk of metal. The man was naked except for a loincloth and leather sandals. His body glistened with sweat as he squatted in front of the fire. "We are continually being tested by God and growing as fast as we can meet our tests," Robert went on.

"That train ride was a pretty good challenge," I confessed. "My Canadian friend is still recovering!" Indeed, Blythe was prone to spend long periods napping and resting in the cool of the youth hostel. We both slept especially well for several nights after the train trip. "I have to admit, though, that I'm already looking back at that train ride with a certain nostalgia."

"It's like," Robert said, "everything has its positive and negative side. It's often easier to see the negative side of things, but we should all try to look at the bright side and consciously control our thinking. Look at events as precious opportunities to learn a little more about yourself and life. You may find that you are really having a great time all the time, but you just don't realize it. We must remember that we are all blessed by Allah."

The philosopher Robert and I smiled at each other. His white teeth and big eyes were a strong contrast to his dark, almost purple skin. Across his forehead were nine or ten thick scar beads, reminiscent of the makeup that Boris Karloff wore as Frankenstein's monster. "Such were the Saints," said Robert, "they must have been able to see the bright side of everything."

Two days later Blythe and I were on top of a gigantic Sudanese truck, plowing our way through the hilly, forested country along the border of the Central African Republic. My Chinese-Canadian companion and I were on our way to Juba, the capital of the Autonomous Southern Region, which was hundreds of miles to the east of us through the dense forests of Central Africa.

I lay back on a big bundle of cotton and stretched out, smiling foolishly at my good fortune. These "lorries," as the Sudanese and British call them, are like big cattle trucks, solid steel cages welded together in Omdurman onto British Leyland truck chassis. Like the old ships of the desert, they ply all the lonely desert tracks and swamp roads to the far reaches of the Sudan. We had organized our lift in the "souk" or market in Wau and were paying about ten dollars for the ride to Yambio.

It was several pleasant days on the road. We'd pick up and leave passengers, and stop for the night in some small village, where everyone would climb out and sleep on the ground. Blythe and I would each tie up a mosquito net to the side of the truck and sleep under it; women would usually sleep on top of the cotton, boxes and other goods in back of the truck; and the men would sleep, without mosquito nets, on blankets by the side of the truck.

Our mid-point was the small town of Yambio near the border of Zaire, a village with a few shops and about thirty grass huts, as well as a police station. We slept in our sleeping bags on the porch of the police station, which is a good alternative in the small villages in Central Africa where there are no hotels.

We waited for two days in Yambio and then picked up the owner of the truck, a Northern Sudanese Arab who wanted to go with his merchandise to Juba. Things were going just fine on that sunny winter day in Central Africa as we bounced and spun our way through the thick forest, all green and wet from the rains the month before, until the owner of the truck brought out a bottle of Sudanese date wine. The biologist and I each had a small taste but didn't like it, so the owner and a friend of his who was along drank the entire bottle in the space of only an hour.

There were eight of us in the back of the truck: the owner and his friend, another dark Arab from the north; Blythe and I; a woman in her late twenties and her ten-year-old son; and a couple of other Sudanese getting a lift to Juba. About half way through the second bottle of date wine, the owner announced in broken

English that he was a Communist and a Russian sympathizer. "Chinee—good!" he declared. "Chinee—good!" He would look at Blythe, very Chinese, and reach over and pat him on the leg. "Good—Chinee, good—Russia!" Then he looked at me, scowled and blurted out, "America bad!" He then flew into a rage and began hitting me very hard on the legs with his flaying fists.

Hurt and shocked, I tried to move away from him, but there was nowhere to go. The owner's friend tried to stop him, but the drunk politician was sitting just opposite me, and though I moved as far away as I could, he still was able to hit me.

"Chinee—good!" he said again as he calmed down. "Merica—bad!" and he flew into a drunken rage again and started hitting me. I tried to stop him grabbing his wrists. He was really hitting hard.

"If he does that one more time," I told Blythe, "I'm going to throw him off this truck!"

"Better not do that," he said, "he owns this truck. He'd probably make you get out and walk. And we're a long way from anywhere!" This was true, and the idea of being dumped out in this forest where there weren't any villages for miles and when the next truck might come by in a couple of weeks wasn't too appealing.

I suddenly remembered Robert telling me back in Wau that you don't always realize what a fun time you're really having....The owner dozed for a while in a fitful sleep, to wake every once in a while and hit me, screaming, "Kill, kill, kill, I kill you!" I'd kick him back, but not too seriously. Later he verbally attacked the lady and her son, calling her a whore and him a bastard, spitting at them and making them cry.

"Not the most uplifting experience I've every had," said the biologist as we got off the truck at Juba the next day. The owner seemed somewhat ashamed of his actions, but made no effort to apologize. Anyway, we were now in Juba and ready for another rest.

We checked into the Africa Hotel, one of the two hotels in Juba, and rested for a couple of days. Mostly Juba is for resting. At the time, there were twenty-odd travelers hanging out at the Africa Hotel, recovering from dysentery or waiting for a truck to Lodwar in Kenya, or waiting for the ferry to come down river from Bor, Renk, or some other swamp town where the ferry stops. As Juba is mainly a commercial center for the Nilotic tribes in the South, it is official policy that everyone must wear clothes in town. When a group of Nuers wanders down the Nile to Juba, they might only have one pair of shorts to share between them. The first puts them on and sprints into town, does his business, and sprints back. He then hands his shorts to the next guy, who sprints into town in this decency relay. All in all, Juba was a lot of fun.

Blythe and I made one side trip south to Nimule Park on the Ugandan border, to check out the elephants and other animals in the park. It is a day's truck ride from Juba, and has a police station and a few shops. However, there was absolutely nothing edible to buy there. Not one banana, watermelon, or anything else. Fortunately, we stayed with the game warden, who was able to feed us from his own private stockpile. Because there had been a sudden change of regulations in Juba, all the truck drivers who regularly drove up from Kenya through Uganda had to return to Nairobi and get their passports which they had never needed before. Consequently there had been no transportation from Nimule to Juba for four days.

Blythe and I spent two days in the park looking for elephants, found a lot of gigantic turds but not much else, and decided to go back to Juba. The next morning we went down to the market in Nimule to catch the first truck in five days to Juba. As I boarded the truck with an Australian couple, who had also been in Nimule at the park for two days, a young man from Uganda said, "Hi," to me.

He was on his way to Juba to visit some relatives, he said in very good English. Short and stocky, he looked to be in good health, and wore a pair of blue slacks and a brown plaid shirt. He had a hearty smile.

"How long have you been waiting here?" I asked him.

"Five days!"

I was genuinely shocked, knowing there was no food in the shops or market. "What did you eat?"

He looked at me with his big, white eyes. "Nothing," he said simply.

Such were those crazy days in Sudan before the war. The Green Monkey Disease had just finished ravaging the south, and Juba had been quarantined some months before. People had been dropping like flies around town, and the Sudanese government had been worried that the disease would spread to the rest of the country. They had no idea how far it would eventually spread. According to the World Health Organization (WHO), the first case of AIDS (Acquired Immune Deficiency Syndrome occurred when a Green Monkey bit some black guy on the ass in Southern Sudan. Yet, according to an article that appeared on the front page of the *London Times* in 1987, there was a strange correlation between smallpox vaccinations in Africa and the spread of AIDS. The article hinted that the Green Monkey Disease was not the cause of the AIDS epidemic, but rather that the virus was genetically engineered in a laboratory and then purposely spread via WHO itself in an effort to wipe out the population of Central Africa. Such an accusation carries with it far-reaching implications of a diabolical plot.

We discussed the Green Monkey Disease while we rode on a truck north up to Malakal and Kosti. It is a couple of days by truck up along the eastern side of the Nile to Malakal, and this can only be done in the dry season, which is winter. Malakal, a swampy little village on the banks of the Nile and the official starting point of the "Black South," has one hotel, the Upper Nile Hotel. Two days before we arrived there, the owner, an Arab, and his friends had been sitting out front having some afternoon tea when a black guy came riding along on a bicycle and tossed a hand grenade into the party, killing everyone, including the owner. This was my first introduction to the tension that exists between the Arabs of the north who control the economy and force their Islamic laws on the non-Muslim blacks of the southern region, who resent it.

Back at the youth hostel in Khartoum, I was shocked to discover that I had scarcely seventy dollars left! I wondered how I was going to make it back to Europe. If I could get to France or Greece I could pick fruit for a while, or ask my parents to send me some money. Then I remembered an Australian guy I'd met in India who told me he had gone from New Delhi to Amsterdam on forty dollars!

On an inspiration, I applied for a job at a geophysical company to work at an oil camp in southern Sudan. They weren't hiring, they said, but they referred me to another company which might be interested. I went there and was hired as camp office manager by the Kenyan-American Catering Company, which was feeding the people who worked on a multi-million dollar oil exploration project in central southern Sudan.

My first job was to work in the office in Khartoum, expediting materials to the camps and managing the huge shipment of frozen meat and fresh vegetables that came in from Kenya twice a week. For this I was given a security clearance for the Khartoum airport. Our office and staff house was the same. We were five of the craziest, wildest guys in Khartoum at the time, and had a reputation that fit the fact. We were three Americans, an English bloke, and an Ethiopian, all travelers who had been hitchhiking around Africa and found work when they ran out of money, except the Ethiopian, Gary, who was a refugee from Eritrea and the war there.

We'd work ten hours a day, seven days a week, but would party it up four or five nights a week with some of our and Gary's friends. Our Eritrean cook would always leave a pizza or some super-spicy Ethiopian curry called zigney, served with a sort of sourdough spongy pancake-flat bread called engera. We'd cool down some of Sudan's local beer, called "Camel Beer," but known popularly as "Camel Piss," to drink and save our tongues as we ate zigney.

I had a habit of stopping at the same ice cream place every day I was downtown, largely because they had the best mango ice cream in Khartoum and the cutest little Sudanese girl at the cash register. Curly black hair was pulled back from her brown skin, and her big gorgeous eyes would open up very wide when I asked her what kinds of ice cream they had that day. Over a period of several months of going there and talking to her, I gradually became bolder and wrote her a note that said, "You are very charming." As my courage increased, I finally asked her out on a date, to have lunch with me and fellow staff members at our house.

Because Sudan is so hot, especially in the spring and autumn, private businesses and government agencies take a midday break of three hours from one to four p.m., when everyone naps. We did this in our office too, and one day during my nap, the charming young Sudanese girl came to visit me. Nazia was her name, and to my surprise, our house boy just let her into my room where I was reading and listening to music.

"Nazia! What are you doing here? This is a pleasant surprise!" She was nineteen and was on summer vacation from a girl's school that was the Sudanese equivalent of a prep school.

"Let's dance," she said, reaching down and turning up my cassette player. Shy and very sweet, her behavior was amazing in a Muslim girl. Even formal dating is not encouraged in Arab countries. We danced in my room for a while, holding each other close and swinging our hips. One thing led to another and soon we were kissing and hugging on my bed. I'm sure she had never done anything like this before; I would have been far too shy ever to have suggested such a thing.

Slowly I undressed her, as we gazed deep into each other's eyes. "My father would kill me if he knew I was here," she said. And she meant it. He would probably have killed me too. The life of a rogue archaeologist on the road can be very dangerous, but sometimes it had its moments.

SYRIA

BENGHAZI
ALEXANDRIA JERUSALEM AMMAN
CAIRO SINAI JORDAN
AGABA
NORTH

LIBYA EGYPT ASYÛT
LUXOR
ASWAN
L. NASSER
WADI HALFA

MEDINA
MECCA
JEDDAH

SAUDI
ARABIA

CHAD

PORT SUDAN RED
SEA ABHA

SUDAN ATBARA MASSAWA SAN'A
OMDURMAN KASSALA ASMARA HODEIDA TA'IZZ
KHARTOUM AXUM ASAB ADEN
ABÉCHE KOSTI ASAB
EL FASHER ELOBEID GONDAR DJIBOUTI
NYALA BERBERA BURAO
BABANUSA RENK T.ANA DESSYE LAS
MALAKAL HARAR HARGEISA ANOD
CENTRAL ADDIS ABABA
AFRICAN BENTIU JIMMA ETHIOPIA GALKAYU
REPUBLIC WAU GOBA BELET
RUMBEK BOR YIRGA-ALAM HUEN
L.ABAYA
JUBA L.CHAMO MEGA
YAMBIO NIMULE L.RUDOLF MOGADISCIO

ZAIRE UGANDA KENYA SOMALIA

NORTH EAST AFRICA

MILES
0 100 200 300
KILOMETERS
0 100 200 300

BORDERS — — · — —
ROADS
RAILROADS +++++++++
SWAMP

A reconstruction by W.B. Emery of the fortress of Buhen near present-day Wadi Halfa in Sudan. It was a huge Nubian fortress as modern as any today, with

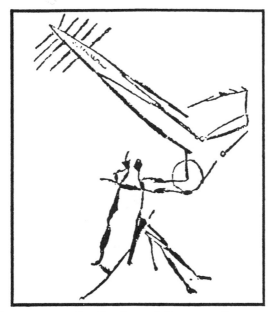

Strange rock glyph at Merowe. It seems to depict some sort of telescope, laser gun, ultra-sound gun, or radiation condenser.

yramid-tombs at Merowe.

Ancient Pyramid in Lake Moeris, Egypt.

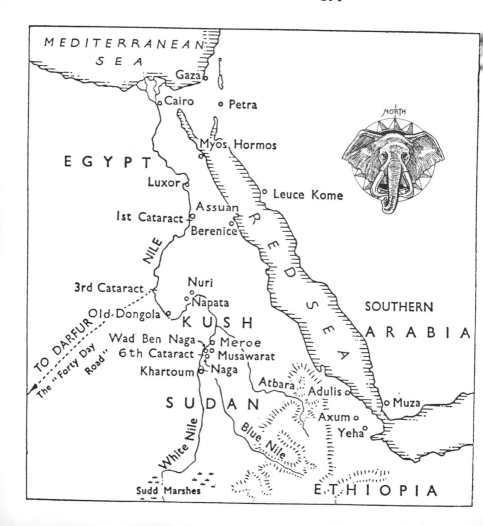

Chapter 7

THE CENTRAL SAHARA:
PORT CITIES & TUNNELS
IN THE DESERT

Hatch Chibango – Trouble shooter
Have Landcruiser, will travel.
–*sign on my tent wall in the oil exploration camp*

That was me, *Hatch Chibango: Troubleshooter.* The office in Khartoum had heard through the grapevine that one of our camp managers, an Ethiopian named Tekle, had gone bonkers while down in the camp of Babanoosa, the same town that I had passed on my way to Wau. They asked me to straighten things out.

This wasn't the first time one of our camp managers had had a nervous breakdown. I had taken one Kenyan guy to Khartoum's only psychiatrist a few month's before. He had been found by the oil company's representative in a pit he had dug beneath his bunk, stark naked and incoherent. Life is difficult out in the swamps; the oildrillers are a strange and rough bunch, and work is 16 hours a day, seven days a week, with little else to do. Many workers were crude, rough, racists who just blew their entire paycheck every two weeks drinking and whoring in Nairobi.

Others were apparently as nutty and crazy as could be. The Vietnam vet helicopter pilots flew rigging in and out of the swamps while often stark naked. They lived in grass huts with native girlfriends who also wore no clothes. Apparently their contract did not require them to be dressed.

At the time I had been transferred down to an oil camp in the swamps of central Sudan. Sudan is a gigantic basin and through it runs the White Nile, the longest river in the world until the construction of the Aswan High Dam in Egypt. From Malakal south to Mongola is the Sudd, a gigantic swamp that covers several hundred square miles, and through it meanders the Nile on its long trip north. French engineers were busy cutting the Junglei Canal through the Sudd, the largest earth-moving project in the world. It was never to be completed because of the impending war.

I worked as a camp manager during the dry season at Bentiu on the western edge of the Sudd. The Nilotic tribes of this area, Nuers, Dinkas, and Shilluks, still live very traditionally, most of them wearing clothes only on rare occasions, and they are always seen with their two spears for protection from possible tribal conflicts or wild animals. Though very friendly to the expatriate oil workers who worked for Chevron and the other companies, they watched approaching natives carefully, noting the tribal marks of deep scars on each others' faces.

One of my duties as manager was to purchase local food for the oil workers. We often had frozen meat flown in, but sometimes we needed local meat for a

169

special Sudanese stew, so about twice a week I'd walk a couple of miles into town to purchase a cow for slaughter at our mess hall. I'd grab some money out of my petty cash box and take my trusted aide and confidant, Kennedy Toot, a local swamp dweller who worked for our company, into town with me. It was company policy that every employee have two names, and since he didn't actually have a first name (or was it his last name that was missing?) Mr. Toot took on the name of Kennedy and became our laundry boy. He'd put on his best T-shirt and, on my insistence, he'd wear a pair of pants. A pair of red plastic Hong Kong sunglasses and a copper pipe topped off his wardrobe. Kennedy's orange hair (a color he achieved by bleaching it with cow urine) was combed straight up from his forehead. All in all, he was sensational, a cool visitor to town, although people might not recognize him with pants on. I had often told him that I'd like to take him to a New Wave bar in London that I know of, but he never really showed much interest.

Once in town, Kennedy went over to talk to some of his friends who were hanging out underneath one of the huge mango trees in the center of town. They were Dwop Thop and Simon Gash, who worked for us as houseboys. Simon, Dwop and Kennedy were all in their late teens, tall but youthful in appearance. I waved to them and walked toward the large open area where some cattle and their owners were starting to gather for the daily cattle auction. Suddenly, as I was checking out the cattle, a slight commotion to my left caused me to turn and look. A naked woman with bone and copper jewelry around her neck and a spear in each hand was singing and dancing, coming straight for me.

Initially shocked, I realized that since I was the only European in town, she was singling me out for a little fun. She danced in a circle around me, waving her spears, singing, and shaking her bouti like a go-go dancer. Everyone had stopped talking and the whole compound waited to see what I would do. I saw Kennedy, Simon and Dwop looking apprehensively in our direction; they worried about me sometimes.

Europeans are known to be weird and stuffy, but I was no tight-ass. With a snap of my fingers I did the strut and joined her in her dance. She was delighted and we both laughed and danced around the park in great rhythm. Everyone was greatly amused and probably a bit surprised at this display of craziness. People laughed and clapped and even joined us in a kind of elephant walk through town. I lost my partner near the crowd that was gathering for the cattle auction and joined my friends for the bidding.

Kennedy looked at me through his red plastic sunglasses, brushed his orange hair back and took his smoldering copper pipe out of his mouth. "You're a pretty good dancer," he said, and Simon Gash, Dwop Thop and I had a good laugh. The new wave clubs of London would never know what they were missing!

Unfortunately, my role as *Hatch Chibango: Troubleshooter* was to take a more serious turn a few months later. Tekle Miriam was a teacher from Ethiopia who had been tortured by the *Dergue*, the ruling military clique of Ethiopia, but had fortunately escaped to Sudan, and was now managing one of our small camps. As I was informed over the short wave radio one day, he had apparently had some sort of strange flashback, and I had to retrieve him from the camp in the field. I would have to take a few days absence from the geophysical camp that I was managing, and head in my Landcruiser across the Sahel to Babanoosa. It was an ugly job, with the civil war starting to heat up, but somebody had to do it.

I went and told Chuck, the oil company rep in the camp, that the office in Khartoum had requested me to go to Babanoosa to see what was going on. On my way out of Chuck's refrigerated trailer, he suddenly handed me a pistol. It was a German semi-automatic, with one fully loaded clip.

"Here, take this," he said. "The war could erupt any day now. You may need it."

I looked at it reluctantly. I didn't want to have to carry a gun, but maybe it was the best thing. "Thanks, Chuck," I said, "I owe you a favor," and turned for the door.

"Buy me a beer when you get back," he said. And with that, I was roaring off across the dry bush on my way west.

As I drove down the desert track that eventually reached the railway line to Wau, I wondered what it was that brought me here to this strange land. My quest was ostensibly for lost cities and ancient mysteries, yet did I hope to find something more? Perhaps it was my elusive self that I searched for. A me that I just didn't quite understand. A person driven to the edge of the world in search of the answers to a myriad of questions.

What did I really hope to learn from the past anyway? Was there any value in my study? As I drove through the bleak scrub brush and desert, I wondered what ancient mysteries there could be here. According to most historians, sub-Saharan African had been a primitive and remote area for all of history, occasionally penetrated by slave caravans. Yet, it seemed to me that the area had long been explored by ancient civilizations, and was, in itself, the land of an old, and vanished civilization.

The ancient Egyptians attempted to explore much of the African country, if not the entire world. They were known to have made exploration and trading journeys across the Sahara westwards to the Kharga Oasis and south-west to Lake Chad. On such trips, because of the fierce desert sun, they chewed a root that they called *ami-majos*. They found that this root gave them extra protection from the Sun by reinforcing their skin pigmentation, and modern research now shows that the root contains the active organic chemical compound called *8-methoxypsorate*, which does indeed reinforce skin pigmentation and prevent the user from sun-burn and after prolonged exposure, skin cancer.[31]

I mused over a fascinating book I had been reading in the camp called *The Ancient Atlantic*.[75] The author, L. Taylor Hansen discusses the ancient Triton Sea and the Antediluvian kingdom of the Tuaregs in the Ahaggar Mountains. According to her, the earliest myths and legends of antiquity say that there was once a lake called Triton or Ticonis where today is the western Sahara Desert. Science does not disagree with this theory. It can easily be seen that the freshwater fish caught in desert wells, or the crocodiles of some remote oasises, must have come from an epoch when the Sahara was a fertile land of swamp and lake. Geologists, however, have maintained that this sea in the Sahara was of a remote time, millions of years ago, and that such a sea would have nothing to do with man. Then the discovery of petroglyphs of water animals on cliffs in central Algeria changed the notion that this sea existed millions of years ago. Instead it was of a quite recent time period.

According to Hansen, the Triton Sea was held by the curve of the Atlas Mountains like the rim of a cup, and the water thus held covered the land from the Gulf of Gabes where it entered the Mediterranean to the mountains south of Lake Chad. Only after the sudden sinking of the southern arm of the Atlas did the Niger river break through these southern mountains and tear its way out to the Atlantic. Lake Chad and the underground lakes of the Sahara were all that was left of the Triton Sea, except for the massive port cities that existed in the Sahara!

Do such megalithic ports really exist in the sands of the Sahara? Hansen assures us that they do. According to her, the Ahaggar Mountains in central Algeria were an island during the time of the Triton Sea, which would be contemporaneous with the civilizations of Osiris and Atlantis.

The area of the Ahaggar is not easy to gather knowledge about, both because of its remote location and especially because the fierce and mysterious Tuaregs, the Veiled People, jealously guard their secrets and heritage. The Arabs are very much afraid of these people, as one Arab told Hansen in her book. Her conversation with an unnamed Arab is quite fascinating. He relates that "Near In-Salah there are three high peaks of the Hoggar. No Arab will go there if he can help it. These peaks touch the sky with claw-like fingers. Once a friend of mine got lost and saw the ruins of one of their cities on the Atlas. It was built of giant stones — each one the size of an Arabian tent. In the front is a great circular wall. But in the desert they live underground. I have heard that under Ah Hoggar are many galleries deep in the earth around an underground lake. These galleries are filled with paintings of the long age."[75]

The Arab qualified his statements by saying that he had traveled to areas of the Tuareg as a messenger. "On one of these times I took a message from my sheik to Tamen-Ra-Set—that is the Tuareg capitol city. That is where Amen-Okhal, their king lives." Hansen comments that these are names of Egyptian gods that were ancient at the dawn of history.

The Arab went on to describe how the original inhabitants of the Sahara were the Tibesti people who were Dravidians of India (or like them), with straight noses and hair and dark skin. The Tuareg tribes invaded from the ocean (Hansen believes from Atlantis) into the Triton Sea and settled in the Ahaggar area and the Atlas. The Tibesti tribe was then driven into the Tibesti Mountains of northern Chad and southern Libya. James Churchward in his book, *The Children of Mu*[76] gives information corresponding to this, that the Ethiopians are also Dravidian stock, coming from Mu to Burma and India many thousands of years ago, and then on to Ethiopia by crossing the Indian Ocean. According to the Arab, they also settled the fertile areas of the Sahara during the present of the Triton Sea.

Hansen asks the Arab how he knows this, and he replies that the Tuaregs told him. Therefore Hansen surmises "...the underlying blood of the Mediterranean is Dravidian and the Tuaregs are invaders.... That must have been what the priest of Volcan in early Egypt meant when he told Solon that the Greeks and Egyptians were really brothers in the days when the Mediterranean was a valley filled with cities!"[75]

The Arab goes on, "The land I speak of is beyond the Mya river—one of the great dead rivers of the Sahara. The terraces rise from its ancient bed in colors of red and white. Once it emptied into the Triton sea and carried ships up to the great cities of the Tuaregs on top of the Atlas and the Hoggar mountains. I believe the Mya emptied where the dry lakes of the Chotts are today. As the sea level of the Triton sank in long dry spells, another lake was where Lake Chad is now with a waterway in between. That land was green then. Ostriches, buffalo, deer, tigers roamed the woodlands and crocodiles slept in the rivers....This place is called the land of the monsters because the cliffs are shaped like monster animals—such as you have never seen. There are enormous shaggy elephants and a giant lizard that sends your hair arising when you first see it. In the heat of the Sahara, they seem to come to life and move....The Tuaregs say that when they first came to the land the living monsters really gathered here to fight over their feeding grounds, but of course, that was untold ages before my people appeared."

The Arab goes on to tell of the Land of Fear, "That lies at the gateway of the Hoggar. There is a narrow gorge one must pass through, and here the rocks explode in the sun with the rattle of machinegun fire. Then from time to time small avalanches tumble down from the peaks above. There is a good reason for that name and it takes a brave man to keep going through the land of fear." The Arab then arrived at the forbidden city of the Tuaregs, and met with the king, Amen-Okhal. He describes a strange scene feasting, drinking palm wine and getting drunk (something not common among Bedouins, though the Tuaregs are not Arabs or Bedouins. Though they say they are Muslims, many Arabs do not believe them). A one point, dancers in strange masks, similar to Kachina-doll dancers of the Navaho, Hopi or Zuni Indians come into the hall, and a poisonous monitor lizard stalks the hall, though the Tuaregs take no notice.

The Arab then speaks of galleries and tunnels in the mountains, and tells how a friend climbed one of the peaks and discovered a shaft that was covered by a metal grate. He then tells of another friend with whom, when they were younger, he went to the Ahaggar mountains in the moonlight out of curiosity. Crouching in the moonlight, they saw a ceremonial war take place between two mounted Tuareg groups who clashed until some of their members were genuinely dead. Then, one group of Tuaregs, while the young Arab boys watched from their hiding place, rode into a cliff wall and disappeared!

The Arab ended his incredible tale to Hansen with the legend that, "Down in the miles and miles of underground galleries, where it is said that they wander about a beautiful artificial lake, and then pass along torchlighted passageways looking at pictures painted of their cities so many thousands of years ago—are their libraries. There are kept the books which are the oldest libraries of the earth. There are the histories which go way beyond the great deluge, to the times when the Tuaregs ruled the seas. How do I know? They told me, that is, the emperor did. But save your next question. I could not get to read them. Neither could you, or anyone else—no one will ever read them except the people of the veil."[75]

§§§

I braked and the Landcruiser skidded to a halt in the dusty nomad village that happened to be along the set of desert tracks that I was following to the western railway tracks near Muglad. An old man in a turban had waved for me to stop, and I quickly brought the four-wheel drive pickup to a halt. The entire nomad clan of twenty or thirty came out of stick huts and ran toward me. The sheik, an old man in turban and heavy mustache, came up to me and asked me where I was going in Arabic. I told him I was going to Muglad. How much would I charge him to go to Muglad, he asked me?

"Nothing," I told him.

"Nothing?" he asked in disbelief. No one had ever offered him a free ride before.

"Nothing," I repeated. "I will take as many as can fit in the back. I am not coming back for many days, though. You will have to find another ride back." And with that, a smiling sheik and about half his clan ran and got their traveling gear and things to trade in the market of Muglad. Once they were settled in the back, I took a few photographs of my captive subjects, and fired up the vehicle. The clan was all smiles as we drove off for through the desert brush to railway town of Muglad, some thirty miles away.

That wasn't Hansen's only strange conversation about prehistoric civilizations in the Sahara, underground tunnels and lakes and mysterious cities. In another chapter of her book, she relates a very interesting conversation she had with a wealthy retired American couple named Johnson in Palm Springs back in the mid-sixties. They showed her their home movies of a very unusual vacation they took in the Sahara desert, apparently some time in the late fifties or early sixties. The Johnsons told Hansen about some of their interesting experiences and conversations while in Algeria, Libya and Egypt. They speak of their guide who told them of the ancient history of the area. This guide, Ahben, told them that when the Triton Sea "sank, many cities went down on it because it was about a hundred miles wide...Ahben said that the Tuaregs say the Triton carried the commerce of the whole world. It was the way from the Atlantic to the Mediterranean valley. This was long before the rise of Egypt. In those days the Atlantic was a double much of its way, and goods could be carried from Africa to England and even to the continent beyond."

Mrs Johnson talked about an old Arab whom they paid to tell them the ancient history. After giving the old man presents, he told them of the three main peaks of the Ahaggar that form a trident. They were hollow, he said, because when climbing them, he would hear voices that seemed to come from inside the mountains. Then he would see riders that would disappear inside the mountains. Mrs. Johnson then read from her diary, "The old man told me that above a mighty gorge in the very heights of the Atlas is the ruins of the city of Khamissa. It has towers of marble which have been deserted for untold thousands of years. Yet at certain times, when the weather is just right and the atmospheric conditions are perfect, especially if one has had a sip of wine and a smoke on the pipe, then the gorge doesn't have the blue-purple veils of evening at all, but is deep purple-blue water upon which are a tumbled mass of shipping going to and from the teeming city. It becomes a kaleidoscope of galleys, triremes and ghost ships from long forgotten nations. And the beauty of the rounded towers are reflected in the sea like those from a never-never land."[75]

The old Arab's tale becomes reminiscent of Hansen's earlier informant on the Triton Sea and the Tuaregs when he says, "In the galleries which are not lakes, or places for the storage of water, the unimagined splendor of the Tuaregs in the full pride of their power is still pictured in endless gallery paintings.

"The city of Khamissa is not the only ruin lost in the vastness of this little-explored land. Between the ranges of the Air and and the Ahaggar, is once proud Tafassaset, and southeast of the Ahaggar range is Essouk, the imperial capital of Heracles, once the most important city on Earth."[75]

The Johnsons also talked of their trip to the Siwa Oasis in the Libyan desert near the border of Egypt and Libya. Mrs. Johnson relates the story of how the Persian king Cambysses had conquered Egypt in 524 B.C. He sent his army to Siwa and gazed upon the temples and palaces of the Ammonian kings. The army went by way of Thebes and they surrounded the city at Siwa and decided to wait until morning to attack. According to legends of the Arabs, she says, the priests at Siwa began to chant. "That night, very suddenly, a sand storm came upon them, and not a man was left alive of the fifty thousand iron clad men their horses. No one ever found them to this day."

Mrs. Johnson continues, in Hansen's book, "Passageways run from below the temple of Agourmi for about two miles to Djebel Muta. The temple was most remarkable, even in ruin. The circular walls had a sheer lustrous red beauty, but it was the galleries which really intrigued us. We had to give the Arabs a large tip, but it certainly was worth it.

"We followed the Arab guide, who carried a torch. The gallery seemed to have a stone floor and smooth walls. Before we had gone over two city blocks, we came to the paintings. I suppose the torches have done them no good, but the colors did seem to be quite brilliant. First there were animals. They must have been from Ice Age times. There was an American buffalo, or something that looked like one. Not far away was a very large elephant—perhaps a mammoth. He was pulling some fruit from a tree. There was water everywhere, and ferny forests beyond the stream. Crocodiles were in the water. Beautiful birds sat in the trees, and strange animals peeped through the shrubbery.

"We went on and came to the paintings of a city on the shore of a sea. The guide told us that this was the Triton. Many square-sailed ships were about. We didn't know whether they were Libyans, Cretan, Pelasgians, Etruscans, or Egyptians because neither Joe nor I was familiar enough with the clothes. The quays were wide and well made – apparently of stone."

Mrs. Johnson also tells of their Arab guide taking them to "ancient stone quays stretching way out into the desert sand. So the picture must have been of the city when it was a trading center of the ancient Triton sea. And the quays proved that there must have been a Triton sea."[75]

L. Taylor Hansen then interjects in their conversation, "During the middle ages a ship was found not too far from the Draa Depression [the present border of Morocco and Algeria, just east of the Atlas Mountains–ed.] in which skeletons of the rowers were lying with the chains still around their bones. The Arabs, I understand, charge a very high fee to take you there. It still must be in existence."[75]

Hansen's book was on my mind when I pulled into the dusty railway town of Muglad, and still on my mind as I drove through the mud wall lined streets on the way to the market. Taylor's book is fascinating, but not without fault. She is the daughter of the co-originator of the Taylor-Wegener theory of continental drift, and she is well versed in geology as well as ancient history and legends, especially American Indian mythology. It is doubtful, though, that the Siwa Oasis was part of the Triton Sea. Rather, the Siwa Oasis was part of an inland lake that eventually dried up and became the oasis it is today, with crocodiles and what not. I recalled the crocodile temples to Sobek, the crocodile god of the Egyptians that I had seen in Alexandrian. Crocodiles don't cross hundreds of miles of desert sands to get to an oasis, that is for sure. Rather than calling this ancient sea the Triton, which was farther west, it would be best to call it the Siwa Sea.

§§§

I dropped off my truck load of desert nomads at the market, and headed for the oil company's staff house in town. There I was met by Eric, a young English lad who worked for the same company as myself. Eric was glad to see me, as there weren't many travelers coming through town, and it was always fun to party with one of the boys. The duty-free Danish and Dutch beer were cracked, and after a great meal, we sat in the back courtyard and talked. I told him of what I had been thinking in regards to Hansen's book and the Triton Sea.

Eric, a history major from Oxford, thought it all made sense. "There are tons of petroglyphs from the cliffs of the Ahaggar. The French archaeologist Henri Lhote discovered many strange drawings on the walls, one he even calls the Martian God! Why, one of the chapters in his book, *The Search of the Tassili Frescoes*[77] was entitled, "Did We Discover Atlantis?" It seems evident that there was some sort of lost world in the Sahara. Scholars just argue how advanced the civilization was. I guess when they see men running around with bows and arrows, as on the paintings, it just plays into their natural bias that mankind was primitive many thousands of years ago."

"That's a good point," I acknowledged. "When one starts with the assumption that civilization began a few thousand years ago, and that everyone else was a cave man before that time, it is hard to develop theories that make sense out of hollowed mountains (cave men run amok?), port cities in the Sahara, and such."

Eric got out a volume that he had been reading, *In Quest of Lost Worlds* by Count Byron de Prorok. Byron de Prorok also believed he had discovered Atlantis. To my surprise, the first chapter of *In Quest of Lost Worlds*[78] was about an expedition into the Ahaggar that described adventures and legends that were a remarkable parallel to L. Taylor Hansen's stories. On their way by specially equipped Citroen cars, they found that the Tuaregs had poisoned an important water hole. This was in the region of Moudir, "where great precipices form a wall of rock, believed by the Tuaregs to be the fortress of the 'Amazons, ruled by a white goddess: the fable of old Atlantis. There was no way up, which perplexed us considerably. Whatever foundation there was for the superstition, it seemed not unlikely that there would be some sort of ruin up there worth investigating; but, though we tried consistently, it was impossible for us to scale that sheer wall."[78]

They then discover an underground lake, much as Hansen describes, though much smaller: "We threaded our way down a narrow corridor, which speedily darkened, so that we had to use our torches, and were surprised to come upon a clear, transparent pool, with a fine sandy beach. The walls around were covered with inscriptions and rock drawings of elephants, buffaloes, antelopes, and ostriches. Not one or two; but scores of drawings were there, and we knew that we were definitely on the trail of the old caravan routes to the gold and ivory lands of the ancients."[78]

De Prorok and his company then continue through the *Bled es Khouf*, the land of fear. It is here that the rocks would explode from the sudden heating and cooling. One of de Prorok's men begins shooting wildly, thinking that they are being attacked by Tuaregs...then they come to the Valley of Giants, giant rock formations, thought to be natural (?) that look like hippopotami, elephants, mushrooms and dinosaurs. When they finally meet some Tuaregs, de Prorok describes them as seven feet tall, with their faces always veiled by their blue or purple robes, much the way the ancient Phoenicians dressed.

De Prorok describes the Tuaregs as "white nobles" and curiously describes a subgroup of the Tuaregs, the Hartani, or slave caste of the Tuaregs as "black, but not Negroid. They have nothing to do ethnically with the Negroes to the south, of Senegal, or the Congo, it would appear; but are a people entirely separate, more like the North American Redskins, with straight noses and thin lips, and are subject to, but now moderately independent of, the white Tuaregs or Targui: the noble caste. Some say they are black Antineans. It is a nice theory."[78]

"What are Antineans?" I asked Eric as he cracked another beer.

"Beats me," he replied, "Do you suppose he means Atlanteans?"

Neither of us could figure it out. De Prorok meets with the Amenokhal Akhamouk, the king of the Tuaregs, though he points out that it is really the women who are the power, and the king is chosen by a queen. The women choose their men, and de Prorok is amazed at the Matrilinear culture of Tuaregs among the patriarchal Moslems. They could not be described as matriarchal, however, because of the their warlike and violent attitudes characteristic of patriarchal cultures. After a feast of dates stuffed with locusts (their legs cut off), swimming in thin honey with the tails of lizards and snakes, washed down by camel's milk, he is given three women who are to cater to his every whim, and are quite forward in their attentions to the Count.

De Prorok finds himself in a sticky position; he does not want to offend his hosts by refusing the sexual attentions of three girls, but he is a faithful married man, and knows that his wife would not approve of this desert menage-a-quatre, even in the advancement of science...

Eric suddenly threw his head back and yelled up at the stars, "Sex! I need sex! That wimp de Prorok! I wouldn't have hesitated for a moment with those girls," he cried.

Eric was drunk, and I could see his months down in the camp were starting to take their toll on him. He was ready for a vacation. "You would have accepted the Amenokhal's gift of the young ladies, Eric?" I kidded him.

"Yer damn right!" he bellowed, spilling beer on his shirt. "I'm no sissy like de Prorok. What did he do anyway?"

"This is your book," I reminded him.

"I don't remember what happened!" he roared and fell over backward on his chair into the dust of the courtyard. He picked himself up and and walked over to the fridge and grabbed another beer. I had a feeling that he would have attempted to take on a whole busfull of nurses on tour of the Sahara at that point, though his motor reflexes didn't seem to be functioning very well.

"De Prorok had his Arab guide get the girl's boyfriends, so that their amorous intentions could be relieved on young Tuaregs, rather than the moral count," I said. "That's about it in the book, except that they find the tomb of the legendary Tuareg Queen Tin Hinan and dig it up. They find cut stone blocks, some gold, and the skeleton of the queen. They have a close call with some Tuareg tribesmen, and hightail it back to Algiers. The queen was only seventeen hundred years old, so its not like she was from Atlantis, or even from the time of the Triton Sea."

"Is that all?" groaned Eric. "Those wimpy archaeologists. I'd show them a thing or too on how to have to adventure. If I ever get out of this rat hole!"

I laughed. It was no wonder that Eric was out here in the middle of Sudan managing a camp, instead of teaching history or something in England. He was a wild sort of guy. Maybe he needed to be out here in the desert, far away from English schoolgirls, and Tuareg maidens as well. At least he let off his steam every once in a while at the camp. Probably his next vacation in Nairobi would set new records for boozing and womanizing. My concern though was my mission as Hatch Chibango Troubleshooter: Tekle was still holed up at one of our other camps, and his problems couldn't be solved by a six pack of beer...

§§§

I left early the next morning, following a track that went north up the railway tracks toward Babanoosa. It was an all day drive to the railway junction, I hoped that I would reach it before it was dark. Dealing with Tekle was something I wanted to do in the daylight.

After an hour's drive, I saw a group of blacks standing by a hut by the tracks. There had been an incident a few months before when guerrillas for the ANANYA II guerrilla group that wants independence for Southern Sudan, had attacked an oil company worker in his landrover and killed him. I didn't see any weapons, but I reached under my seat and put the German pistol on my lap.

I drove by and waved without stopping. They waved back, and I passed without incident. It seems that they were waiting for the train, and probably going the other way.

To keep my mind off the unpleasant task that awaited me up ahead, I thought more about the ancient mysteries of Africa. One ancient mystery that has received a lot of attention is that of the Dogon near Timbuktu in Niger. In the late 1940s two French anthropologist went to study the Dogon, much of whose traditional culture has survived the impact of Christianity and Islam. The anthropologists discovered that the Dogon priests had an astonishing knowledge of the star Sirius, which played an important part in their religion. They knew that Sirius A, which is the brightest star in the sky, has a satellite, Sirius B, which revolves around it.

Sirius B is invisible to the naked eye and the Dogon priests had no telescopes, but they knew that it exits and that its orbit round Sirius A is elliptical, not circular. They knew the position of Sirius A within the orbit and that the orbit takes fifty years. Sirius B was not discovered in the West by telescope until 1862 and was not photographed until 1970. The Dogon were also aware that Saturn has rings, and these rings can only be seen through a telescope.

The Dogon priests' knowledge of astronomy is too old to be attributed to Western influence, so where did it come from? The ancient Egyptians had an advanced knowledge of astronomy, and may have known about Sirius B, but according to the Dogon, they got their knowledge from space travelers who traveled from a planet attached to Sirius B and who landed on earth.

In his book, *The Sirius Mystery*,[73] Robert Temple explores the subject, and finds that the only credible answer to the strange knowledge of the Dogon is exactly what they say, that their knowledge came from visitors from the planet Sirius circa 3000 B.C. In his book he traces these extraterrestrials, which the Dogon call Nommos, to ancient Sumeria and Egypt as well. According to the Dogon, the visitors from Sirius B were a type of fish-man. Apparently, on their planet at Sirius B, the Nommo are aquatic beings who have perfected interstellar space flight, and have visited our planet. Temple relates the Dogon tradition of the Nommo to the Sumerian tradition of Oannes, a fish-tailed god who came from the sea (or sky) to give civilization to the Sumerians.[73]

Another fascinating mystery of ancient Africa is that of a prehistoric nuclear reaction that scientists say took place in Gabon 1,700,000 years ago. In a report made by Dr. F. Perrin former chairman of the Commissariat `a l'Energie Atomique in 1972 and published in *New Scientist*, November 9, 1972, a "natural" water-cooled nuclear reactor had operated intermittently at Oklo, Gabon in West Africa for as long as 1,000,000 years. A nuclear chain reaction occurs when neutrons created by the breakup of uranium bombard other uranium atoms, thereby producing more neutrons. If there are insufficient "moderators" to gobble up some of the excess neutrons, an atomic explosion results. If there is an excess of moderators, the process dies out quickly.

They key evidence that made French researchers adopt the natural reactor hypothesis was a puzzling reduction in the uranium isotope U-235. Normally, uranium deposits contain 0.72 percent of U-235. The deposits at Oklo contain less of the isotope, and, strangely, some material actually contained more of the isotope. Scientists concluded that the only possible explanation was the occurrence of a "spontaneous nuclear reaction thousands of millions of years ago."

It was concluded that, just like a man-made nuclear reactor, some plutonium was formed, which then decayed back to U-235, which was then younger and thus richer in U-235 than the primordial uranium. An Oklo sample of ore showed the presence of four rare elements, neodymium, samarium, europium, and cerium, with proportions of isotopes previously found only in man-made reactors. In the same article it is said that because of continued radioactive decay "no such fantastic combination of circumstances can ever occur on our planet again."

As I turned up onto the embankment of the railway line to avoid a potentially dangerous water hole, I wondered how such a reaction could ever take place naturally. It didn't seem possible. One of the common methods of dating is to measure the U-235 content of an article, yet it is impossible to use the same method on U-235 itself. The thought occurred to me that this spontaneous, "natural" nuclear reaction, was not a natural explosion at all, but rather man-made. A man-made nuclear explosion millions of years ago? I doubt it. What if scientists went to Hiroshima, or better yet, the glazed sand of Alamagordo, New Mexico where the first atomic bomb was tested, and tried to date it via the U-235 that was left over. Would they conclude that it had happened forty or fifty years ago, if they knew nothing of the test? I would venture to guess that we would get a similar date of many millions of years for the radioactive decay and similar theories on how it happened, after all, isn't it scientific fact that no one was using or testing atomic weapons in prehistory? Such is the twisted logic of the experts.

I arrived in Babanoosa just before sunset. Somehow the small railway junction seemed more ominous than it had the time I had passed through on my way to Wau. I drove straight to the camp, and went to see the British supervisor of the camp. I introduced myself and asked him where Tekle was.

"Haven't seen him in three days. He's barricaded himself in the storeroom out back. Plenty of food in there, at least. Better be careful though, he may be dangerous. Want a beer?"

I declined the beer, it wasn't Miller Time yet. I took a deep breath to calm my anxiety, and then walked around back to where the storeroom was. I had never met Tekle before, so I had no idea what kind of person he was. He could be some tall, powerful Ethiopian with a fist that could wring my neck in an instant. He might be armed as well. All I knew was that he had been tortured by the Military in Addis Ababa before he had escaped from the country. It was enough to make any one crack, plus, the strain of managing these camps was great. I imagined that the British seismic workers that stopped in at the camp weren't much for socializing with an Ethiopian refugee.

"Tekle!" I called from the doorway. "Are you in there?" I looked inside the long, trailer that had been turned into a storeroom. It was also his office, I could see by the desk about midway down the long, dark building. "Tekle, this is David from the office in Khartoum!"

"Stay where you are!" came a voice from the back of the trailer.

"Tekle, I'm from the main office. We're sending you back to Khartoum."

"I'm not going back!" he suddenly screamed. "I'm never going back!"

Apparently he thought I was trying to send him back to Ethiopia. Indeed, that would be his death sentence. I held the semi-automatic lightly in my right hand. The cold metal warmed gently with my body heat. It felt heavy, yet light at the same time. Was I in danger? "Tekle, I'm coming inside. You're not going to Ethiopia. We're only sending you back to Khartoum. I'm your friend. Don't be afraid."

There was silence. Then I heard a gentle sobbing from the back of the trailer. I put the gun in my pocket and walked forward. I walked past a desk with a mess of strange Amharic writing and delivery orders from Khartoum. In the back I found Tekle, a small, meek former schoolteacher from Asmara. He was weeping, his head in his hands. I put my hand on his shoulder to comfort him while tears streamed down my face. It was a strange world, and I was a stranger in a strange land. And I don't just mean Africa. It is a strange and cruel world, where torture, death, slavery and horrifying savagery are everyday occurrences. All I wanted to do was help Tekle. The war in Ethiopia and elsewhere was beyond my control. I sat down next to him, and we both cried for what seemed the whole night.

The next day I put Tekle on an oil company flight back to Khartoum. One of the gang from the office would meet him at the airport. What would happen to him after that, I didn't know. I was sure that he wouldn't have a job with us anymore. I drove back to Muglad and then back to Bentui to continue my tour of duty at the camp, but I was never quite the same after that.

A year or two after I left Bentiu, guerrillas of either ANANYA II or the

Sudanese Peoples Liberation Army (SPLA) came into the camp and killed everyone there, including all the oil company employees, and one of the men who replaced me. The reason was because the oil finds in the swamps of Sudan were an extremely important factor in the Civil War in Sudan, and since the oil was located in what is considered the "south," the liberation armies of the south felt it necessary to stop all oil drilling and exploration activities. Fortunately, I had had enough of troubleshooting in central Sudan, and had left the country many months before.

Prehistoric natural nuclear reactions occurred in zones 1–6 in this uranium mine at Oklo, Gabon; part of zone 2 will be preserved.

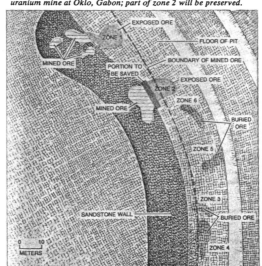

A rock painting from Sefar in the Tassili Mountains of Algeria. Said to be of "The Great God of the Decadent 'Round Heads'." Surrounded by naked women, it is assumed he is part of a fertility cult. The grid on his face could be a mask or possibly writing.

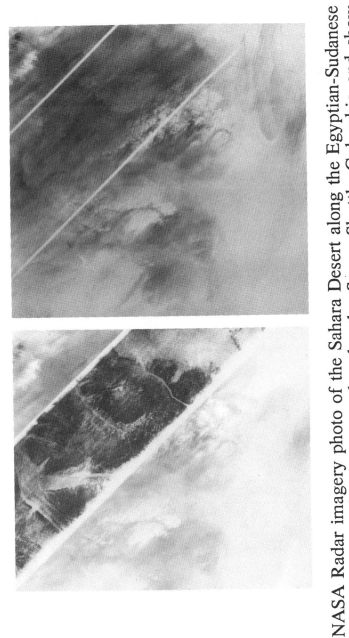

NASA Radar imagery photo of the Sahara Desert along the Egyptian-Sudanese border. These photos were taken by the Space Shuttle Columbia and show riverbeds and tributaries beneath the desert. It is estimated that these rivers existed 10,000 years ago when the Sahara was fertile, and much of it a sea.

Group of triliths from Ahaggar

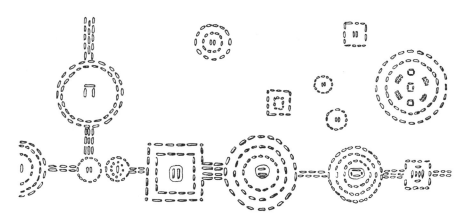

A group of megalithic structures in Algeria

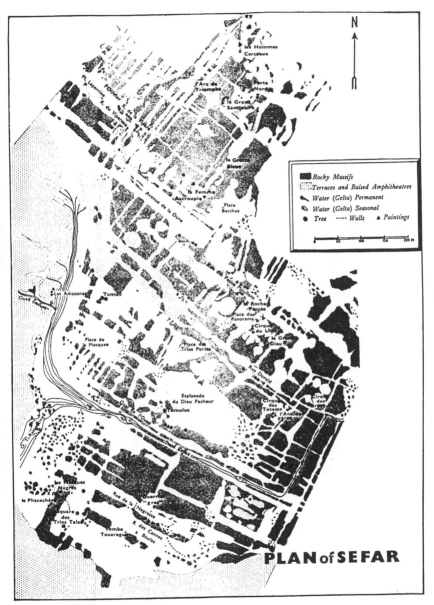

The mountain area of Sefar in the Algerian desert. Bizarre rock formations, thought to be natural, but possibly artificial, extend through the area and are in squares and right angles like a city. An ancient Atlantean city in the desert?

The area of Sefar in the Algerian desert near Djanet. The ancient stone formations make one wonder if they were not once part of a megalithic city built many thousands of years before the last cataclysmic poleshift.

Controversial "Martian God" called Jabbaren, said to be "from the decadent period of round heads". Photo by Henri Lhote, *The Search for the Tassili Frescoes,* 1959. A strange figure by any explanation, it may be a man in robes and turban, though some claim that it is an ancient astronaut (from Mars?).

Strange Bird-Headed women with Egyptian-like headdresses. Inhabitants of the fertile Sahara, sylized ancient gods of Osiris, or what? From *The Search for the Tassili Frescoes,* 1959.

Sirius, the brightest star in the sky and its white dwarf companion. Though only photographed in 1970, the Dogon of Mali have made drawings of it for centuries.

Top right: Germaine Diertalen and (bottom) Marcel Griaule, the two French anthropologists who lived with the Dogon tribe for over 20 years and to whose careful study we owe much of our knowledge of Dogon Mythology.

The Egyptian goddess Isis, showing the small fishtail in her headdress

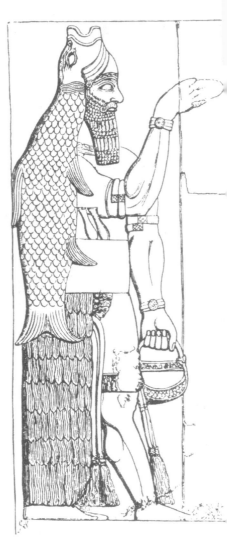

The Babylonian semi-daemon Oannes, a fishtailed amphibious being from the heavens who, according to the Babylonians, founded civilization on earth. *From Nimrud*

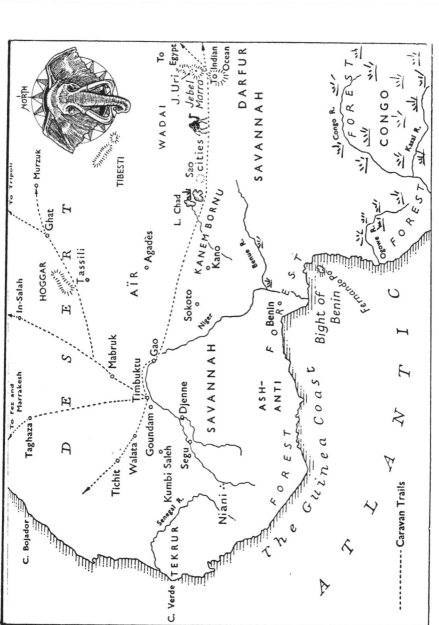

Ancient caravan routes through the Sahara.

Chapter 8

SAUDI ARABIA & THE GULF STATES:
LOST CITIES OF THE EMPTY QUARTER

Do they not travel through the land,
so that their minds and hearts may learn wisdom
and their ears may thus learn to hear?
Truly, it is not their eyes that are blind,
but their hearts that are in their breasts.

–The Holy Quran, Surat al Haji

During my year working for the catering company, I was able to save up more money than I had spent in two and a half years traveling in Asia and Africa, and then some. With a passport pouch full of traveler's cheques and some money in the bank back in Montana, I decided to see the rest of Africa, first going over to Arabia, and then traveling south to Capetown. I said good-bye to the gang at the Khartoum staff house, and left for the airport, where I would fly to Abu Dhabi in the Persian Gulf.

That evening I was on a Gulf Air jet flying over the Empty Quarter, which was very exciting except it was dark. Like the Sahara, it appears that the vast, largely unexplored interior of the Arabian desert was once an inland sea with large fertile areas teaming with wildlife around it. In theory, there were lost cities to explore and discover, and with a little luck I hoped to do just that.

It was after midnight when we arrived in Abu Dhabi, and a horde of Pakistani taxi drivers descended on me as I stepped through immigration. I had gotten a one month tourist visa in Khartoum and breezed through customs and immigration. Hotels are extremely expensive all over the Persian Gulf, and it was out of the question for me to check into one at this late hour. I had to spend my money more wisely than that. I decided to sleep in the airport with a hundred or so other people, putting a foam camping mat on the floor of the waiting lounge and crashing out.

I walked around Abu Dhabi the next morning. I had to get some mail there, so I tripped down to the General Post Office and walked around the downtown. Abu Dhabi is a city of 200,000 and is the capital of the United Arab Emirates, a cluster of small sheikdoms and sultanates comprised of Abu Dhabi, Dubai, Sharjah, Fujairah, Ras al-Khaimah, and Umm al-Qaiwain.

Abu Dhabi is the richest of the Sultanates. The main oil fields of the Emirates are concentrated in the area. It is a modern cement, steel and glass city that has sprung out of the dust in the last twenty years. Without even spending a night in Abu Dhabi I got picked up by a van of Pakistani workers on their way out of town to a factory. I sat in the back and took a swig of some lemon-flavored dehydration salts in my plastic canteen.

The Pakistanis looked at me and with disdain said, "Whiskey!"

I laughed. "No, it's not whiskey," I told them in Arabic. "It's water. Here, try some."

"Oh, no, it's whiskey!" they said. "All Americans drink whiskey!" This was the familiar "James Bond" syndrome: Americans (or Europeans in general) are all men who are government spies, drink whiskey like water and sleep with a a different woman every night. There is little you can do to convince them otherwise, after all, *they've* seen those movies. *They* know! By the way, James Bond films are X-rated in Muslim counties.

"No, really, it's not whiskey, it's just water. Here, smell it. I'm not drinking whiskey, really!" But no matter what I did, I could not convince them that I was not drinking whiskey, and they wouldn't try it themselves. They let me off on the main highway north up the coast to Dubai. Lying next to me in the sand was a dead camel, run over by a truck. It was like a huge camel balloon, inflated by the decomposing gasses as it rotted. It was twice the size of an ordinary camel, and the stench of decay was terrible. I was horrified that it might suddenly burst, exploding gore all over me. I knew that I would then spend the rest of the day throwing up. To forestall that likely event, I walked down the road a way and soon got a lift with a Turkish guy on his way to Dubai.

He let me off near the Gold Market on the Dera Dubai side of Dubai Creek. Here there is a nic-free cameras and cassette players. Dubai is sort of the Hong Kong of the Persian Gulf.

I got a bed at the cheapest hotel that I could find in the market. At ten dollars a night in a dormitory, it was the best deal that I could find. I spent the next few days wandering the market, the only tourist, it seemed, in the country. There was a good Arabian Nights feel which clashed strongly with the constant development and new steel and glass buildings going up everywhere.

I was particularly interested in the Persian Gulf region because of a mysterious lost civilization that is only just now coming to light. Known in ancient times as Dilmun, the remains of the civilization were first unearthed on the small Persian Gulf island of Bahrain. Sometimes called the "island of the dead", Bahrain is almost entirely covered by some 100,000 burial mounds, both large and small, dating back to prehistory. Until the 1960's, it was thought that the pre-Islamic population of the eastern coast of Arabia went to the island to bury their dead and that no prehistoric city, temple or palace existed on the island, which is mostly desert. The island was essentially one huge graveyard.

Excavations conducted by Geoffry Bibby for the Prehistoric Museum of Aarhus, Denmark, have now shown that this idea is incorrect. They discovered walls and houses of a town, a palace of unknown style and a sanctuary. It is now clear that Bahrain was the center of a vanished empire which extended to present-day Saudi Arabia, a country ancient to the Sumerians and referred to them as the realm of Dilmun, country of the Rising Sun. It was the center of an earthly paradise, and the only region to have survived the flood of Middle Eastern and Biblical legend.

It was at Dilmun that Gilgamesh sought the only survivor of the flood, Utu-nipishtim in Babylonian texts or Ziusudra in Sumerian texts. Reads the Sumerian version, "Anu and Enlil cherished Ziusudra, life like a god they give him, breath eternal like a god they bind down for him. Then Ziusudra the king, the preserver of the name of vegetation and of the seed of mankind, in the land of crossing, the land of Dilmun, the place where the sun rises, they caused to dwell."[83]

Bibby, in his book, *Looking for Dilmun*,[83] puzzles a bit at this statement since Bahrain is southeast of Sumer, rather than east, where the sun should logically rise. He can only conclude that the Persian Gulf was called the Sea of the Rising Sun. What does not occur to Bibby is that it is possible that in Sumerian times, Bahrain was possibly directly east of Sumer, and that a slippage of the earth's crust, or slight change in the earth's axis, has since shifted Bahrain so that it is no longer to the east. Egyptian records talk of the sun rising in the west and setting in the east as if the earth's orientation to the sun had completely changed. Many researchers, and particularly cataclysmists, believe that this is a referral to a "pole shift".

Indeed, the more discovered about Dilmun, the stranger the scenario looks. For instance, it is known that the Sumerian port city of Ur is now many miles inland in the desert. Yet, in 4,000 B.C. Ur was a major port city. Obviously, the shores of the Persian Gulf must have come farther inland, yet, if we assume that this was just because of higher sea levels, Bahrain would probably have been underwater! Rather, it seems we are talking about a tectonic shift of some kind, that not only changed sea levels, but shifted the relationship between Bahrain and Sumer, as well as lifting up the entire Arabian plate, thereby draining the Arabian Sea and leaving behind it the vast sands of the Empty Quarter.

Objects dug up in Bahrain indicate the existence of commercial activity oriented towards both India and Mesopotamia. At one time subterranean canals piped spring water to irrigate gardens. There are so many mounds in Bahrain, the excavations are still in initial stages. Yet enough is known to provide overwhelming evidence that an extensive civilization even older than Sumeria existed in the Persian Gulf many thousands of years ago. Bahrain was apparently the center of a vast and busy trading empire. Sumerian texts record loads of 17 or 18 tons of copper coming on one ship through Dilmun. Dilmun/Bahrain was the Singapore or Hong Kong of the ancient Middle East, largely trading with the advanced civilizations of ancient India, and transshipping the material to the fertile crescent area of pre-flood Sumeria.

Even more fascinating than the discoveries on Bahrain are the discoveries in Arabia that are linked to Dilmun. These other traces of civilization are even older than that on Bahrainl. Some are found in the interior of Arabia, though they were port cities just like Ur. Remains of a port city were found at Thaj, a small town ten miles inland from the Saudi Arabian coast. As in other towns linked to Dilmun, some of the stonework is of well-cut, massive stone. Unfortunately, at Thaj, excavation is very difficult because the key site is located at the women's bathing place, making it strictly out of bounds according to Arabian custom.[83]

At another site 200 miles in the interior of Saudi Arabia at Jabrin, there are extensive fields of grave-mounds identical to those in Bahrain. They lie on the edge of the great desert of the Empty Quarter, and this was the farthest point penetrated by Bibby's expedition.[83]

It appears that Arabia was a fertile area with some sort of inland sea at some time in prehistory, and the strange desert port cities of Dilmun, legendary land of immortality, were there on the coast of this inland sea that was part of the Persian Gulf. During one of the catastrophic tectonic shifts that rock our planet every few thousand years, probably the Biblical and Sumerian Flood itself, Arabia was drained of its shallow sea and became a dust bowl.

Bibby describes exactly such an event himself, though naturally he places the event millions of years ago and describes it as gradual in nature. It is not fashionable for respectable archaeologists such as Bibby to embrace such myths as the Flood in explaining anomalies of archaeology. This was even more true in the pre-tectonic plate era of the fifties and sixties. Nevertheless, his account seems fairly accurate if speeded up in time: "...during the last great mountain-building period of the world, the Persian massif had lunged southward, tipping the whole slab of Arabia. In the east Arabia had been pressed down below sea level, forming the Arabian Gulf. And in the west the slab had been cracked off from Africa, forming the deep chasm of the Red Sea, the Rift Valley of East Africa, and the crack which was now the Gulf of Aqaba and the Jordan Valley. It was not unlikely that a recovery had been going on ever since, that Arabia was gradually returning to the horizontal.

"It would explain many things in the historical record. Such a rise in east Arabia would reduce the flow of underground water from the high land to the west.... The exposed sea bottom would dry out and blow away as sand and dust, which would choke the vegetation on the land, already threatened by the diminishing water supply. Dust-bowl conditions would result, adding more sand to the dunes. The supply of pasture for grazing animals would diminish, and what there was would be overgrazed, giving more denuded areas, and more sand."[83]

It was still the popular belief among Europeans at the turn of the century that the interior of Arabia held strange civilizations, lost cities and hidden dangers. In 1916 a pulp novel named *The Bowl of Baal* by Robert Ames Bennet ran as a serial in *All-Around* magazine. The story was about a British aviator named Larry O'Brian who ventures into the unknown Arabian desert during World War I. His discoveries are epic; an ancient hidden race with a conflict between two beautiful priestesses; a barbaric tribe of cave dwellers; and a gigantic saurian still surviving in the dreaded swamps that are all that exist from the ancient inland sea that is now the Empty Quarter of Arabia. *The Bowl of Baal* is a large helping of adventure fantasy, yet it is curious how close to the truth of ancient Arabia it might be!

§§§

I met a Pakistani clerk one day in a cafe in Dubai. He was on holiday and just passing the time. We talked at great length about Pakistan, America, and the world. He had been a sailor and had sailed the seven seas several times. He also spoke English quite well, and we often sat around in the bazaar drinking tea. We would meet at the same cafe every day, down in the central market, sit at a table on the street and watch the bazaar pass by. We'd have a good strong fan aimed at us and drink cup after cup of tea and eat snacks. He'd have on his immaculate white shirt that covered his brown, trim body. He was in his late thirties, had dark skin and a thick black mustache.

One day I was sitting in the cafe with my Pakistani friend reading the Dubai paper, in English. One headline was about a Genie queen who was getting married. She was going to grant mortals a wish with the only stipulation that they not ask for immortality. Otherwise, they need only make a wish at precisely 7 P.M. her time on the day of her marriage. Anyone who mocked her, however, would be jinxed for life.

Suddenly we were joined by an Indian friend of his, a doctor by the name of Mr. Kintia. We sat on a hot afternoon drinking tea in the central market and Mr. Kintia and I discussed the Middle East from a geographical point of view.

Mr. Kintia was 45, rather overweight, and very hairy. He had thick black hair on his shoulders, back, and ears and sported a hearty laugh.

"It only makes sense that the major religions of the world stem from the Middle East," he said. "It is the crossroads of the world! Why the only other major religions, Hinduism and Buddhism, stem from my country, India! So it is only natural that Jesus was born in Israel, the crossroads of the entire civilized world at the time.

"Why, two thousand years ago, the caravan trade between civilized Egypt, Southern Arabia, Greece, Rome, Persia, India and China all centered around Israel, a rich cultural area. And those areas, too, inherited their civilization from those countries before them, the Chinese from the Uigers; the Egyptians from the Osirian Civilization and the Indians from the Rama Empire."

"Why, did you know that Jesus as a young man traveled on a caravan through Persia to India?" Mr. Kintia asked.

"Is that right?" I said, taking a sip of tea.

"Indeed the Bible mentions nothing about Jesus' life between the years of eleven and thirty, but there are records that indicate what became of Jesus during those years. They say he traveled to the Jaganath Temple in Orissa, on the Bay of Bengal in India, and studied for four years. He was a good student, but preached against the false Hindu doctrines of caste and transmigration, and the Hindu priests in Nepal, the place where Buddha was born some 500 years before in the foothills of the Himalayas. From Nepal, he traveled to Lhasa in Tibet and there he met the Chinese sage Meng-tse, who was very old already. Jesus studied the ancient records kept in Tibet by the Ancient Masters and began his return journey at the age of twenty-three.

"He traveled and studied in Persepolis, where the three wise men, who were Zoroastrians, and who had visited him in Bethlehem as a child, were living. He went on to Athens and Zoan, Egypt where he studied for some years and was initiated in the Great Pyramid before starting his ministry at the age of thirty in En-Gedi, in Israel!"

"That's amazing!" I said. "Is there any proof of that?"

"Plenty," said Mr. Kintia, sitting back in his chair and lighting a cigarette. "It is mentioned in several Indian texts as well as in the so-called "lost books" of the Bible, which were excluded from the Bible in the fourth century A.D. Even the Holy Quran mentions these facts. St. Thomas, one of Jesus' disciples, is buried in South India; I myself have been to the tomb. It is really not so incredible, you see. Many people traveled widely in those days on the caravans. The trade caravans and ships daily plied the well-worn and established trade routes of Asia."

"I can buy that," I said, "why not spend fifteen years traveling around on caravans? Lots of other people did."

"Sure," said Mr. Kintia, "Jesus was an especially knowledgeable person. He received as fine an education as anyone could have possibly had at this time. It is evident by his words and actions that he was a highly evolved Ego."

§§§

I was especially keen on going to Oman, a country straight out of *The Arabian Nights* where people still live as they did five hundred years ago. Slavery is rumored to still exist in Oman, a country completely cut off from the outside world until 1970. At that time, there were a handful of automobiles on some one and a half miles of paved road! Sunglasses were banned as western decadence. Oman is officially known as the Sultanate of Oman, and is ruled by a absolute monarch, Sultan Qabus ibn Said al Bu Said, a young man in his twenties who overthrew his father in 1970. His father is known as one of the most tyrannical despots in this century; sanitary facilities and stoves, as well as movies, drums, tobacco and cultivating new land, were all banned. The Sultan did his best to keep Oman in the Middle Ages.

During the 17th century, Oman was the most powerful country in Arabia, controlling such far-off places as Zanzibar, much of the Persian Coast and parts of Pakistan from its capital of Muscat. The Omani sailors were some of the best in Arabia, and their ships plied the Indian Ocean with each monsoon. It is essentially from the stories of the Omani sailors that we get our tales of Sinbad and the many strange places that he visited.

That Omani sailors must have sailed to remote islands far to the South in the Indian Ocean is evidenced by the Sinbad tales of the giant Rok, a huge bird that carries away sailors that lives on a remote island. Both the elephant bird of

Madagascar, and possibly the Dodo of Mauritius, would have filled the bill. The elephant bird of Madagascar was one of the last descendants of giant prehistoric birds called Ratites whose relatives can be found in New Zealand, Australia, Islands of the Indian Ocean, Africa and South America.

That Omani sailors went as far as Indonesia and Australia is evidenced by the spread of Islam into these areas. Islam spread into wild island areas of Borneo and the southern Philippines, as well as through all of Indonesia. While the Omani sailors of yore may have sailed parts of the Australian coast, there was probably little of interest there for them, because the gold mines of the Egyptians, Phoenicians, Persians and King Solomon (Ophir?) had probably been long closed down. The stranded miners were either killed, merged with the local populace, or took off to colonize remote Pacific Islands. New Zealand itself may have been settled this way. This is essentially the scenario given by Barry Fell in his books.[20,21]

An exciting, romantic place, Oman offered a very good opportunity to step back in history a few hundred years. I caught a ride out of Dubai to a place called Al Ain, near the border of Oman, and a university town. I couldn't get a visa to enter Oman, but I thought I would at least go to the border and try to get a transit visa. But in spite of my eagerness I was refused entry into Oman. I hitchhiked back to Dubai, which was only a few hours away, and wondered what to do next.

I decided to fly to Jeddah, Saudi Arabia. Saudi Arabia is a strange country by anyone's reckoning. It is even stranger when one realizes its bizarre ancient history. A mysterious land of lost cities in the desert, virtually uninhabited desert in the dreaded "Empty Quarter", hostile bedouin, and of course the holy shrine at Mecca, forbidden to non-Muslims under pain of death.

Traveling to Saudi Arabia can be very difficult, simply because they just don't want travelers or any other tourists. Saudi Arabia will issue no tourist visas, and therefore has no tourists. They do allow pilgrims, however, and still get a few travelers: those intrepid adventurers who don't let small details like passports and visas stop them from exploring a country. As an old Tibetan horse trader in the Karakorams once told me, "Where there's a will, there's a way."

I was determined to go to Saudi Arabia and had gotten a visa at the embassy in Sudan. The fact that Saudi Arabia didn't want me made me want to go even more. It was a major challenge, and I was determined to meet it. All the oil-producing Arab countries are difficult to get into because they earn all their foreign exchange in oil and therefore do not need tourists and their dollars, nor do they desire any more foreign influence than they already have. They are highly conservative countries resistant to change.

It had been a major hassle just getting a visa in Khartoum. Three types of visitors are allowed into Saudi Arabia: businessmen working in the oil fields or some other development project; transit passengers who are driving across the vast desert expanses on their way to the Persian Gulf or Yemen; and pilgrims, or *hajis* as they are known in Arabic, on their way to Mecca, the holy city of the Muslims. I spoke Arabic fairly well after living in Sudan for a year, so at first I didn't see why I couldn't qualify as a genuine pilgrim to Mecca. However, at the Visa Office I discovered that Arabs take their religion a lot more seriously than I. The pilgrimage requires a lot of study, and you must be an orthodox Muslim, able to recite verses from the Quran in Arabic, and to follow the prescribed rituals on the pilgrimage exactly. Failure can mean death! I wasn't going to risk everything to get to Mecca, but nothing would prevent me from seeing something of Saudi Arabia.

Determined to get a visa by hook or by crook, I told the Saudi Arabian Consulate in Khartoum that I was a medical student who had been working in

Central Africa and I wanted to visit my father, who worked in Saudi Arabia for the Saudi Arabian Oil Company, ARAMCO. Furthermore, I said, I hadn't seen my father for two years (a clever device which preys on Arabians' strong paternal instincts) and I was flying to Jeddah on the coast of Saudi Arabia in two days.

"Have your father send us a telegram," said the Consul, "and I'll give you a visa."

"Impossible," I countered, "I'm leaving for Jeddah in just two days. There's not enough time. My father will meet me at the airport and get me a month's visa while I'm in the country. Just give me a transit visa so I can get in."

After some discussion with the Vice Consul, he gave me a transit visa. This was apparently completely against the rules, but I have heard it said that rules were made to be broken. Certainly, no Saudi Arabian, official or not, would keep a son from his father.

A couple of days later at the airport, as I was standing by the duty-free counter looking at cameras, what few they did have, when a tall, thin Canadian stepped up next to me and ordered two bottles of Dewar's White Label Whiskey.

"Aren't you going to buy some duty-free whiskey?" he asked, turning to me.

"Hmm," I thought, "that might be a good idea." One traveler's trick is to buy duty-free whiskey and cigarettes in the airport as you are flying into some Asian or African town and then sell them to your taxi driver, since imported luxury goods such as Scotch and English cigarettes are quite expensive. I could certainly use the money to help me pay expenses while I was in Saudi Arabia. I knew it was a terribly expensive country. What the heck, he was buying whiskey so I bought a bottle as well, and stuffed it way down in my pack.

As it turned out, the traveler wasn't going to Saudi Arabia. He was going somewhere else, perhaps to Cairo. On the plane I spoke to the guy next to me, a swarthy, middle-aged, olive-skinned Arab gentleman who looked to be of good education.

"Tell me," I asked him, leaning out from my seat and sticking my head as far out into the aisle as I could, "what can I get for a bottle of Dewar's White Label in Saudi Arabia?"

He thought for a moment and then replied, "Oh, a good whipping and one to three years in prison, I suppose."

"Wha-what?" I croaked, my throat suddenly getting rather dry.

"It's a very serious offense to drink alcohol or have any in your possession in Saudi Arabia," he said very calmly. After all, he didn't have a bottle of whiskey in his luggage. He took a sip of his orange juice. "Certainly you would get a public flogging, and then possibly a prison term—why do you ask?"

"Uh," I could hardly speak. My mouth was totally devoid of moisture and my stomach had dropped beneath my seat. My life seemed to be flashing before my very eyes. A public flogging and a prison term in Saudi Arabia were not what I had in mind. How would I ever explain this to my parents? "I, oh, I was just curious, that's all. It was nothing really, I mean...will you excuse me for a moment while I go to the toilet?"

It's really funny how these things happen. In retrospect I can sit back and have a good laugh. At the time though, I thought I was going to throw up. What could I do? Because of the strange loading set-up at the airport, my bag with the bottle of whiskey was in the hold of the airplane and I wouldn't be able to get it until I was in Saudi Arabia, and through immigration, which would be too late. Gritting my teeth and fighting the sinking feeling in my stomach, I decided to go though with the whole thing and actually sneak the bottle past the customs officers. Now I knew how smugglers must feel.

I walked up confidently to the immigration officer who was looking pretty sharp at the immigration booth, one of many, at the new Jeddah International

Airport. I could tell just by looking at him that he was no Saudi Arabian. Egyptian, I guessed. The Saudis hardly do any menial work, importing laborers from all over the world. The airport was crawling with Koreans, Philippinos, Egyptians and Pakistanis as well as Sudanese, Jordanians, Palestinians, Yemenis and workers from all over the world who were attracted to jobs in Saudi Arabia by the high pay.

"Salaam Aleikum," I said, stepping up to his booth and handing him my passport with my visa all ready.

"Where are you going?" he said without looking at me.

"Mecca," I declared while out of the corner of my eye I noticed my bag coming out of a conveyor belt behind the officer.

He took a quick look at me; I was no pilgrim to Mecca.

"What?" he said.

"Did I say Mecca?" I answered. "Just stopping in Jeddah on my way to Abu Dhabi." As an afterthought I added, "I'm here to see my father."

"Let's see your airline ticket," he demanded. After inspecting it momentarily he saw I did not have an onward reservation. "Make a reservation for your flight out," he said, indicating a Saudia airline reservation booth behind him. Putting my passport on the bottom of his little stand, he turned to the next passenger behind me.

I made a reservation for exactly three days later to Abu Dhabi, the very limit of my transit visa. Actually, I had no intention of flying to Abu Dhabi at that time. My plan was to hitchhike south, but I couldn't tell them that. I showed the immigration officer my new reservation.

"Good," he said, glancing at it and handing me my passport. "You may wait in the transit lounge until the flight."

"You mean wait three days in the transit lounge until my flight to Abu Dhabi?"

"Yes, that is correct."

I was already a little tense about having the whiskey in my luggage, but this was too, too much. It looked like the best thing to do at this point was to throw a scene right there in the airport. I simply had to get into Saudi Arabia...

"Are you crazy?" I said. "My father is waiting for me right now outside the airport, I've got a perfectly valid visa and airline reservation, and you want me to go to the transit lounge for three days. Are you crazy?"

I felt somewhat better after letting off steam, but this poor immigration officer, who was merely following orders, just turned away and continued stamping in other people. In desperation, I turned to the airline officer who had just seen my tantrum and was standing there watching me sympathetically. He went to talk to the chief immigration officer, who had a lot of clout and was a real Saudi Arabian. After explaining that I was here to see my father, whom I hadn't seen in two years, and who was waiting outside for me at this very moment, the chief officer took pity on this hitchhiker and wrote me a note, allowing me to actually enter Saudi Arabia. I felt better already.

I was stamped into the country and walked over to my pack which was circling around a luggage carousel. By now it was the only one, but fortunately more passengers were arriving all the time. I had, also fortunately, packed the whiskey inside my sleeping bag so it wouldn't get broken along the way. As calmly as possible, I picked out the busiest customs counter I could find, walked up, and flopped my bag on top.

"Open your luggage!" commanded the customs officer. "Take everything out!"

Thank Allah he was a man in a hurry. I took out all the stuff in my pack, which was quite a lot of junk, and put most of it on the counter. A few things, including my tent and sleeping bag, I laid on the floor. He passed by me, glanced at my stuff and scrawled a line on my pack with a piece of chalk. "OK," he said. Whew! I had made it! I packed up my pack faster than you could milk a camel and was out on the street. "What a hassle getting into this crazy country," I thought. "But, after all, what is life without its challenges?"

The airport is near the central bazaar and I walked into the main souk checking out the hotels. Everything in Saudi Arabia is expensive, except gasoline. The cheapest hotels ran $15 a night for a single! A better option was staying at one of the tea shops that abound throughout Saudi Arabia, called *garuwaz* in Arabic. In these *garuwaz*, I slept on a rope bed, usually under the stars, and it cost a mere $2 a night.

I selected a *garuwaz* on the roof of a building, and after leaving my luggage there, I headed down to the main market. Jeddah is the main port on the west coast of Saudi Arabia and has a nice bazaar within walking distance of the hotels, just down by the docks. There are vendors selling gum arabic out of big baskets, unflavored sap straight from the tree. Arabians like to chew it as bubble gum. There are shops galore, selling gold, cameras, watches, cassette players, and such. All of Saudi Arabia is duty free. There are sandwich shops and some sidewalk cafes, plenty of women on the streets, too, to my surprise. There were Saudi Arabian women there, and a lot of Orientals, wives of expatriot workers, Americans mostly, who had married while in Thailand, the Philippines or Korea. Very few of the women wore veils, but all women were dressed conservatively, in long sleeves, long dresses, or pantsuits, and often with scarves over their heads.

Lots of foreigners in this city, I noticed, looking around at the sailors and other "ex-pats" from the cafe where I was drinking tea. I started up a conversation with three young Scottish doctors who were working at a hospital in Jeddah.

After a cup of tea, I inquired quietly, "Would any of you be interested in buying a bottle of whiskey?"

They were suspicious at first; after all, alcohol in any form was forbidden in Saudi. But their Scotch taste for some Dewar's White Label won them over and I was able to unload my bottle on them, to my great relief. The whole episode was decidedly not worth the risk, and I swore I would never try anything like that again. The tension in my back from anticipating the flogging quietly melted away and I felt I could stand a little straighter.

I walked back to the *garuwaz*. Old men were sitting on their rope beds watching TV and smoking tobacco mixed with honey from water pipes. I nodded to the watchman, lay on my sleeping bag, and gazed up at the stars until I fell asleep. It was good to be in Saudi Arabia, a pilgrim even if I couldn't get into Mecca.

§§§

The next morning I was awakened early by the sun warming my face and sleeping bag until I couldn't comfortably sleep any longer. Around me the many robed and turbaned pilgrims or wayfaring merchants were starting to stir from their beds as the bright Arabian sun began to beat down. I packed up and left. I had only three days in Saudi, and I had to make good use of my time. Today was the first day of Ramadan, when Muslims fast to commemorate the Prophet's flight from Mecca. During daylight hours, Muslims cannot eat, drink, smoke, or even swallow their own spit. In most Muslim countries, this is all done purely by choice, but in Saudi Arabia, it is mandatory whether you are a Muslim or not. If you are seen in public eating, drinking or smoking, you may be put in jail or publicly whipped. Most Muslims know better than to do this, so it is usually foreigners who get into trouble.

I was sitting in the lobby of the Allharamein Hotel talking with a young

Sudanese kid, tall and dark, who was the receptionist. I told him I was just coming from Sudan and had lived and worked there for a year. "What are you going to do here in Saudi?" he asked.

"I'm planning to hitchhike south to Yemen," I told him. "I think I'll go to Taif, in the mountains, today."

"Today is the first day of Ramadan," he said. "The royal family will be going to Taif today. It is better that you take a bus because if they see you on the road, you could be in trouble."

That was good advice, so I went down to the bus station and bought a bus ticket to Taif. I left my luggage in a locker and walked around the port and market for awhile; I had three hours before the next bus. The road to Taif, the summer capital, was high and windy. Twisting up into the Hejaz Mountains, which run down the length of the western coast, the road runs right past Mecca (less than an hour from Jeddah) and on to the cool mountain city of Taif, a three hour bus trip.

At the bus station in Taif, I asked an American working at the counter what there was to do there. He was in his late thirties, going gray and bald, and he had a slight curvature of the spine, causing him to hunch over slightly. He looked at me sharply. "What are you doing here in Saudi Arabia, anyway?" he snipped.

"Oh, I'm just hitchhiking around. I'm going to Mecca," I said casually.

"Mecca!" he snorted. "You'll never get to Mecca! Mecca is for Muslims. They would kill you if you tried to get into Mecca!"

I did know that the Kaaba is guarded by guards with machine guns who watch the pilgrims very carefully. Any person who does not follow the rituals, and appears to be a non-believer trying to sneak into the courtyard around the Kaaba, would be in big trouble.

Travel to Mecca has always been dangerous, and not just because it is forbidden to infidels. The first report of a pilgrimage to Mecca is from the astounding Sheik Mohammed ibn Abdulla Ibn-Batuta, born in Tangier in 1304 to a wealthy family. Ibn-Batuta was very well educated, adventurous, and was probably the best-traveled person of his time. At 21 years of age, he left Tangier for Egypt, went to Damascus, made the pilgrimage to Mecca, and went on to Damascus, Persia, back to Mecca, and even traveled to East Africa, Yemen, Oman, India and China.

The first non-Muslim to visit Mecca was the Italian adventurer Ludovico di Varthema who published a book about his travels in Arabia in 1510. He disguised himself as a Mameluke, one of the Christian prisoners of war who converted to Islam and were now mercenaries for the Egyptian and Syrian caliphates. After many adventures, and foolishly deserting the Mameluke guards, he returned to Italy and published his book.

It was nearly 200 years later that the next westerner was to travel to Mecca. This was the English seaman Joseph Pitts, who had been captured by Barbary pirates in 1678 and then sold as a slave to a Muslim cavalry officer. He accompanied his master to Mecca in 1685. Later he escaped, returned to England, and wrote a book about his adventures.

The next European was a mysterious Spaniard who called himself Ali Bey. He made the pilgrimage in 1801 with a number of scientific instruments and was the first to fix the position of Mecca by astronomical observation. The German Ulrich Jaspar Seetzen made the journey in 1809 disguised as a Muslim, but was unfortunately murdered two years later near Ta'iz in Yemen.

The Swiss explorer, John Lewis Burckhardt, who had discovered Petra in Jordan, made the haj in 1815, but died a year later of dysentery in Cairo while preparing to cross the Sahara. In 1853, the most celebrated western explorer to visit Mecca, Sir Richard Burton, an army lieutenant who had served in India for seven years, managed to sneak into Mecca disguised as an Afghan pilgrim.

Because he had stepped on a sea urchin while wading ashore from a pilgrim ship at Jeddah, he had to make the journey in a hammock slung to a camel, and this kept him from his ambitious plan of crossing Arabia. Another English adventurer named A.J.B. Wavell repeated Burton's dangerous voyage disguised as a Zanzibari pilgrim in 1908. He made the pilgrimage on the newly completed Hejaz Railway, built by the Ottoman Empire to help control Arabia and its warlike Bedouin.

The fact that Mecca was a forbidden city to westerners probably helped fuel the nearly impossible task of journeying to the city. What was it that travelers saw when they reached the holy spot? In the center of Mecca is Islam's holiest shrine, the Great Mosque. It stands loftily in the white-hot desert sun, lifting its seven tall minarets up to heaven. In the center of the mosque is the fifty-foot high, cube-shaped Kaaba, draped in black silk, embroidered with verses of the Quran in gold thread. It is toward this building, the Kaaba, that some 800 Muslims pray five times a day. Tradition has it that the Kaaba was built by Abraham, the Biblical progenitor of the Hebrew peoples, in the second millennium B.C. Abraham built the Kaaba as a House of God in which to worship the monotheistic deity of the Hebrews. For thousands of years the desert dwellers of Arabia made an annual pilgrimage to this shrine, which netted a tidy profit for the inhabitants of Mecca who created a yearly festival for the event. The desert tribesmen would often bring their own idols or deified gods and over the years the Kaaba was filled with such pagan deities as Awf, the great bird, and Uzza, the goddess of the morning star, as well as some 360 other images.

Sometime in the late sixth century A. D. the elders decided to renovate the Kaaba, but could not decide who should have the honor of resetting the most holy object in the Kaaba, the Black Stone, in its place back in the masonry after it had been cleaned.

According to Islamic tradition, in 1892 B.C. God ordered the Biblical patriarch Abraham to emigrate with his son, Ismail, and Ismail's mother, Hagar, to the valley at Mecca. Here Hagar lived with her son and herself a house, the Patriarch coming from Palestine to visit her on occasion. He was then ordered by God to make his house into a temple where people prayed. Therefore, he demolished his house, and built the Kaaba. According to Arabs, Mecca is the Babylonian word for house.

The Black Stone, considered to be the only thing remaining that was originally placed in the Kaaba by Abraham himself, was then and now, the central article of the pilgrimage to Mecca. Believed to be a meteorite, some 12 inches in diameter, the Black Stone plays a very important part in Islam.

It was agreed by the elders that the next man to enter the sacred precincts would judge the dispute of resetting the sacred stone. The first man to appear was a young Meccan merchant and former caravan guide, whose wisdom and honesty had earned him the nickname of El Amin, "the trustworthy." A lean man, strong of body and character, his thick black beard framed his tanned face and dark eyes. He was said to be so serious that his laugh was rarely more than a mere smile.

To settle the dispute, he spread his cloak on the ground and placed the stone on it, then had two nobles grasp a corner of the cloak and together raise the stone. Then with his own hands, the young merchant, El Amin, placed the stone into the niche, where it remains to this day. That merchant's name was Mohammed!

Born north of Mecca in Yathrib, now renamed Medina, in A.D. 570, Mohammed was orphaned at six and raised by an uncle. He traveled widely on the caravans as a conductor, carrying frankincense and silk through Mecca north to Damascus. Mohammed has been said to have been illiterate, but many scholars feel that is unlikely, for he was very familiar with sixth century Jewish and Christian thought.

Mohammed devoted much time to contemplation and often climbed to a small cave among the rocks of Mount Hira, just north of Mecca, to meditate and fast. In the year 610, Mohammed had a vision. In his vision, the Angel Gabriel told him to clear out the false deities in the Kaaba. The leading elders of Mecca, however, did not like this at all, as the annual pilgrimage was a major event. They threatened and scorned Mohammed.

Mohammed and several trusted friends and family members slipped out of Mecca one night and went to the oasis of Yathrib, about 200 miles from Mecca to the north. The year was 622, and the Muslim calendar of 354 days begins on the date of this flight, or as it is known in Arabic, *hegira.*

In Yathrib, Mohammed and his followers built the first mosque and established the first Islamic community in what would eventually become Islam's second holiest city after Mecca, Medina. Mohammed and his ever-increasing followers waged a superficial war with the elders of Mecca for twenty years. During one battle, Mohammed himself was almost killed and companions dragged his unconscious body to safety.

By the year 630 Mohammed had converted enough Bedouin tribesmen from the surrounding desert that he was able to reenter Mecca with an army of 10,000, a considerable army in the desert expanses of Arabia at that time, and Mecca surrendered without a fight. Mohammed walked to the Kaaba, touched the Black Stone and made the prescribed seven circuits around the building, a ritual still followed today. Within two years most of Arabia was under his sway, and within 100 years, shouting the name of Allah, Arab armies seized a realm that stretched from Spain to Central Asia, and came within 100 miles of Paris.

Mohammed did his best to civilize the desert dwellers of Arabia. They had fallen into barbaric practices; Mohammed abolished idol worship and the killing of baby girls, he limited the number of wives a man may have to four, and forbade gambling and intoxicating liquors. His most important reform was a ban on violence and war, except in self defense and for the Islamic cause. It is this last clause, that war can be waged for Islam, that has fueled the many violent wars that continue to this very day. It was the first caliph of Islam, Abu Bakr, one of Mohammed's followers, who was not content to merely protect the faith in Arabia, but was moved to organize a *holy war,* or *jihad,* that would forcefully and violently spread Islam throughout the world. And so the early Muslims were able to continue indulging in mankind's favorite sport, war, even though it was officially forbidden by Mohammed.

There is a certain fanaticism about these dwellers in one of the largest and most desolate deserts in the world that enabled Mohammed's followers to weld such a huge empire out of a bunch of camel herders in such a short time. Believing themselves to be divinely inspired, they waged holy war on the rest of the world, their battle cry, "There is no God but Allah, and Mohammed is his prophet!"

§§§

It appeared that my dream of going to Mecca was going to be shattered. Mecca was a fascinating place, and I had always wanted to go there, planet-earth pilgrim that I am, but it wasn't worth risking being riddled by bullets from the guards who were bound to spot me. I decided to keep heading south to Yemen.

It wasn't difficult to part from the American in Taif. The guy hadn't been too helpful but he told me about his two wives, one in Philadelphia and one in Bangkok, Thailand.

"Isn't it illegal in America to have more than one wife?" I asked.

"I don't give a damn!" he snarled. He seemed bitter, perhaps because he had to work in this bus station in Saudi Arabia while both his wives lived in foreign countries. "Besides," he said, "I live in Saudi Arabia, I can have as many wives as

I want!" He didn't seem to think that there were many things to do in Taif. But to me it apeared to be a pretty nice place to live, fresher and cooler than the hot humid coast. With rugged purple mountains that shot straight up into the cloudless sky, a pleasant little bazaar to wander in, plus the summer palace of the royal family, it was Saudi's version of a resort town.

I was heading for Abha, straight south along the mountains. I caught several short lifts out of Taif and south down the mountain highway. My best ride was with a young Saudi Arabian named Saeed. He was taking a new Dihatsu truck to Khamis Mushad, an air base that the Saudis had in the south. He was neat, cleanshaven and courteous. Friendly enough, he didn't speak English, so we conversed as best as we could in Arabic. I told him about life in America, and he was astonished at the ease of getting married.

"A wife in Saudi Arabia costs between 50,000 and 100,000 rials in dowry money," he said. I figured this out to be about $16,000 to $33,000. In the Middle East, a young man must save all his money for years and then pay a huge dowry to the family of the bride for his wife, a women he is unlikely to even see before his wedding day, as women are kept in their homes most of the time and wear veils on the rare occasions that they go out.

He was amazed, as are all Arabs when they hear it for the first time, that you do not have to give a big dowry in America to your wife's family when you get married. This dowry is a heavy financial and psychological burden for young Arab men, it being almost impossible to satisfy any sexual urges until they can afford to get married. Prostitutes, while available in other Arab countries to a limited extent, are virtually unknown in Saudi Arabia, and therefore many men turn to homosexuality out of desperation. Saeed was fine, and at no time in Saudi Arabia was I ever even teased. Homosexuality is also illegal in Saudi, but there is still some prostitution and abuse of young boys.

In the east part of Saudi Arabia, where all the oil fields are, ultra-modern cities are springing up out of the dust and are manned by virtual armies of American, European and Asian workers. Damman, on the Persian Gulf, is a miniature American city, complete with American TV, Dairy Queens, pizza parlors, and American schools. While in Saudi Arabia I was told by an American that a teenage daughter of an American family living in Dammam had been selling her favors to wealthy Arabs in the area, and by the time it was discovered and the family expelled, the teenager had collected about $50,000!

I stared out the window and looked at the mountains as Saeed and I continued on the long drive south. I thought of the many mysteries of the vast, unpopulated and unknown land. Early travelers had dispelled some myths of the country. For instance, Di Varthema was able to disprove the popular Medieval legend that Mohammed's coffin in Medina was suspended in the air by giant magnets. Perhaps a more real sight reported by Di Varthema was a supernatural light that traditionally shown over the tomb of Mohammed. It is possible that some light did shine over the tomb for some period, though its source is not known.

In the northwest of Saudi Arabia lies the mysterious ruins of Mada'in Salih, carved into solid cliffs, which is somehow connected with the meteorite now kept in the Kaaba. Tradition claims that it fell from the Paradise with Adam and was given by an angel to Abraham. This legend is apparently the Islamization of a cult of stone practiced by the Semites of the desert.

The Nabataeans were stone worshipers who carved out buildings in solid rock, and at Mada'in Salih there are a number of sanctuaries carved out of solid rock. Corridors and rooms lead to the end of one temple, where there is an enormous block of block of stone which has been removed from the wall. On this stone the Nabateans are said to have make made sacrifices to the divinity of the place. The surrounding cliffs from a vast necropolis of family tombs, and in many of these sanctuaries there are similar sacred stones, some of which bear people's names and inscriptions. These stones were thought to be the residences of a god and bore the name of *baetyl,* meaning "house of the god". The Black Stone of Mecca is believed to be connected with these stones.[23]

Curiously, Ibn-Batuta relates that when his caravan stopped at Mada'in Salih in 1326, the pilgrims would not drink there, even though there was plenty of water and they were virtually dying of thirst. According to the Tunisian traveler, because they believed that Mohammed had once ridden his camel through the ancient place without stopping and had forbidden his followers to waste time by stopping to drink.[90]

The book, *Citta Perdute Nel Deserto* (Lost City of the Desert) by Cino Boccazzi[81] is, to my knowledge, the only book ever written about Mada'in Salih (or Madain Saleh, as Boccazzi spells it in his book). Boccazzi points out that the complex of buildings is virtually identical to Petra, and seems to harken back to a time when the Arabian desert was fertile. Boccazzi even thinks that the structures are so old that he likens them to Atlantis and other vanished cultures. That some of the doorways, carved into solid cliffs, are so worn and chipped away at the bottom that there is little left, and this makes Boccazzi think that the city was built before the Biblical flood! As incredible as this supposition may seem, the structures cannot actually be dated, as they are solid rock, and there are no inscriptions.

Most interesting to me is the common motif of stairsteps in the city. This same motif is found at the megalithic buildings in the high Andes, and an identical motif can be found at the megalithic, ancient, Temple of the Sun at Ollantaytambo on the Urubamba River just north of Cuzco (see my book *Lost Cities & Ancient Mysteries of South America*[22] for more information).

Was Mada'in Salih an ancient city at the time of Atlantis, Rama and Osiris, when Arabia was a fertile area teeming with wildlife and the Empty Quarter a sea? Who were the mysterious Nabateans, said to have built the cities? Was Mada'in Salih even a city? Many researchers believe that the strange rooms are only tombs, pointing out that no other structures can be found. Were they swept away in a cataclysmic flood?

Certain clues to the above questions can be found at the cities. Besides the stairstep motif, there are a few carvings of pumas, eagles and one of the head of a man. This head is unusual, fat, with wide lips, a broad nose and seemingly sunken, bug-eyes. This man hardly looks like the Semitic Bedouin of the desert, more like the strange Olmec heads in the Yucatan, or the weird statues at Tiahuanaco.

There are also some inscriptions, probably Sabean, but undeciphered. It is an interesting thought that these inscriptions, ancient themselves, were carved into the walls long after the city was created, the original builders having vanished many millennia before.

Whether Mada'in Salih was a lost outpost of Atlantis was not a question that Saeed was particularly interested in. We stopped frequently at gas stations, and munched on apples, bananas and meat pies while we talked intermittently.

"If it is Ramadan, Saeed," I asked him, "how is it that you are eating and drinking during the day?"

"Because," he answered, "it is all right for travelers to break their fast while they are traveling. They must have sustenance. Still, you should not eat or drink while in public." There was a beautiful sunset that night, a deep set of rainbow colors that sandwiched along the rows of jagged bare mountains to the west. We stopped at a tea shop for a while in the evening and ate dinner. Saeed ordered a water pipe and puffed on it for a while. I had some shish kebab mutton and whole wheat pita bread, a flat, round loaf. Restaurants are quite expensive, but street vendors, selling mutton sandwiches and sesame snacks, are quite cheap. Water is still precious all over the country. In some places, a glass of water may cost a dollar or two.

Saeed finally let me off at a *garuwaz* on the outskirts of Abha, a town of a few thousand in the mountains near North Yemen. It was about midnight and I was tired. Saeed drove on to the east and I went to sleep on one of the rope-strung beds in the tent-like frame and canvas structure that was the tea shop. Because of Ramadan, old men were still up, watching TV and smoking their water pipes all night, then sleeping most of the day.

The next morning I walked through town looking for the road out on the way to Jizan, on the coast and closer to the border of Yemen. Because they were just building this part of the road down from the mountains to the coast, the way was closed until three o'clock while they did some blasting in the steep slopes. There was a police station and road block just at the end of town, and the police seemed to think I was a bit of a strange character. I guess they had never seen a hitchhiker before. They utterly tore my luggage apart. Again I was grateful that I had sold my whiskey in Jeddah.

They were a nice bunch of guys, these policemen. They were amused and delighted to find a frisbee in my pack. I showed them how it worked and then three of us, including the chief of police, played frisbee for a while in the back of the police station. I bought some apricots, almonds and a jar of American peanut butter at a shop and walked back to the police station. It was about time for the road to open.

I got a ride with a Lebanese engineer who was in charge of the construction site, down the switchback road to where there were cranes and bulldozers working. He was thirty and quite sophisticated. Lebanon is the most cosmopolitan country in Arabia, often called the Switzerland of the Middle East because of its mountains. As we neared the construction site I could see that a crane had driven off the edge of the road and down the steep mountainside. As we neared the wreckage we saw a worker pulling a lifeless body out of the twisted mass of steel that was once a crane.

As the worker laid the body down, he took a deep drink from a water bottle. "When you touch a dead person during Ramadan, you may break your fast," the Lebanese engineer told me.

He stopped at the wreckage site, and I began walking down the road. I got a ride in a share-taxi, the big Mercedes Benz sedans that drive long routes and pick up passengers along the way, charging them accordingly. They are not marked, but they usually have quite a few people in them.

"Are you a taxi?" I asked them in Arabic when they stopped for me.

"Yes, where are you going?" replied the driver.

"To the coast," I replied. "But I only can take free rides. Thank you for stopping anyway."

"That's all right," they said. "We won't charge you."

"Thank you very much," I said, and got in back. I was sitting next to a guy I had spotted earlier at the police roadblock. He was a wild-looking fellow with fierce dark eyes and long braided black hair with a beaded headband. He wore a sort of black cotton pajama outfit and a cartridge belt with some very large shells in it for what must have been a powerful rifle. Tucked into his belt was a large curved Arabian dagger which I imagined he was quite an expert with—not your typical oil-soaked, Mercedes-driving Saudi! He was rugged, handsome and silent, the type who might throw a blonde woman over his horse and ride off into the desert with her while the movie cameras kept on rolling. The taxi let this guy off in the middle of the mountains, not a human habitation to be seen anywhere, and it was difficult to imagine people living out there in those steep, dry, barren slopes.

He paid with some old, tatty notes that he carefully pulled out of his belt. The other guys told me as we drove off that he was a Bedouin and had probably stashed his rifle out there in the desert somewhere.

The share-taxi took me down to the coastal plain, to a place called Ad Darb, where, after standing in the dust by the side of the road in this tiny town of a few hundred people, I got a ride with two Saudis in white robes driving a Toyota Landcruiser. We traveled down the road for about ten minutes while I told them who I was and where I was going in Arabic, and that I was hitchhiking. They did not look like a share-taxi, and I don't think they were, but often in Arabia, just as in Africa, people will charge for rides. They said I would have to pay, only 100 rials they said, to the next town which was about sixty miles away. This was about $33, which was way too much, I felt, although transportation is expensive in Saudi and they certainly didn't think it was too much. In Africa I probably would have paid for a lift like this, but they really wanted way too much, and there are so many vehicles in Saudi Arabia that I knew I would get a free ride without too much waiting. I nicely told them that I didn't have that kind of money, especially for such a short ride. They stopped the truck right there in the middle of the desert and told me to get out. They seemed a little irritated; there is a general impression among Asians and Africans that Americans and Europeans are made of money, so why shouldn't I be able to pay?

"All right," I said in Arabic as I got out of the truck. I thanked them and they sped away in the dust, with me standing alone on the highway.

I confess I was worried as I took a drink of the precious water in my canteen and surveyed the road. A dust storm was threatening in the distance. It could be that the next car to come along might be a police car, a thought that was decidedly unpleasant. I had already been searched by the police once today; they are a suspicious lot, and I feared that if they found me out here in the middle of nowhere, I would be in trouble. My visa was expiring today and furthermore it said right on it that I was on my way to Abu Dhabi, while here I was 300 miles south of Jeddah near the North Yemen border. What could I say? "Gee, officer, I must have taken a wrong turn."

To take my mind off my plight, I thought about the strange book written by an Arabic scholar named Kamal Salibi. Salibi's book is entitled *The Bible Came From Arabia*[89] and it is full of lost cities, to say the least. A professor at the American University of Beirut, Salibi reveals startling linguistic evidence which controversially suggests that the Promised Land, and the place where most of the Old Testament takes place, was not in present day Palestine/Israel but in west Arabia, in the very area that I was now traveling through, the Jizan mountains of the Asir in Saudi Arabia!

Whilst looking at a gazeteer of Saudi place names, he noticed a remarkable concentration of Biblical place names in an area 600 km long by 200 km wide, the Asir. Ancient Hebrew, like Arabic, was written without vowels. Salibi believed that scholars of the sixth century might have added the vowels wrongly when standardizing texts, and so he went back to the original unvowelled Old Testament to prove his theory. What Salibi believes he has discovered is that the geography of Palestine has never corresponded in any way to the apparently specific stories in the Bible. He believes that the ancient towns of the Asir do correspond with the Bible.

Salibi says that the Jordan River was really the Asir Ridge, or the Jizan or Hijaz Mountains and that King David's Zion was the present day town of Siyan in Asir. He maintains that ancient Jerusalem lies beneath the present day village of Al-Sharim (or Arwa Salam as it is also called). Just about every town in Palestine can be found in the Jizan area, and even towns mentioned in the Bible which can't be found in Palestine.

Salibi's book has caused quite a political and academic storm, as one might imagine. However, it is entirely based on deciphering small words with a few consonants and no vowels, to prove his point. My own objections to his theory would be such things as the Wailing Wall in Jerusalem, the remains of King Solomon's Temple, and Ezion Geber at Eilat where King Solomon's ships sailed for Ophir. Salibi's one point that is most probably true is that this is the area where Abraham had once lived, as in the legend of the Kaaba and Black Stone at Mecca. This is almost certainly true.

In the distance I could see a car emerging from a dust cloud to the south. I prayed it wasn't a police car and started to wave. Attracting a ride in Arabia or Africa is a lot different than in other countries. You don't just stand there on the road with your thumb out, you take a far more active stance.

Setting down my pack, I waved with both my arms over my head and jumped up and down a few times, just to make sure I had gotten the driver's attention. Then, using my whole right arm, I motioned down the road in several long sweeping gestures and pointed to the road where I was standing, thus indicating I wanted him to stop for me. I repeated this several times until he was so close it was superfluous for me to continue. I stood and waved, smiling, all the time confident he would stop, and sure enough he did. It's not every day that you see a blond kid standing in the middle of the Arabian desert waving his arms like a madman. Who wouldn't stop?

To my great relief, it wasn't a police car but a friendly truck driver who immediately offered me a drink of water, a handful of grapes and a gleaming smile. "What in the name of Allah are you doing out here?" he quizzed me in Arabic, obviously confused by finding me in the middle of nowhere.

I explained about my former ride, that I couldn't pay the fare the gentlemen were asking and so they dropped me off right there in the desert. I told him that I was going to Yemen. He shook his in wonder. It was an unusual day for him, I guess.

He took me pretty close to Yemen, to a place called Jizan, and let me off at the crossroads just outside of town, the road to the Yemen border. I stood there for a while at dusk, the sun gradually sinking into the grey-brown dust of the atmosphere to the west. Just as it got dark, I got one last ride out to a *garuwaz* about twenty miles out of Jizan toward Yemen. Right in the middle of the desert, it was full of Yemenis who worked at the local cement factory near Jizan; this was like their dormitory. They were fun and talkative, told me stories of Yemen and drank tea with me all night. I finally crashed out under the stars in one of those comfortable rope-strung beds and had a good night's sleep.

The next day, a young Yemeni boy, hardly fifteen, gave me a lift on his motorcycle, a Honda 125, painted a bright orange and yellow with six reflector-studded rearview mirrors on the handlebars and multi-colored streamers flowing out of every orifice on the bike. We were like a two-man parade bombing south through the desert.

It wasn't long before we were at the border, where I woke up the immigration officer. I thanked the kid for the ride and he took off back to his cement factory. The immigration officer looked at my passport and noticed that my visa said that I was going to Abu Dhabi. He didn't notice that I was one day over my transit allowance, however.

"This visa says Transit to Abu Dhabi," he said, wiggling his mustache in order to scratch his upper lip. He was thin and had the faded scars of smallpox all over his face.

Hoping I looked more confident than I felt, I leaned back in my chair and responded in my best Arabic, "Yes, I'm on my way to Abu Dhabi, but I'm going to Yemen first." This sounded like it made some sense, and the immigration officer nodded silently.

The border post was in the middle of the desert, just a couple of small cement buildings and lots of sand. Looking out the window, the immigration officer thought this over, wiggled his mustache and talked with another man in the office, a customs officer.

"Will I ever get out of this country? It's as hard getting out as it is getting in! Well, if I can bullshit my way into this crazy country, then I can bullshit my way out!" I thought.

"Yep, on my way to Abu Dhabi," I said, "Just have to go to Yemen first." We then talked about the weather and had a few laughs about this or that. My Arabic was getting better all the time.

Finally, the officer looked at me with large dark eyes, wiggled his mustache once and said, "Well, O.K." He put a big exit stamp on the page that had lain open on his desk so long.

I gripped the edge of my chair to keep from sliding off from relief and tried to suppress a grin of self-satisfaction. As I picked up my pack and headed out the door, I couldn't help but feel like some sort of pilot being ejected out of his plane into space. Saudi Arabia had been interesting as hell, but I was glad to be leaving. I turned to wave goodbye to the immigration officer who kindly let me walk away. He wiggled his mustache one last time and a gleam in his eye seemed to say, "Crazy kid."

One of the sights on the way to Mecca was the supernatural light traditionally believed to shine over the tomb of the Prophet Mohammed. From an old wood print.

THE BOWL OF BAAL
Robert Ames Bennet

The Bowl of Baal is a period piece; a nostalgic, half-forgotten survival of the year 1916 when it appeared as a long serial in *All-Around* magazine. Larry O'Brien ventures into the unknown Arabian desert during the days of World War I. His discoveries are epic; an ancient hidden race with a conflict between two beautiful priestesses; a barbaric tribe of cave dwellers; and a monster saurian survival.

THE BOWL OF BAAL

A small section of one of the six major moundfields of Bahrain. Every mound is the work of man, and every mound covers a stone-built burial chamber. From *Looking for Dilmun* .

At Buraimi, a hundred miles inland in Abu Dhabi, lie tomb-chambers built of megalithic stone blocks. From *Looking for Dilmun* .

ARABIA

MILES
KILOMETERS

BORDERS
ROADS
RAILROADS

View of the magnificent gateway, high and narrow, at the "palace" at Qala'at al-Bahrain. Note the large cut stones. From *Looking for Dilmun* .

Atleast four thousand years old (and probably older) this wall a
Qala'at al-Bahrain still survives to an appreciable hight. From
Looking for Dilmun .

Above: A smaller tomb-chamber on Umm an-Nar, where one of the two triangular entrances still survives. *Below:* Among the stones fallen from the ring wall of one of the Umm an-Nar tombs were two bearing reliefs of animals, to the left a camel and an oryx, to the right a bull. The stones originally flanked the entrance to the tomb, and the stone between the two reliefs was the capstone of this entrance.

Finely cut blocks at the lost city of Umm an-Nar in the Empty Quarter of Saudi Arabia. Were these cities when the Empty Quarter was fertile and parts of it a sea?

Monolithic rock-cut cistern at Dedan in Saudi Arabia. The existence of these gigantic cisterns cut out of one solid rock is an engineering wonder! From *Looking for Dilmun* .

The south front of the Barbar temple on Bahrain showing terraced walled of finely cut stones. From *Looking for Dilmun.*

An early "Dilmun" settlement on Failaka island, Kuwait. The ruins contained quite a number of stamp-seals similar to those found at Mohenjo-Daro. Dilmun was apparently connected with the Rama Empire of ancient India. Dilmun was said to be the only area to survive the Biblical "Flood".

One of the facades at Mada'in Salih, Saudi Arabia, cut into the rock face. Like Petra in Jordan, it is attributed to the Nabateans. Note the severe weathering, as if by water, on the lower part.

Another view of one of the facades at Mada'in Salih. The stairstep motif is common in megalithic structures, and appears at the massive sun temple of Ollantaytambo in the high Andes.

trange Negroid head cut into the rock at one structure, the only
igure represented anywhere. Is he representative of the
nhabitants? Nabateans are supposed to be semites of the desert.

Another cut stone building at Mada'in Salih. This was obviousl
part of a greater structure which interlocked with other wall
now gone. Vanished in a cataclysmic upheavel?

Chapter 9

YEMEN & THE HADRAMUT:
A MEETING WITH THE QUEEN OF SHEBA

For God's sake give me a young man who has
brains enough to make a fool of himself.
—*Robert Louis Stevenson*

He who stands with both feet on the
ground cannot take a step forward.
—*Arab proverb*

In the distance, through the waving wall of heat, I could see the outline of a city, walls rising six stories above the desert, small black windows masking each floor. Bleary-eyed, I stared off into the desert. It was hot; I had to be careful of touching the side of the Landcruiser, lest I burn my arm as I leaned on the door. The city in the distance was a mirage. I had seen half a dozen already that day. Through the heat, my sweat-stinging eyes saw great lakes of water, endless beaches and waves, and even huge cities and tall, brown, adobe skyscrapers in the distance, like the misty veil of some long-lost city out of a prehistoric past. Each time I thought it was real, but each was a mirage, an illusion of the oven-hot desert along the North Yemen coast.

Suddenly the Landcruiser stopped and I wiped the sweat that was beading up on my forehead with a faded blue bandana. We actually were in a town! It wasn't a mirage, but for once a real sight of adobe skyscrapers, small shops and, I hoped, water.

Back at the Saudi Arabian border, I had met a taxi driver who was on his way to Hodeida as I stepped out of the immigration-customs house. "Would you like a ride?" he asked me in Arabic, looking me and my backpack over.

"Sure," I said, though I suspected he was a taxi. "I'm only taking free rides, though."

"That's OK, I'm happy to give you a ride. Where are you from?"

"America," I said, throwing my pack in the back seat of the four-wheel-drive Japanese jeep that is so popular in Arabia. "Thanks for the ride."

"Think nothing of it. God provides for us all."

Speaking only Arabic, he was a dark man in his thirties of medium height with a thick mustache. He seemed a very serious person with a lot on his mind. He had been working in Saudi Arabia for a year, driving his Toyota Landcruiser around from place to place, saving his money; now he was returning to Yemen. Things are not so serious in Yemen, he told me, forcing a smile.

He was very friendly, and bought me one of the small 8-ounce Japanese Pepsis that you see all over the deserts of Arabia. The Japanese are so clever, I thought, they have even made a million-dollar exporting business out of little half-sized cans of Pepsi. They were definitely the rage in Saudi Arabia and Yemen when I was there. He kept a big bottle of mineral water in the jeep which we gulped from every now and then as we followed the two-wheel track through the gentle rolling desert plain, steering around sage brush and spiky aloe plants.

With a grunt, the taxi driver jumped out of the Landcruiser. Feeling greatly lethargic, I managed to stagger out after him, still stunned by the fact that we were actually at a real town and not just another mirage. I aimed for a bit of shade across the street, a garuwaz/tea shop.

I collapsed on a rope bed, next to an old man, thin and naked except for a loin cloth, who was sleeping noisily by the water urn. After a drink of fresh, cool water and a Japanese Pepsi brought to me by the taxi driver, I lay back on the bed and almost fainted.

God, it was hot, I thought, there weren't even any flies out! Too hot for flies: now, that was hot! Too hot for flies...flies...hot... I was sinking down into a state of heat exhaustion. My head was aching, my mouth was dry; I dozed fitfully.

I dreamed of cities in the desert,
tall and ancient
the wind blowing sand through empty streets;
I dreamed of lost continents
rising from the ocean,
and flying machines hovering
above the Great Pyramid.
Of caravans, a long line of dusty camels
making their way to India and the far east,
I dreamed of a planet
in the birth throes of a new age.

"Let's go," said the taxi driver, kicking me gently on the bottom of my foot. Still dazed, I staggered over to the Toyota and got in. I don't know how long I had been sleeping, but the sun was still high, and it was as hot as ever. Once again we were bumping and jumping through the dirt and low brush toward Hodeida. I glanced at the book in my lap, *The Ultimate Frontier*, by Eklal Kueshana. It had something to say about illusions. We know that matter is made up of molecules and atoms, it said; what most people don't realize is that atoms are made of almost entirely empty space—nothing! An atom is empty space consisting of a central blob of whirling energy sheathed by the orbits of electrons. The orbiting electrons describe a more or less spherical shells of energy. Imagine an atom—if the nucleus were enlarged to the size of our sun, the shell described by the electrons would be twelve times greater than the diameter of our solar system. A pound of iron, so concrete and solid, is just a mass of energy swirls separated by comparatively vast reaches of emptiness. Our bodies are just the same, space and vibration!

It's probably a cop-out to say that the whole world is just a mirage, I thought, staring out at the shimmering glare of a huge mirage lake, complete with sandy beaches, palm trees and surface waves. I tore open a package of United Nations dehydration salts that a friend in Khartoum had given me.

I remembered the last time I had had lunch with him, at the Greek Club in an upperclass district of Khartoum. "Life is illusory," he said, dipping a stalk of celery into a bowl of humus, a chickpea paste. "Einstein proved time is a function of matter and vice versa. Without matter there can be no time; without time there can be no matter." After a gulp of Camel Beer he went on, "When you're working with a lot of vibrations and space, well, anything's possible. These miracles we read about occasionally, stuff the prophets did in the Bible, things these yogis do in India and Tibet, it's no big deal in the world of the void. What physicists are just now proving, the mystics of the east and west have known for thousands of years. The first precept of Buddhism is that the world is an illusion, and that's also the Indian concept of Maya."

Recalling my conversation in Khartoum and the beers we'd had for lunch made me awfully thirsty. I finished off most of my dehydration salts and put my feet up on the dash board. What I would give for a cold beer right now! But, as in Saudi Arabia, alcoholic beverages were illegal. In heat like this all a person could really do was sleep....

§§§

We eventually made it to Yemen's chief port, Hodeida, a medium-sized town of 80,000 and the most modern of all towns in Yemen. It was dark when we finally got there, and I made the mistake of buying my taxi driver dinner as thanks for the long ride from Saudi Arabia. We had a bowl of mutton stew and some bread with salad in a thoroughly dumpy little cafe-cum-greasy spoon, and to my horror, it cost me twenty dollars for the two of us! I was expecting maybe three. That was the last time I ever ate anything in Yemen without first checking the price. I found a downtown *garuwaz*, one of those outdoor dormitories of the Arabian desert, on the roof of a six-story building, the tallest in Hodeida. They charged five Yemeni rials a night for a rope-strung bed and mattress, and another five rials (about one U.S. dollar) for a shower.

I wandered around Hodeida for two days. The heat was stifling. I was awakened in the morning by a score of flies that happily buzzed around me and crawled on my hands, legs, and face until I got up. I could only afford one shower a day, so I waited until sunset to wash the day's sticky sweat off. I walked down to the port to check out ships to Djibouti or the Persian Gulf and India. I wasn't too fussy about where I was going; this was the Zen of travel. There wasn't too much in Hodeida, except flies, but Yemen becomes one of the most fascinating countries in Arabia once you head up into the mountains.

Known in Roman times as Arabia Felix, Latin for "Happy Arabia," "Al Yemen" means happiness and prosperity. A series of rich agricultural and trading kingdoms flourished in Yemen in the first million B.C., the Kingdom of Ma'in was here in 1300 B.C., but the Kingdom of Sheba in 1000 B. C. is the best known. The spice caravans from Oman skirted the dreaded Empty Quarter, traveled through the ancient Hadramut valley and came through Yemen. In the middle ages, because of the fertile mountain areas, Yemen became a major coffee grower, its old port of Mocca lending its name to a certain variety of the bean.

Yemen is said to be one of the oldest inhabited areas on earth, though little is known about it. This may be largely due to the official discouragement of research into pre-Islamic civilizations in Arabia.[82] What little (and it is precious little) is known about ancient Yemen comes from the Bible. The tenth chapter of I Kings records the visit of the Queen of Sheba to King Solomon's court at Jerusalem. This would have been around 940 B.C. It was during this time that King Solomon, in partnership with the Phoenician King Hyram of Tyre, was engaged in extensive gold trade somewhere in the Indian Ocean from the port of Ezion Geber, next to Eilat. Their ships were making the three-year journey to Ophir, the mysterious land of gold somewhere south of India, either in Sumatra, Indonesia, Australia or a combination of all three. It is interesting to note here the petroglyphs on the Finke River in Australia which show no less than ten traditional Jewish candle sticks or menorah.[92] Did these petroglyphs show the way to King Solomon's Ophir?

The Queen of Sheba's visit to King Solomon was something of a trade trip, as the Queen brought with her some hundred and twenty talents of gold plus a multitude of spices and precious stones. As the Bible records, "there came no more such abundance of spices as these which the Queen of Sheba gave to King Solomon." The Queen also wished to test the wisdom of Solomon, and see for herself his magnificent court. "And King Solomon gave unto the Queen of Sheba all her desire, whatsoever she asked, beside that which Solomon gave her of his

223

royal bounty. So she turned and went to her own country, she and her servants."
(I Kings)

So, the 120 talents of gold were largely for trade, with the Queen taking the "royal bounty" in return. Yet, scholars to this day are divided as to who the Queen of Sheba (she is never given a name) was and where she came from.

The famous Arabian explorer H. St. John Philby, in his book *The Queen of Sheba* [88] maintains the popular Arab legend that the Queen of Sheba was a Sabean queen named Bilqis is false. He points out, as does Wendell Philips, that the name of Bilqis does not appear in any Sabean records, which go back as far as 800 B.C.[87,88] However, this proves nothing, since the Queen must have lived several hundred years earlier. Philby believes the real Queen of Sheba came from Northern Arabia and is somehow confused with the legends of Zenobia, last queen of Palmyra in Syria. This also seems unlikely, since Palmyra was supposedly founded by King Solomon! However, the very last statement in Philby's book does shed some light onto the identity of the Queen of Sheba, stating "strong circumstantial support to the claim of the Ethiopian dynasty and its last representative, Haile Selassie, to have been descended directly from Bilqis of the land of Sheba, if not from her utterly apocryphal union with King Solomon in a far earlier age."[88]

Indeed, the brief story of the visit of the Queen of Sheba gives us, three thousand years later, clues to life in the South of Arabia and the Horn of Africa in the first and second millennium B.C. It seems that the land of Sheba was far more extensive than most scholars will dare to admit, not because there is not evidence to support it, but just because it would be impossible given the preconceived notions of the ancient world. That the Queen's caravan was loaded with spices is only natural. Not only would her realm have included the Dhofar region of the Hadramut, virtually the only source of frankincense in the world, but it controlled the thriving trade with India and the spice islands of Indonesia. As the German scholar Joachim Leithauser points out in his book, *Worlds Beyond the Horizon*, [91] if it were not for the incredible demand for spices, almost all of which came from Southeast Asia or India, the world would never have been explored. Nutmegs, cloves, cinnamon and other spices were literally worth their weight in gold. Some spices, such as saffron, are still that expensive today.

Trade in spices had been going on for thousands of years before King Solomon's time and it seems only natural that the empire of Sheba was in control of a portion of that trade during the Queen's time. The Babylonians and Persians had their own trading vessels that made the easy voyage, but it was "Sheba" that controlled the trade to the trading cities in Egypt and the eastern Mediterranean.

It has been suggested that the 18th-dynasty queen of Egypt, Queen Hatshepsut, may have been the Biblical Queen of Sheba. At least the time frame is correct, though it seems unlikely that Biblical authors, so familiar with Egypt, would not have identified her country correctly. The evidence (much of it to be related in the next chapter) is overwhelming that the land of Sheba included most of Southern Arabia and Ethiopia as well as Somalia at the time of the first millennium B.C. I think that, with the capital probably in Axum, and other major cities such as Marib in Yemen as part of the empire, the Queen of Sheba, whose real name was not Bilqis, but Makeda, journeyed from Axum to the Red Sea, crossed the sea to her cities in Yemen, and then journeyed by caravan up one of the major routes, probably via Mecca, to Jerusalem. Therefore her journey was both by camel and ship. It would have been important for Makeda to have traveled via the Arabian route on the east side of the Red Sea, as otherwise she would have had to transit Egypt, a country with which Sheba (Saba) was frequently at war, particularly in this later period.

The land of Saba or Sheba begins to look like the remnant of an ancient empire that existed at the time of Rama in India and Atlantis and Osiris to the west. As Churchward points out in *The Children of Mu,* [76] the cultural history of Ethiopia and Southern Arabia is connected with ancient India. India was ruled at the time by Priest Kings, much like other ancient civilizations circa 8,000 B.C. It is interesting to note here that according to Wendell Phillips, in his book *Qataban and Sheba,* [87] the earliest rulers of southern Arabia "of which we have knowledge were mukarribs, priestly rulers, such as were found in most South Arabian kingdoms in their early days."

Eventually, the Sabeans of the Kingdom of Sheba in Arabia were superseded by the Himyarites under whom Judaism and Christianity took root. Later their rule was eclipsed by a force of invading Christian Ethiopians in the fifth century A.D. In effect, the Sabeans were reconquering old territory lost to them a thousand years before. It didn't last long, though, for within a few centuries, Mohammed's *mujaheedin* arrived, forcefully converting the world to Islam.

The Ottoman Empire controlled Yemen and the rest of Arabia from the sixteenth century until 1918, when a theocratic dynasty founded in Yemen as far back as the ninth century regained full power over the country. In the late fifties, Yemen joined the United Arab Republic (UAR), forming a federation of Arab states with Egypt and Syria. These three countries found it difficult to stay together, all being geographically separated by other countries who had no interest in joining the federation. Shortly after the dissolution in 1961, a long civil war broke out in Yemen between Republican and Royalist forces. Egypt was supporting the Republican forces and sent 40,000 troops and an air force to Yemen to fight the Saudi Arabian-backed Royalist forces, who were forced to withdraw to Marib, their headquarters on the edge of the Empty Quarter in western Yemen. The Egyptians remained in Yemen until 1967, but never completely subdued the Royalist forces, even after bombing Marib.

The country continues to be a land of violent change and uncertainty. In June 1978, the Presidents of both North and South Yemen were assassinated in a 48-hour period. One was blown to smithereens when a briefcase exploded; the other was shot in the head by a hired assassin. There have been border clashes between the two Yemens over the years, as well as unification talks. Saudi Arabia does its best to thwart any progress, fearful that a unified Yemen eight million strong would overrun their country.

One of the most untouched and traditional Arab countries, Yemen still operates on a tribal law system much as it did two hundred years ago. In the interior, it is common to see Yemeni tribesmen with cartridge belts on their chests and bolt-action rifles on their shoulders, though it's becoming more and more common to see automatic rifles and four-wheel-drive Toyota Landcruisers.

The interior is ruggedly beautiful, with mountains and green cultivated fields. Here the rainfall is much higher than anywhere else on the Arabian Peninsula, and Yemen was once a major grower of coffee. Now the narcotic drug qat has taken the place of coffee. East of the mountains is the Empty Quarter, a huge sea of sand that lies mostly in Saudi Arabia. There is virtually no plant or animal life, and water is extremely scarce. Even the Bedouins rarely cross this inhospitable area of Arabia. To the west is the Tihama, the coastal desert plain where there are many African strains mixed with Arab blood, and the villages have a distinct African flavor. It is here in the Tihama that Hodeida is located, and I was determined to escape the heat and hitchhike east into the mountains and San'a, the capital of Yemen.

At the crack of dawn one morning, after the flies woke me up, I decided to get an early start hitching out on the road to San'a. No sooner had I crossed the street, which was the main road out of town, than an Arab man in his forties picked me up. He was on his way to a construction site about thirty miles out of town, where he was an engineer. He spoke perfect English and had traveled in Europe quite a bit. We had a nice talk; he was very helpful and concerned about there not being too many travelers who make it to Yemen. As I got out of the small jeep he was driving, he asked, "Do you need any money?"

I chuckled quietly, genuinely touched by his offer. "No thank you, sir, I'm fine. I have enough money for the time being, I just have to spend it wisely."

"Take care, then," he said.

"I will, and thank you again for the ride." And he was off and I was once again left standing in the road.

I caught a ride the rest of the way to San'a. It was two hours in a truck with three Yemeni guys, winding up the steep road up into the cool and green mountains of the central, most populated area. It was mid-afternoon when they let me off near the main bazaar of the capital. San'a is a city of a hundred and fifty thousand or so, located in the heart of the agricultural area. Yemen's architecture is famous; its brown mud buildings are square and tall, sometimes rising ten stories above the street, with lots of dark, narrow windows.

San'a has some fantastic markets that invite endless wandering. Rarely in my travels have I seen bazaars as exotic and fascinating as those in the major cities of the Yemen mountains. Small shops lit with kerosene lanterns in twisting narrow streets offer everything the Arabian wanderer could ever want, from silver jewelry and amber to jewel-encrusted daggers and exotic fruits and sweets. You'd think you were in the sixteenth century.

On my second night in San'a, I met a young Yemeni kid, Ali, about eighteen, who spoke especially good English. He was tall and slender, and wore blue jeans and a T-shirt that said "University of California" on it. He was cleanshaven and had big dark brown eyes. They got bigger as he asked me questions about America.

"I've always wanted to go to America," he said. "I like to watch American movies at the theater, whenever they have one, which is rarely." He showed me around the Bab el Yemen Souk, one of the bazaars, while we talked about Yemen and America, I trading him questions on Yemen for his about my country. The streets were narrow and winding, barely big enough for two donkeys to pass. After rain, which is frequent in this part of Arabia, the streets are muddy and slick. The shops are boxes with large wooden doors that will swing up and back in place to make a shade. A vendor, often dressed in a striped robe, or baggy pants, shirt and a colorful vest, will sit cross-legged about chest height on a platform with his wares. There were knife shops, dried fruit sellers, cloth sellers, jewelry shops, sweet shops, knickknack stores, and much more.

"The souk stays open until very late," said Ali as we strolled through the endless maze of booths, tea shops and stores. We stopped at one street cafe, which was a couple of tables on the street and a few kerosene burners with some pots of tea, coffee and stew in a shack off the street. We each had a cup of coffee while we watched an old man at the next table. He was haggard and thin, dunking some brown flat bread into his mutton stew. His gaze was vacant and bare, and he felt his way with his bread to the bowl of stew. After he had slurped up the last of the juice, he produced a worn and tattered note that looked like it had been around as long as the old man. It was a one rial note.

"How much is this?" he asked another man at a table sitting to the other side of him.

"One rial," said the man.

"No, it's five!" said the old man, who then left it on the table to pay for his dinner. He slowly got up, leaning heavily on one ragged, greasy sleeve and grabbing a cane that was propped on the table. As he shuffled away down the street, I looked at Ali, but he shrugged his shoulders.

"Just another blind beggar," he said, "on the streets of desire."

After finishing our coffee, we paid and walked in silence for a while. Then came to a strange alley with many carts and vendors, each with a number of bundles of small green twigs. The sellers sat on the street with their twigs in front of them on a cloth, or in a permanent stall, or on a movable cart, selling their wares. It was the busiest street in the market, full of men, and one or two women, dressed in black dresses but without veils (most women in Yemen seemed to go without them) selling some of the bundles.

As we watched the men picking their way carefully through the bundles and bargaining with the vendors I asked Ali, "What is this stuff?"

"Qat!" exclaimed Ali. "This is what everyone, at least the men, and many of the older women, do for recreation. Qat is a bush that is cultivated by farmers in the mountains here, and the young reddish green shoots are snipped off and taken to the market and sold.

"People then chew the tips of a twig, mixing the juice with their saliva and swallowing it. It makes you...feel good!"

We walked slowly along the twisting, muddy streets. Gas lamps were glowing on the many tables, casting flickering shadows on old men who walked about, browsing through the many vendors and their carts full of qat. In the booths, I could see men reclining on a mountain of pillows with huge piles of qat next to them. They had huge wads of qat in their cheeks, like cows chewing on cud, or pandas with fresh stalks of bamboo. Next to them were Pepsis or bottles of water they would occasionally sip on.

A couple of young women were selling qat from a cloth in the middle of the street, enticing buyers with their fresh green bundles. One girl, young and pretty, with dark eyes and long black hair, only sixteen or so, smiled a big golden smile of metallic teeth and offered me a few stalks of qat for free.

"Go ahead," said Ali, "just bite off the ends, the reddish part, and chew it slowly, don't eat it."

I popped a few twigs in my mouth and bit off the tips. A few short moments of chewing told me that this was pretty bitter, awful-tasting stuff.

"Yuck," I said to Ali.

"Just keep on chewing," said Ali, who bought a bundle from the girl. Like everything in Yemen, it was rather expensive, five dollars or so for a small bundle. He began chewing on it. "Keep the qat in your mouth and just swallow your saliva. Keep biting off new pieces as you go. Eventually you will build up a large wad of qat in the back of the mouth, between your teeth and gums."

I thanked the girl, who flashed me her golden smile again. Ali and I continued to stroll down the street past the many piles of qat and people chewing it.

"In order to relieve the bitter taste in your mouth," Ali went on, "we like to have a sip of Pepsi Cola or ice water, or even a little mint candy and sweet tea."

As we walked down the street, an old man saw me chewing and the stalks in my hand. In surprise he asked, "Do you chew this qat?"

I replied in Arabic, "Yes, I guess so."

"Here, have some more," he smiled, revealing teeth with bits of green qat stuck between them.

After chewing for a while, I started to feel light-headed and energetic. Ali took me to a room just off the market where a lot of men were sitting around on pillows, bundles of qat next to them, thermoses of water, bottles of Pepsi; you could call it a "qat den." We found a little spot in the corner and started to chew our qat in earnest. So many people had given me qat as I walked through the market that I now had a pretty good-sized bundle, which I was chewing slowly. Still, it was nothing compared to the piles that the Yemenis themselves had.

This qat den was called a *muffrage*, and I learned it was somewhat the social center for men, at least the ones who chewed qat, and that was a very large portion of the population.

"It's great that an American is here chewing qat with us," said one man.

"Do you have qat in America?" asked another.

"When you chew qat," one old, grizzled man told me in Arabic, leaning toward me with his right elbow resting on a pillow, "You go to America!" He meant, I suppose, that you spaced-out on qat and daydreamed.

He was right. I was indeed "going to America." I was alert and coherent, but my mind seemed to be working overtime, thinking of all kinds of pleasant things. There was a certain numbness in the back of my head and I didn't feel tired at all, even though it was getting late.

I found myself back in the Landcruiser, coming from the Saudi Arabian border, the heat, the mirages, the fine line between reality and a vague world of dreams. But instead of being tired and uncomfortable, I was alert and fascinated. I realized what a great trip it had been so far and the incredible wealth of experiences I had had.

Still, I never lost contact with the men in the muffrage. "How are you enjoying Yemen?" one man asked.

"Yemen is great," I told them. "I like it very much."

It seemed these guys would go on all night chewing qat, watching TV, drinking tea and daydreaming. They probably did. I left at some point; I don't know what time it was. Ali showed me the way back to my hotel. I was coherent, not staggering or anything. My mechanical functioning, like walking, seemed utterly unimpaired. But my mind was racing, active and inventive. There was a terrible price to pay, however, as I found it very difficult to get to sleep. I lay in my bed, next to an old man who snored noisily and talked in his sleep, wishing my mind would slow down off its racetrack. Who was I anyway? What was it that had led me here to the hinterlands of Arabia? What was it that I wanted out of life? As I finally dozed off, I realized that I was just like that old man in the tea shop: another blind beggar on the streets of desire.

§§§

I awoke with a very dry mouth. It was as if I had been sucking on cotton all night and the cotton was still in my mouth. My head ached as well, and I wondered if everyone in Yemen woke up like this, with a qat hangover? After a vigorous brushing of my teeth and tongue, I was ready to hit the street. Qat was an interesting social drug, but once was enough, I felt. But it wasn'tt so easy not to chew qat, as there were not many other ways to be sociable.

I decided to hitch out to Marib, an ancient and still barely-living city on the edge of the Empty Quarter in eastern Yemen. Although it was only a hundred miles or so, I knew it would take me the good part of a day to get there. According to legend, it was supposed to be the capital of the Sabeans, where the Queen of Sheba sat on her throne. However, no inscriptions have ever been found. Because it had been the site of the Royalist Forces' headquarters during the civil war in Yemen and was still a very lawless place, both the Tourist Office and the American Embassy in San'a had told me not to go there.

"Off limits to tourists," said the rather thin and busy man who ran the tourist office. Yemen could probably count all its tourists on one hand, I figured, but he didn't have much time for me. "Marib is out of the question, why don't you go down to Hodeida?"

I was polite and friendly but determined to go to Marib, though I didn't tell him that. I knew I could just go. My Arabic was getting pretty good, and I felt I could take care of myself. It might be a little dangerous; all those guys out there carried rifles and knives as women carry purses, and if someone decided he didn't like me and wanted to use me for target practice, well, he damn well would, and nobody would try and stop him! I have faith in my positive thinking, but I must admit that my sense of adventure does get out of hand sometimes. Nevertheless, though I get into a sticky situation now and then, this was one adventure I felt I could handle.

I left San'a in the late afternoon, hitchhiking out of town just before it got dark in an effort to miss any road blocks that the Yemen government might have set up, not so much to stop crazy travelers like me, but to search people in and out of San'a for weapons. I'd just stay in some roadside funduk/dormitory or muffrage/qat den. I got a ride out of town with an unshaven young trucker in his dump truck. He was friendly and rowdy and understood perfectly why I wanted to go to Marib. "Great place," he said.

He let me off at a road camp full of Egyptian road construction workers who wore striped pajamas all the time and were interested to see me. I had dinner with them, played pingpong and watched TV with them in their little tent city out in the desert. I thought it was funny that they wore their pajamas all the time, looking very unmacho in a macho country. Certainly Yemenis wouldn't wear pajamas like that all the time. They were very nice, a generous, worldly bunch; we talked about Egypt and how they missed the more sophisticated atmosphere of Cairo. Yemen was truly the backwater of the Arab world.

They found me a spare bed, and I spent the night underneath the cool and cloudless starry Arabian sky. After breakfast with them, I was off again, hitchhiking over a mountain range of black barren crags. There weren't too many vehicles on this road either, I noticed. Eventually, after I had been walking down the road for forty-five minutes, two guys in a pickup truck stopped and gave me a ride in the back. They took me over the mountains and then let me off at a dusty, barren crossroads. I waved goodbye to them as they drove into the desert toward the south.

I sat down on my pack and took a drink from my canteen. I had prepared for the trip by bringing a change of clothes, my sleeping bag, some food and a liter canteen, plus my usual odd assortment of books and writing stuff. Sigh. This was what hitchhiking was all about, I thought, sitting on the side of a road in a desert without a car, building or person in sight. On my way out into the forbidden territory of Marib and a taste of the Empty Quarter. Ah, the dreaded Empty Quarter! I liked the sound of that. The size of Texas, the Empty Quarter is known for its infrequent rains that set off the hatching of grasshopper eggs and trigger the vast plagues of locusts that cause the crop destruction for which Africa is famous. Once it may have been a great sea, with port cities and commerce...

As I was staring dreamily into the east, from behind a couple of empty oil drums beside the road an old lady emerged, a Bedouin in black embroidered robes and sandals, accompanied by a little boy and a young girl of about sixteen. Both women were veiled, but as the older woman approached me, she took her veil off. The little boy and the young woman remained by the oil drums.

We greeted each other in traditional fashion, "Salaam Aleikum," and I asked the woman if I could take a photo of her. The old lady declined but kept coming closer and began saying something in Arabic which I didn't fully understand. It had to do with going to her home, which was not visible but which she indicated was to the north of the road somewhere. We could have some tea with her daughter, she said, whose name, I gathered, was Sheba. She was very aggressive about this proposition. Everything centered around her daughter, who was very attractive, I thought, and alluring with her thin veil on. I presumed that this lady was offering me her daughter, Sheba, for sex.

This was the last thing I expected to happen to me on my way to Marib, and not the sort of situation I wanted to get involved in. It was my belief at the time, and generally still is, that women in Arab countries do not just come up to strangers and talk to them, much less offer them sex. Furthermore, I was in some of the wildest country west of Afghanistan, where everybody, except me, carries a gun, and uses it quite frequently! Shootings and feuds generally center on women in some fashion or another. If some Yemeni guy were to come along and find me just talking with these women, heaven forbid making love with one, I could be in big trouble. A bullet in the head or a knife in the chest is pretty common. I had no intention of going with this lady and her kids to her house somewhere in the desert; that seemed like suicide...

"Uh, lady, your daughter is very beautiful, but I really don't think I have time to go to your house right now. You see, I'm just traveling to Marib and..." I managed to get this out in my best Arabic.

"Come to my house and drink tea with my daughter," said the old lady.

"Look, lady, " I tried to say, "just get away from me." As soon as I said this, I saw a jeep coming down the road from the same direction I had just come.

"Now I'm in big trouble," I said aloud in English. "Lady, please leave me alone." I imagined the tough Yemeni tribesmen in the trucks cocking their rifles, ready to shoot this infidel who was pestering a female desert dweller. I took several big steps backward and picked up my pack, trying to put as much distance between myself and the lady as possible.

The jeep pulled up to the lady, who was a few feet in front of me now. In it was a Yemeni in desert garb: robe, headcloth, cartridge belt across his chest, an automatic rifle on the seat next to him, a large curved dagger in his belt. The old lady was talking to him and he cast a few glances my way and argued with her.

Eventually I got up my courage to walk over to him and stick my head in his truck.

"Are you going to Marib by any chance?" I asked in Arabic.

"What are you doing here?" he asked, ignoring my question.

"Oh, I was just on my way to Marib, you see, and I was let off here..."

"I'm not going to Marib," he said rather gruffly, and then the old woman got in the car with him and they drove off in the direction she had also indicated to me. The little boy and veiled girl remained standing by the empty oil drums along the side of the road.

I was relieved that there hadn't been too big a fuss, as it was a sticky situation. The tribesman must surely have thought it strange...not too many backpackers out in the desert talking with women just like that. I started walking down the road and fortunately another jeep came along before too long. Looking over my shoulder as we drew away, I heaved a great sigh of relief. That was one set of desert crossroads I was glad to be leaving.

An early morning sunrise
A dusty desert ride
We came upon a pickup
Abandoned by the roadside.

The wind howls around the truck;
The only sound to hear.
A Bedouin smiles up at us
His throat cut ear to ear.

– A Hitchhiker's Guide to Armageddon

I never made it to Marib that day, but I did get to Al Hasool, a smugglers' town near Marib. It was one strange place, full of tall, tan adobe skyscrapers and kids as young as eight years old running around fully armed with automatic weapons and driving Toyota Landcruisers, and sometimes even large Dihatsu trucks. I had to look twice to see if it was all real: a camel caravan coming into town from the Empty Quarter, a sesame mill using a blindfolded camel to drive the grain wheel, and the tall, adobe buildings, just as they were five hundred years ago. Then the Toyota Landcruisers, adult desert dwellers and their young sons, daddy's pride and joy, all armed to the teeth with cartridge belts across their chests, automatic rifles slung over their shoulders and their pickups full of smuggled goods from Saudi Arabia; cases of Pepsi, tunafish, canned meat; nothing really contraband like alcohol. They met every day there for a smugglers' market, the Landcruisers coming in from Saudi Arabia, and people selling their goods in bulk.

It was about noon when I got there, and the market didn't open until later in the afternoon. I found a shady porch, an old abandoned storefront with a couple of other guys lounging around, their rifles on their laps. Leaning back on my pack, I looked over the town. It was run down, with thirty or forty tall mud skyscrapers. A few stores here and there, but it was still Ramadan, the Muslim month of fasting during the day, so most of the places would be closed until dusk. The center of town was a large parking lot which was where the smugglers' market took place. There were quite a few people and trucks there already, but they kept their goods covered.

I fell asleep in the hot sun. Not even the flies could wake me up. When I did wake, I was looking down the barrel of a Soviet AK-47 assault rifle with a bearded and rather scruffy-looking tribesman standing behind it, gazing down at me.

I took a rather dry gulp and blinked real hard, thinking this was probably a dream. I wasn't really sure where I was at that stage, as my dreams had been pretty wild of late. It wasn't a dream, nor was I in any danger, actually. Some desert dweller had just been curious about me, and was checking me out, his rifle pointing in my face as he looked me over.

He sat down next to me. "Salaam Aleikum (peace unto you)," he said.

"Aleikum a Salaam," I returned, rubbing the sleep out of my eyes. He was tall and had a rich olive-colored skin. His face was badly scarred and even though his general appearance was tough, with his knife, cartridge belt, and black robe, he looked like a nice guy.

We talked for a little while. Like most people around here he had a little farm, which he cultivated from an artesian well and where he grew some subsistence vegetables. He had gone to Saudi Arabia to work for a while, where he earned the money to buy a jeep. He already owned a gun: that was about the first thing he ever owned. He was in town for the market, and said he would give me a ride to Marib, which was only a couple of miles away, after the market.

I went around taking pictures of people. Everyone loved it, I was the instant celebrity of the market. A group of kids, eight to twelve, would line up and hike their rifles up proudly, show a gold-capped tooth and try to be as tough as possible. I walked up to one old man and asked him if I could take his photo.

"How much?" he asked in Arabic.

"Nothing, it's free," I told him.

"Oh! All right, then," he chirped with a smile, nestling his rifle on his shoulder. He stood straight and proud, straightening his cartridge belt and brushing the dust from his white, neat beard. His head was wrapped in a black and orange turban; his clothes were a deep but fading blue. He stood very seriously as I took my photo. I felt like the official photographer to these modern-day desert smugglers, recording the scene for the Queen of Sheba herself.

For all their knives, guns and bullets, they were very nice people. They kindly gave me advice, invited me to eat with them, and offered me rides back to San'a. Later, just after dusk, I left these dusty warriors, and got a ride to Marib with the fellow who had greeted me with his rifle after my nap. In Marib, an ancient but now small town, I was shown to the leftovers of a hotel by a shop keeper. An old guy ran what little was left of it. He showed me up to the very top floor where there were some mattresses on the floor, and I promptly fell asleep.

Marib was pretty bombed out, as I discovered the next morning. After the Egyptian Air Force bombed it a couple of times during the civil war in the sixties, it was never rebuilt. Today there are just a few shops and the one hotel, which I thought was cheap at fifteen rials (three dollars). The ruins of Marib's Temple can still be seen, as well as the ruins of the ancient dam that was built to water the farmland. Now it has been reclaimed by the desert, but the glory of the Sabean culture, gone for three thousand years, is still visible about a mile out of town.

I hitched back to San'a that same day, and spent several weeks hitching and even trekking around in the mountains of the central part of Yemen. Near the northern border by Saudi Arabia is Sadah, a walled city in one of the wildest parts of Yemen.

Just nearby, between Sadah and San'a, is the road to Wadi Jawf, and in Wadi Jawf is the ancient lost city of Barakesh. Once an important walled city on the caravan routes in Arabia, it died out several hundred years ago. According to some sources, Barakesh was an ancient city, even in ancient times! Did this mean that it went back to the days of the Rama Empire, Dilmun and Atlantis? It was situated in the correct area, geologically, to have been active while the Empty Quarter was still a sea. What had caused the ancient city to die? Possibly the eventual lack of water supply, I guessed, but I never learned the real reason. I was determined to go there just as I had been determined to go to Marib. But Barakesh is another story. Several Peace Corps volunteers had been killed there the year before, and unlike Marib, it was a dead city; no one lived there, just ghosts.

I thought I would give it a try, and caught a ride to the town where I might get a jeep out to Wadi Jawf. I had two canteens of water and a couple of days' worth of food. Now if I could just get a ride...

"Where are you going?" asked a merchant who saw me standing by the road.

"To Wadi Jawf," I told him.

He thought for a minute and looked at me. "I would advise you not to go to Wadi Jawf," he said carefully. "The people there are not good. They do not keep their word. The law of the desert is that if a man asks you for shelter you must give it to him. If a man is staying with you, even if he is your enemy, you are honored to take care of him and do him no harm. But in Wadi Jawf, they do not honor this code." That was what had happened to some Peace Corp volunteers: they had been killed in the night by their hosts. "Don't go to Barakesh," the man stated again.

I know when to quit. I hitchhiked back to San'a and headed south the next day. About a hundred miles from San'a, I stayed in Ibb for a couple of days. A quaint town of 20,000, it has a neat bazaar that winds up a hill through the twisting, narrow streets. I hitched out of Ibb, and stopped at the old capital city of Yemen on my way to Taizz. Jibla, the old capital, is a picturesque small city built around a hill in a green, mountainous valley, and is a perfect place to wander in for an afternoon. The entire road between Ibb and Taizz, the other major city in the mountains after San'a, is very scenic.

It was a couple of hours' easy hitching to Taizz from Jibla, a city of 79,000 with probably the largest and most interesting markets in Yemen. The market in Taizz is full of character. It's full of characters too, all kinds of merchants, beggars, and wild, red-eyed old men who have been chewing qat for fifty years and are out of their trees, and even prostitutes (a cloth merchant I was sitting with pointed them out to me).

I traveled around some small towns and walked through the mountains for a few days, staying in *funduks* and eating beans and bread in the little cafes. Eventually I ended up hitching down to the coast to Mokka, the ancient port of Yemen. Old and dilapidated now, there wasn't a lot to do there, but I had to see it. With only a few thousand people there now, it is another smugglers' town, with

Arab dhows sneaking in from Djibouti, across the Red Sea, with holds full of whiskey.

Hot and humid, Mokka seemed like the epitome of the washed-out, rundown town. I was sitting in a tea shop/fruit stand on the main street, which was no more than a block long. I could see that I was in the middle of the weirdest collection of characters ever to assemble at a tea shop/fruit stand in Mokka during the month of Ramadan, or at any time for that matter.

To my right was a brown, shirtless, handsome but retarded beggar, who stood there and looked through everyone, muttering to himself and occasionally breaking out into an enigmatic grin of realization. The keeper of the shop gave him dates to eat, and water.

To my left was a black kid with a cowboy shirt and a red bandana around his neck, looking like an escapee from Wagon Train. He too was begging, rather silently and awkwardly, picking through the dates the other had left behind, and then taking some chewing tobacco from the display of fruit and consumables in front of the shop. Neither he nor the retarded beggar seemed able to fathom their own existence.

Then there was the old red-haired Haji, who must have been to Mecca, as it is customary to dye your hair and beard red when you have made your pilgrimage. He wandered about silently in a daze. And there was a wide-eyed kid in jeans with his hot-rod motorcycle taxi, looking like a prepubescent James Dean of Arabia.

Last was a fat, pot-bellied old man with one blind eye, his cheeks bulging with qat, moving in and out in front of the counter of his shop. And strangest of the lot was me, sitting amongst the dregs of Mokka, myself some sort of freak. I must have been crazy—I guess it was the incessant heat—to stay in Mokka for even one day. Still, I had to see what it was like....

Economically, North Yemen is a strange place. They have absolutely no exports, not even coffee anymore, as all the farmers have switched production to qat, which is consumed in the country. Yemen's only export is laborers, who go to Saudi Arabia or the Gulf States. Recently oil was discovered in Yemen, though they have yet to export it, so why is the exchange rate so inflated against Western currency? The reason is that Saudi Arabia backs the Yemeni rial with its own currency, which is in turn backed by its huge petroleum deposits, giving the currency value. With Saudi Arabia backing the Yemeni rial, Yemen can then import all the goodies it wants, export nothing, print up all the paper money it wants, and its currency can be used by Japan, America and Europe to buy Saudi Arabian oil. This all makes Yemen a false economic area.

A strange lost city of sorts is to be found just a few miles to the north of Mokka off the coastal road that goes to Hodeida. This is a group of basalt crystals, naturally formed into long, six, seven or eight sided "logs" which are stacked up and placed on end over an area one and half kilometers long, forming what may have been some kind of strange, ancient temple. Basalt crystals are an unusual building material; they are extremely heavy, naturally form into blocks, and are generally magnetized. This ability to magnetize naturally lends them to levitation and other strange sciences. I am reminded of the massive, basalt crystal city built into the ocean just off of the island of Pohnpei in Micronesia. According to island legend, the stones were levitated into place (for more information on this mysterious city, see *Lost Cities of Ancient Lemuria & the Pacific* [35]).

I caught several long lifts out of Mokka, one to Taizz and then another to Ibb, and a third into San'a, getting me there just at dusk. I was ready to leave North Yemen now, and decided to fly to South Yemen, where I thought I could get a transit visa at the airport, and then on to Djibouti and Somalia. I had my ticket changed to go on to Aden, the capital of South Yemen, and flew there a few days later.

South Yemen, known officially as the People's Democratic Republic of Yemen, is well entrenched in the Soviet political sphere. It is difficult to get into; they do not encourage visitors from western countries and overland travel from North Yemen or Oman into the country is impossible. I flew in one sunny afternoon and figured I'd just breeze through immigration and customs, but this was not to be the case. Back at the airport in San'a, I had written in my journal: "Here I am taking a plane to Aden, no visa, no Onward Ticket, no morals, and in a whole lot of trouble!"

At the time I wrote that, it was just a joke. It was no joke to the army officer who was in charge of immigration, however.

"How did you get to Aden?" he demanded. "They should never have let you on the plane. You have no visa." He called me an American spy and said I would have to fly back to San'a, but the plane had already left. "You'll go to Moscow then."

"Moscow! I want to go to Djibouti," I said. But there were no flights to Djibouti for several days, and they wouldn't allow me into the country. I spent the night in the transit lounge with a bunch of Somali guys returning from the University of Moscow, where they had been studying for several years. They were friendly and bought cases of duty-free Japanese beer in gigantic liter cans and got drunk. I'd have a beer with them and talk about Russia or Africa or America, and read occasionally from Basho's *Narrow Road To the Deep North* or *The Ultimate Frontier.*

South Yemen is a pretty cheap country, they told me, compared to other Arabian countries. A socially progressive country in many ways, it made women stop wearing veils in 1967, took away all the guns from the tribesmen, which cemented the rule of the Marxist-dominated military, and made the chewing of qat illegal except on two specific days of the week. The entire eastern part of South Yemen is called the Hadramut area and is especially wild and mountainous. It is in this area that the frankincense tree grows.

There are lost cities and gold treasure out in the remote mountains of Southern Arabia, or at least there were. In a personal letter to me, Maurice Egan claimed that in 1961 he accompanied a university group to Aden and from there went by helicopter to a small oasis where a "trickle of water seeped. The whole area looked as if it were used to test H-bombs, it was so bleak." They traveled through the desert for a week, until they came to a small range of mountains.

"We finally came to a plateau where we made camp...At a place that seemed a natural harbor we began our dig, coming across petrified piling driven into the earth. Sometimes we even found a stone roadway, following it as it walked up the mountainside. We came at last to what must have been a fortified position, the walls in places about twelve feet thick. Then a place of worship, of Baal and one probably of Kemosh with a firebowl still gray-black, its face with those huge eyes and pendulous ears. By this time we were finding bowls of better quality, the building of well dressed stone. It was a palace of some kind, the walls had paintings of a people that hunted in wetlands.

While they were black men and women, there were paintings of fair-complexioned peoples as well, that threw wide nets.

"We followed wide steps down into the underground and discovered an enormous treasury. Ivory, gold, jewels, every manner of wealth that is sought by man. More so were stacks and stacks of baked clay tablets, no doubt of records. We just sat down there completely stunned by what we'd found, then our trouble really started. They may have been bandits, but whomever they were, they began to show themselves, but only on the fringes of where we were camped. Each day they constricted themselves closer, so we decided that we'd best get out. Each of us took a few of the stones we'd found in the underground vault, we stripped the camels of everything except water and food, and our weapons. We made as if we were again settling in and built a large fire. When it got dark we sneaked out, but left presents for our guests. Small chests of stones we'd brought up from that vault, along with trip wires that ran to pineapple grenades. While they might want to harm us, greed is a more powerful motivation. The short of it was that we got back to where we started, called in for a chopper pick up, and were well out of Yemen in just under a month a half. We never had time to really explore, other than what we had done, so we never found out for sure whether or not it really was the land from which the Queen of Sheba came from, but it must have been a most prosperous state until the weather changed, plunging it into the inferno which we found it in."

Had these adventurers discovered one of the port cities of the Empty Quarter? We may never know. South Yemen is pretty much closed to foreigners, and Arab countries actively discourage investigation into their ancient history. Yet, assuming that this story is true (it is a genuine letter that I have received in the mail, though I cannot vouch for its authenticity), it would seem that Arabia has many mysteries still to divulge.

The next day, as I was hanging out around the airport lounge, the immigration officer, a general in the army, ordered me to get my bag and come out onto the blacktop strip where a small Canadian Twin Otter plane was waiting to fly to North Yemen. He handed me my passport which he had been keeping. "Get on!" he said.

I looked at him squarely. "I don't want to go back to North Yemen." It was a small plane that only held about fifteen people and it was already full. "Furthermore, I won't pay for a ticket."

The pilot, a Frenchman, said, "You'll have to pay for a ticket, it's about a hundred and fifty dollars."

"Well, I'm not going," I said flatly.

"You will go!" shouted the general, turning red in the face.

He spoke amazingly good English, I thought. This was natural really, as Aden was a British naval base and colony until 1967 when the British pulled out. Since then the economy of South Yemen has suffered greatly, and the port is largely inactive despite Soviet help. The closing of the Suez canal ruined it; and it has not regained much since the Suez was reopened.

The general was tall and muscular, clean-shaven and handsome, with very angular features, just like a career soldier should have, I thought. He wore an old but neatly pressed tan uniform with just the right number of medals on it.

"You are going!" he said again.

"No, really, I'm not going. I don't want to pay to go back to North Yemen. I've already been there," I explained as nicely as I could. "I'm going to Djibouti."

"He can't make you go," said the French pilot. "Besides, we're full anyway... .OK, that's it, we're going without you." And he shut the door and started the engine on the other side of the plane.

I'd never seen a general cry before, but Arabs can be very emotional. He stood on the airstrip and cried unabashedly, tears streaming down his face and onto his medals. Really, Europeans, and Americans especially, should learn to let their emotions out like Arabs. Eventually, the general turned his back and walked fuming into the terminal. I was left standing there by myself on the tarmac as the Twin Otter taxied down the runway. I picked up my bag and walked slowly back to the transit lounge.

I spent another night on a chair in the lounge, and in the late afternoon the general came to me and said, "There is a flight to Djibouti today. You will be on it."

"Great," I said, and apologized for all the inconvenience I had caused him. Within two hours I was in the air over the Red Sea on my way to Djibouti. I made a mental note to remind myself about little things like visas; they can complicate life sometimes for the rogue archaeologist.

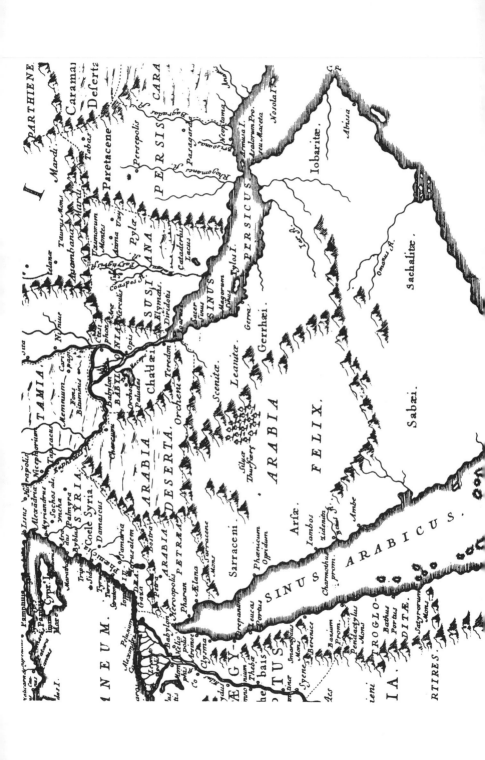

Sir Edwin Poynter's painting of the Queen of Sheba visiting King Solomon. This is perhaps the most romantic of Biblical stories, with a mysterious Queen which the Bible has very little to say about, visiting King Solomon with a great treasure as a gift.

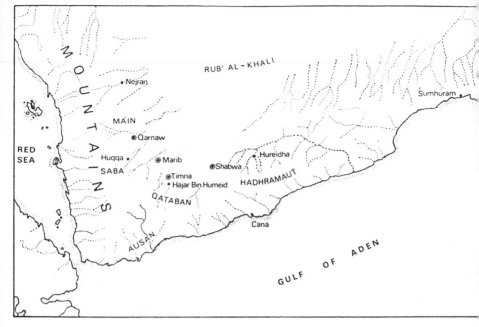

Sketch plan of southern Arabia showing major archaeological sites and approximate locations of the ancient kingdoms

George Sprod: 'Solomon Visiting Sheba', cartoon (from Sir Leonard Woolley, *As I Seem to Remember*, George Allen & Unwin Ltd, London 1962)

Solomon visited Sheba in a flying machine according to Ethiopia's most ancient text, the *Kebra Negast*.

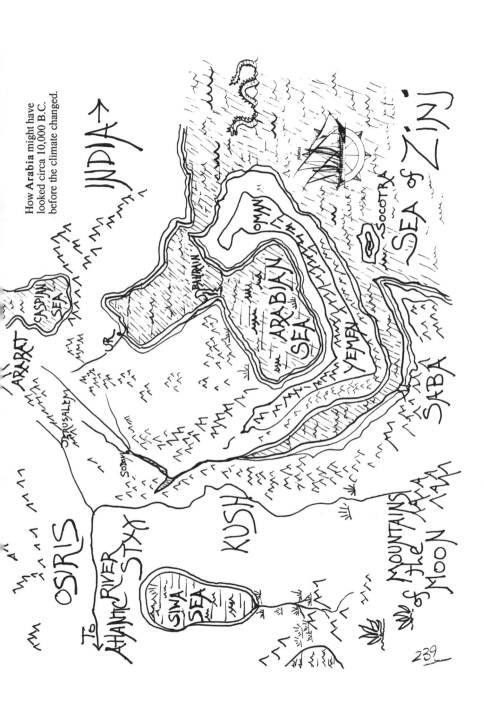

How Arabia might have looked circa 10,000 B.C. before the climate changed.

239

South Gate of Timna (probably from the sixth century B.C.) after excavation. Before the expedition arrived this entire city area containing important buildings, streets, inscriptions, and art objects lay buried under the sand.

The ancient caravan city of Marib on the edge of the Empty Quarter. Near by is the great dam built many thousands of years ago to irrigate the desert. The Egyptian airforce bombed the city during a civil war in the early 1960s.

One of the two towers of the
Great Dam of Marib, a huge
and sophisticated construction
built about the ninth century
BC to irrigate tens of thousands
of acres of desert

Stones carved with Himyarite
inscriptions still abound among
the sand-dunes and the rubble
of Marib today

From *Qataban and Sheba* by Wendell Phillips.

Near these thirty-foot Marib pillars, called by the Arabs *el-amayid* (the pillars), a broken inscription was discovered giving the name of this temple as "Bara'am, dedicated to the god Ilumquh," the moon.

From *Qataban and Sheba* by Wendell Phillips.

The first pre-Islamic inscription discovered in Dhofar Province, Oman, this bronze plaque, deciphered by Dr. Albert Jamme, dates from about the second century A.D. and gives the name of the Hadhramaut moon god Sin and the name Sumhuram, a long-lost city. The lines read from right to left. Line six can partly be made out on the plaque's lower edge: 1) SHAFSAY AND HIS MOTHER, 2) NADRAT DEDICATED TO, 3) THEIR LORD SIN, HE OF (THE TEMPLE OF), 4) 'ILUM, IN (THE CITY OF) SUMHURAM, 5) FOR THE PROTECTION OF THEIR PERSONS, 6) AND OF THEIR KING

From *Qataban and Sheba* by Wendell Phillips.

Shibam—South Arabia's most spectacular skyscraper city—whose many-storied buildings constructed mainly of mud were built over the centuries for defense against attack by surrounding Bedu tribes

Chapter 10

ETHIOPIA, DJIBOUTI, & SOMALIA: PRESTER JOHN AND THE HIDDEN EMPIRE

He who does not really feel himself lost,
is lost beyond remission. He never finds
himself, never comes up against his own reality.
–Ortega

I was hit by a blast of hot air as I stepped out of the airport in Djibouti. It was even hotter than Aden! I didn't have the trouble getting into Djibouti that I had in Aden, although they did charge me twenty dollars for a visa, the most I have ever paid in my whole life. I knew Djibouti was going to be expensive from the very beginning.

Djibouti is mostly famous for its bars and the fact that it is a duty-free port. Coming from Arabia, it was amazing to walk into the go-go bars that abound all over Djibouti. They are filled to the brim with tall, beautiful Somali and Ethiopian women who play for pay. The bars are rather seedy, filled with French Foreign Legionnaires with lots of money to spend and a few skeletons in their closets, which they try to forget by running away to foreign countries and drinking, or so one legionnaire told me.

I must admit, I didn't really feel like spending a lot of time in Djibouti. Aside from the fact that it is one of the hottest places in the world, it is just too expensive. Almost the last remnant, of France's once vast African Empire, it only achieved independence in June, 1977. A tiny little enclave on the north of Africa, the main reason for its existence is the 486-mile long railway to Addis Ababa in central Ethiopia, and its French naval base. More than half of Ethiopia's exports go out through Djibouti, making it a major port along the Red Sea. Of course, its thriving trade in booze, duty-free goods and hot night life also make Djibouti a popular spot for sailors and other travelers in the Horn and Arabia.

It's a city of a hundred and twenty-five thousand people or so, but you would never know it unless you drove around the sheet metal shantytowns that lie to the south, west, and north of the main port. The rest of the population, another 125,000, live out there in the desert. To the north is the Danakil Desert, the hottest geographical area in the world. Ethnically, Djibouti is divided into two groups; the Afars, or Danakils, and the Assis, a Somali tribe. From these two tribes comes the old name for Djibouti, the French Territory of Afars and Assis.

"They're pretty savage," said the French lieutenant about the Danakils. We were sitting together at a sidewalk cafe in downtown Djibouti. He was drinking beer, and I was drinking cold mango juice. It was a hot, humid afternoon, as usual, and our clothes were soaked in the perspiration that gushed from our pores and dripped off our noses from our foreheads.

He had the look of a true legionnaire, with a hardened and tanned face, tough as leather, a few scars, a trim, brown mustache and dark aviator sunglasses. He spoke English in a thick French accent. "You know," he went on, "those Danakils have a real curious custom. They must present their brides with a pair of human testicles before they can get married. Human testicles! They have to cut them off somebody! God! They're savage!

"The first thing a Danakil boy owns is a knife, then a gun, then a shirt! They're rough. Don't get caught out there in the desert by yourself...or schlept!" and he made a cutting motion onto the table. I gulped down the last of my mango juice and ordered another, hoping it would dispel a slight tingly feeling between my legs.

"One night I was driving from the very north part of the country to a small army camp in the middle of the desert, right in Danakil country, when my jeep broke down. I didn't have a gun or anything. Whew, that was scary! If a couple of unmarried Danakils came along..."

"I know, I know...schlept!"

"Oui! Well, fortunately, another car happened along just after me. That was fortunate, very fortunate! I was a nervous wreck! I just abandoned the jeep. Never even went back for it; probably wasn't there anyway." He finished off his other beer, and I could see him shudder. I suddenly felt a little chilly myself, a rather nice feeling on a sweltering hot Djibouti afternoon.

He wasn't kidding, either. The most famous artist in Ethiopia, who did the well-known stained glass windows at the Addis Ababa airport, was castrated as a young kid in Ethiopia in Danakil country, which runs for hundreds of miles north along the coast. He and his father had been walking and his father saw some tribesmen coming toward them, and in a panic, left his son, fleeing himself. The son was castrated and left to die, but survived. He went on to become a famous artist, but never spoke to his father again!

I wanted to go to Ethiopia, and spent some time in Djibouti trying to get a visa. At the Embassy, the Ethiopian consulate told me that I would have to fly in to Addis Ababa, remain in the capital for my entire stay in Ethiopia and then fly out. I wanted to take the train up from Djibouti and stop in the towns of Harar and Dire Dawa. This wasn't possible at the present time, they told me, and refused me a visa.

I went back downtown, somewhat depressed, as I had always wanted to go to Ethiopia, and now it didn't seem that I would be able to go. Sitting in another cafe in the late afternoon, I met an American named Chuck, who was just coming from Ethiopia. Chuck who was in his thirties, had worked in Saudi Arabia for two years and had gotten to Djibouti a few days ago from Addis Ababa.

"It was pretty crazy there in Ethiopia," he told me over an imported French beer. The beer was cool and refreshing. It is the drink of African travelers from Cairo to Capetown, and the customary exchanging place for information is over a beer on the terrace of a cafe. "The government wants you to stay in these expensive tourist hotels, like in them commie countries. But I wasn't about to pay twenty dollars a night when there are hundreds of African hotels around for two dollars a night. So I just stayed in the African hotels.

"I've been around Africa for a while and I know all the tricks," he said, tipping his baseball hat back from his dark sunglasses. "You see, to get out of Addis, you have to have a special pass to show at the road blocks that are all over the country. I acted like I knew just what I was doing, and showed them my International Student Card like it was some special pass or something. You see, it has my photo in it, and writing in all these different languages, including Russian, which is what the Ethiopians care about, since the commies are running the country. They looked at my student card and let me pass. I took a bus to Bahir Dar on Lake Tana, the source of the Blue Nile. But wouldn't you know it," he went on, wiping his forehead with a handkerchief, "the commies were having some kind of convention there in the town. There were Russians all over. I checked into a cheap hotel and went down to the lake.

"You see, I travel with this here little mini-recorder," had he showed me a miniature cassette tape. I tape what I'm thinking and conversations with people as I travel around. When I saw all those Russians down by the lake, I went behind a bush and dictated into my tape recorder, 'Jeez, this place is crawling with commies. They're everywhere!' Stuff like that. Then that night I was lying in my bed, in my sleeping bag, listening to what I had taped earlier that day through an earphone, when suddenly the door to my hotel was busted open! The hotel manager knew that foreigners were not supposed to be staying in the cheap African hotels.

"So, the police just burst into my room. Kicked the door in and jumped inside, pistols on me the whole time. They thought I was James Bond!"

"That is quite an experience," I put in, and ordered us each another beer. This was as good a story as any African traveler could tell, and deserved another round. "What did you do?"

"Well, without getting up, just lying still in my sleeping bag—my tape recorder was underneath me and I didn't want them to see it; I knew it was rather incriminating—I talked to them. 'OK, I told them, 'I'm coming out. Just go back outside, close the door, and let me put my clothes on, and I'll let you in.' And you know what? They did! They actually left the room and closed the door. I got up and put my cassette recorder way down in the bottom of my sleeping bag, got dressed and let them in.

"Of course, they arrested me as a spy, and I was escorted by bus back to Addis, which took a day. I was put under house arrest in the Hilton Hotel, and they developed my film, expecting it to be full of photos of military installations, bridges, etcetera. They were sure I was a spy, kept my passport and everything to do with me. And do you know what happened? Well, they found my tape recorder, the only incriminating evidence I had—I mean, had they listened to that, I would have been in real trouble! But when the detective was going through my stuff in front of me, he turned on the recorder, which still had the earphone plugged into it, and therefore you couldn't hear anything. 'Broken?' he said. Boy, did I breathe a sigh of relief. 'Yeah,' I said. 'That's right.'

"They eventually gave me my passport back, but by then I had been staying in the Hilton for two weeks, and had racked up a hotel bill of a few hundred dollars. I wasn't going to pay that! I booked two flights out of Addis to Djibouti, and slowly moved most of my luggage out of the Hilton into a cheap African hotel. At the Hilton they had room checks four times a day, and you didn't even have a key to your own room! On the morning of my plane flight, I left the hotel just as if I was going out for a walk, took a taxi straight to the other hotel, paid my bill and took all my stuff straight to the airport.

"I was afraid they would look in my room and find I was gone, although I tried to make it look like I was still staying there by leaving a few things lying around. The first plane that I had a reservation on never showed up, and the second plane was two hours late. As I sat in the departure lounge, I cringed every time someone walked in, thinking they were coming for me. Finally I got on the plane and flew here to Djibouti. Thank God! When I got off the plane, I was so pissed off at the Ethiopians that I wanted to just yell at some Ethiopians and tell them what a lousy country I thought Ethiopia was. I walked up to this guy who was on the same plane as I was at the Djibouti airport. 'Are you from Ethiopia?' I asked him. 'Yes,' he said, 'and am I glad to be out!' Both of us heaved a big sigh of relief. I couldn't yell at that guy!"

Ethiopia isn't such a bad place really, though I didn't get there myself. Known as the Hidden Empire, its official name, until recently, was Abyssinia. Its history fades back into the mists of time. The Empire was founded, according to tradition, by Menelik I, Solomon's son, born to Makeda, the Queen of Sheba in about one thousand B.C. According to ancient Ethiopian tradition, recorded in the *Kebra Negast* ("Glory of Kings" a sort of Ethiopian Old Testament that is the most important document to all Ethiopians) the reigning Queen, Makeda, left Axum, then the capital of Saba, and journeyed across the Red Sea to present day Yemen and up the Hijaz to Jerusalem to visit the court of King Solomon. Seeing the important Ark of the Covenant was also a key goal of her visit. She made love with the King, and essentially they fell in love. She had to return to her own kingdom, however, and there she had King Solomon's son. He was named Menelik I, and it was with this child, later to become king, that Solomonic line of rulership over Ethiopia was begun. This line was unbroken for three thousand years until the death of Haile Selassie (born Ras Tafari, 225th Solomonic ruler) in August of 1975 under house arrest by the Military Council which then usurped power, and rules Ethiopia to this day.

According to the *Kebra Negast,* [93] when Menelik I was a young man, he went to visit his father, King Solomon, in Jerusalem. When he was to return to Axum, King Solomon invited him to ask one supreme favor as a token of his affection. The prince asked that he be given the Ark of the Covenant. This, of course, was refused. He then asked if it would be possible for a replica of the ark to be made, and that he be permitted to take it with him to Abyssinia. This was granted, but Menelik I decided to steal the real Ark of the Covenant and, accompanied by Israelite noble youths to serve the country as priests, transported the Ark his homeland, where many Ethiopians believe it is to this day.

Assuming this story is true, it is most likely the real Ark did in fact remain at the temple in Jerusalem, and that a duplicate Ark was taken to Axum. One begins to appreciate the importance of the Ark of the Covenant, again, when one learns of the Ethiopian traditions. Given that the Ark was an extremely important ancient relic, said to have been stored for centuries in the Great Pyramid and then taken out of Egypt by Moses, the mystery of where it is today gains weight. In this theory, the Ark is such an important relic that it surely would not be stored casually or left accidentally in some cave or monastic cellar.

The Ethiopians revere their Ark as their most sacred relic. One of the few people to ever see the Ethiopian Ark was Dr. J.O. Kinnaman, who organized the National Museum of Ethiopia at the request of Haile Selassie more than twenty years ago. As a mark of special respect and appreciation, Dr. Kinnaman was permitted to spend many hours in the immediate presence of the Ark at the Coptic Cathedral in Axum. His request to photograph the Ark was denied, but he was permitted to take exact measurements and make sketches of the details. The Ark is apparently kept in secret vaults somewhere in Axum, and the Coptic Church has taken great care to see that the relic does not fall into the hands of the ruling military council.

Ethiopia has plenty of mysteries, and its history goes back into the mists of time. James Churchward believes that the original Ethiopians were of the same race as the ancient Dravidians of India who had come from Mu, the lost continent in the Pacific.[76] While this can never be proven, it is interesting to reflect on Ethiopia's ancient connections with Indian and the Rama Empire. The ancient kingdom of Saba was quite probably the remnant of some state that was affiliated or within the Rama Empire. According to the *Kebra Negast,* King Solomon would visit Makeda and his son Menelik by flying in a "heavenly car". "The king...and all who obeyed his word, flew on the wagon without pain and suffering, and without sweat or exhaustion, and traveled in one day a distance which took three months to traverse (on foot)."[93]

I would surmise that King Solomon was in possession of a vimana left over from the horrific wars that came at the end of the age of Rama and Atlantis. Like the Ark of the Covenant, certain high-tech relics still existed in the world of 1,000 B.C. (For more information on King Solomon and his heavenly car which he reportedly flew into central Asia as well as Ethiopia, see *Lost Cities of China, Central Asia & India*).

In Axum, relics from this ancient period of history can still be seen. Sabean script can be found in Tigre and the gigantic obelisk in Axum is 22 meters high (about 69 feet high) and weighs more than 200 tons. It is carved out of single piece of granite, and is as impressive as any Egyptian obelisk. In Lalibela are monolithic rock-cut churches built in the thirteenth century A.D. It seems that, like the Nabateans, the Ethiopians were fond of carving structures into solid rock.

The rise of Islam in the seventh century caused Ethiopia to withdraw into itself and move the capital from Axum and lose control of its coastal areas, the inner highlands being too rugged for the invading Muslim armies to conquer. Ethiopia entered a long period of isolation which was well summed up by the seventeenth-century historian Gibbon, who said, "Encompassed on all sides by the enemies of the religion...Ethiopia slept near a thousand years, forgetful of the world by whom they were forgotten."

Even though Ethiopia was completely surrounded by the enemies of its religion, it clung to the Coptic Christian faith, a religion that teaches reincarnation, karma and that Christ was an Archangelic being from the sixth plane of existence. Like Nestorian Christians and Gnostic Christians, the Coptics did not sign the Nicene Creed in the 4th century A.D., and were essentially banished from further participation in the church councils at the time. They neither recognize the Pope nor the editing and censorship of the New Testament that was taking place at the time under the orders of Emperor Constantine. What eventually grew out of Ethiopia's Christian isolation was the legend of a Christian king who battled the Muslims and wanted to free Jerusalem from their grasp.

Around 1165 A.D. a mysterious letter addressed to Manuel Comnenus, Emperor of Byzantium, began circulating around Europe. It was from a Prester (Priest) John who claimed to "exceed in riches, virtue and power all creatures who dwell under heaven. Seventy-two kings pay tribute to me. I am a devout Christian and everywhere protect the Christians of our empire...Our magnificence dominates the Three Indias, and extends to Farther India, where the body of St. Thomas the Apostle rests. It reaches through the desert toward the place of the rising sun, and continues through the valley of deserted Babylon close by the Tower of Babel... "

The three Indias are, apparently, Ethiopia, India proper, and Indonesia (or rather South-east Asia in general, and the spice islands in particular). Thomas the Apostle is indeed buried in India, in a small church just outside of Madras. That trade was going on between Ethiopia and the Indies is evident. Europeans took great heart from this letter. The took encouragement in their struggle with the Muslims from the knowledge that they were not alone in that struggle. While some scholars believe the whole Prester John affair to be a hoax, there seems little doubt that the letter referred to Ethiopia and had been written by some traveler who had been there. The search for the kingdom of Prester John was on its way, yet the Europeans knowledge of geography was so poor that they hardly knew where to begin. Many believed erroneously that Prester John's kingdom was in Central Asia. Even Marco Polo hoped to find the Christian kingdom of Prester John, but it wasn't until a Portuguese expedition in 1520 managed to reach the interior of Ethiopia that Prester John's kingdom was found. Even though the Ethiopians had never heard of Prester John, the Portuguese were content that they had found his fabled kingdom, and probably they were right.

Ethiopia is full of strange sites that testify to the antiquity of the country. At Tiya, near Soddu, 156 miles south-west of Addis Ababa are huge stone slabs, or stelae. They stand 16 feet high and are large, flat slabs with one side bearing reliefs consisting of uninterpreted marks and the outlines of a varying number of daggers, sometimes arranged in two rows with their points facing inwards. The significance of these daggers and arrangement is not known, nor do the locals know who built them. They are similar to menhirs found on Corsica, but the markings are unique. Perhaps it was part of an ancient weapons manufacturing site, which might explain the daggers.[23]

Not too far away is the deep gorge of Chabbé. Ranging from 15 to 30 feet wide, the local people say that for a long time it was a cave or tunnel, and that in the remote past, the roof collapsed. The walls are covered with about 50 elegantly stylized cattle carvings, whose bodies appear in profile, with the horns viewed from the front. However, the cattle have no heads, the horns spring directly from the heads. Above the horns, an elongated triangle pointing at each beast can be distinguished. The cattle are each following each other as though they were all leaving the tunnel.

The carvings are similar to other carvings at Harar, and a layer of residue which covers the carvings is proof that they are of great age. The locals know nothing of their construction, and even though there are other gorges and rocks in the vicinity, none have any rock carvings.[2] One wonders if the tunnel at one time extended farther then it does now? Why did the roof collapse? Was this some sort of anti-diluvian slaughter house? We will probably never known.

Closer to Addis Ababa, on the River Awash, 30 miles south from the capital, an ancient settlement was discovered in 1963. Known as Melka Kontourea, the area is littered in stone debris and tools. Stone tools are found all around a completely bare oval platform 30 feet square, except on one side where five small circles formed by six to ten stones are found. Most incredibly, the site is dated geologically as more than one million years old![23] This is of course in uniformitarian dating, where it takes many thousands of years to accumulate a few feet of clay or dirt. Still, the evidence is there for a settlement of great antiquity, no matter what form of dating is being used. Similarly, it is proof that Ethiopia and the surrounding area were inhabited, and apparently civilized, in the dim past.

Rugged and mountainous, Ethiopia is for the adventurer. In 1930, the Lion of Judah, Haile Selassie, 225th Solomonic ruler and central figure in the Jamaican Rastafarian cult, ascended the throne of Abyssinia. He was destined to be the last of a continuous line of Solomomic rule, when he was deposed in 1974 by Military Council and died a year later. The Military Council of "Dergue," as they are known in Ethiopia, chose to ally themselves with the Soviet Union, who gives them military, technological and financial aid in combating the several sessionist wars that it has been fighting for the last fifteen years. Because of these wars, the unstable nature of the Ethiopian government and its xenophobic paranoia about the capitalist western states, travel in Ethiopia is severely restricted.

§§§

"Talk little, listen plenty," was a Somali saying an old man had repeated to me as I walked away from the border of Djibouti into Somalia. It made sense to me, and made me think of the Chinese philosopher Lao Tzu, who said, "Those who speak do not know, those who know do not speak."

Buttoning my lip, I prepared to travel out of the little border village where I found myself. Since I couldn't get into Ethiopia, I had decided to head south into Somalia. It was only an hour or so by public bus to the border, and it took only a few minutes to get through immigration and be on my way.

It was good to be back on the road again. Traveling can be a form of meditation. It's a focusing of the mind upon the present, a quietness of thinking. Meeting people and taking the time to be part of them for that moment, without being critical or judging them, is all part of the pleasure of hitchhiking. Just relating to different people around the world on a one-to-one basis, letting them be themselves and appreciating them for what they are, is what every traveler should try to do. One hitchhiker I met in Afghanistan, a Canadian named Winston Whittaker, said, "In every person is an inner core of perfect energy, a light that is their inner self. We can, and should, love each person unconditionally. The less we know someone, the easier it is to do this." Winston was a die-hard traveler, and a philosopher of sorts.

I caught a ride to Hargeisa, the main town in northern Somalia and a good day's drive from the Djibouti border, in a big two-ton cattle truck with a score of traveling Somalis in the back. I climbed in back with everyone else, after negotiating a price for the ride (about a dollar) with the driver. After a while I sat up on a box above the cab of the truck as we voomed along in the vague wheel tracks of this remote desert road. Suddenly another truck passed us, and our driver was so incensed at his daring that we were treated to the maddest, wildest, craziest truck race across the Ogaden Desert that any adventurer ever had. At first I enjoyed it, leaning back with the sun in my face, singing aloud a Somali tune: "The desert was hot, but my baby was not..."

We shot off the road into the bushes and roadless wasteland at full speed in an effort to catch and pass the truck ahead of us. I was too stunned to breathe as we dodged bushes and roared down ravines and over flat spots trying to get to the next set of definite tire tracks and pass the truck in front of us.

My teeth were clenched and gritty with dust as I hung on for dear life, absolutely horrified and frozen by this display of kamikazi truck driving. Eventually we came to a win-or-lose situation where we raced along parallel to the other truck on a flat section toward a steep ravine and a narrow set of tracks cutting to the left. If we missed the road, and we weren't even on it at the time, we would surely run down into the ravine and turn over. On this final leg, both trucks gunned it, side by side, neck and neck, each determined to make it to the "pass," so to speak, before the other. Our driver would not quit, and while I suppose the Somali in the truck were used to this kind of driving, I was not, and was practically shitting in my pants. In desperation I began stamping with both feet on the top of the cab, which I was sitting directly over, as hard as I could in order to make him slow down. To my great relief, he did, and even came to a stop, while the other truck zoomed ahead.

The driver stuck his head out the window and gave me a look as if to say, "What's the matter with you?" Maybe he just wanted to give me my dollar's worth.

Later, when my heartbeat had returned to normal, we were barreling off through the desert with no other trucks in sight, into the low afternoon sun. I thought about how difficult is is to break down the barriers between people. It is easier to label people than to be non-judgemental. Oh, he's a Foreign Legionnaire, or a policeman; write him off. Or he's a fanatical Muslim nomad, can't talk sense with him; or a racist white living in Africa, I refuse to even speak with him...ah, if we could all just reach out, accept each other as we are and love.

Fat chance, it seemed. With one third of the nations on our planet at war at any one time, it would take a miracle to straighten out the mess the world is in right now. As Winston Whittaker had told me in Afghanistan, "We have to change people's attitudes toward each other before we can make any real progress."

In the distance, the sun was setting in the west, over the Ogaden. The sky was slowly turning from yellow to orange and red. The sun was a white orb floating above a dusty, musty, stark desert planet of mountains. It seemed like some other world. A cool evening breeze was just starting to pick up. Sitting next to me was a young, dusty Somali kid, with a small suitcase, a comb in his fuzzy hair, and a wistful look on his innocent face. I put my arm around his shoulders. As least I could make an effort to straighten out my corner of the world. We have to start somewhere...

We continued on to Borama, a small town in the mountains near the Ethiopian border, where we stopped for dinner. After an hour break, a stop for the other driver and his greasy, fuzzy-haired teenage mechanic to do a little work on the truck, we were off again, driving through the crisp night air toward Hargeisa. We eventually stopped at around two in the morning and everyone slept. Most of the people in the back were already asleep anyway. At the crack of dawn, we all jumped, sleepy-eyed, back into the truck from our blankets and mats on the ground.

We were in Hargeisa by mid-morning, pulling right into the central market where everyone piled out. Hargeisa is the main town in northern Somalia, an area of arid mountains. Northern Somalia was once part of British Somaliland, and therefore English is widely spoken. Southern Somalia was an Italian colony, and Italian is more commonly spoken; of course, Somali is the official language and Arabic is widely understood.

The southern part of Somalia is situated on the dry coastal plain that eventually becomes the Ogaden Desert. Believed to be the land of Punt of the ancient Egyptians, the capital Mogadiscio was known to sailors as early as 1500 B.C. Punt was the only other source of frankincense and myrrh than the Hadramut area of Yemen. Here we see the connection with the ancient Sabean kingdom of 1000 B.C. which stretched across both sides of the Red Sea. Hot and arid, its people are mostly nomadic camel herders, Muslims who are dark-skinned but Semitic rather than Negroid in appearance. The country has been virtually untouched by travelers and tourists; it was closed to foreigners until a few years ago and is still well off the major travel routes for hitchhikers. In the three weeks I spent in Somalia, I never met one other traveler. Currently, the country is having a civil war.

§§§

Everyone will be mobilized, and all boys old enough to carry a spear
will be sent to Addis Ababa. Married men will take their wives
to carry food and cook. Those without wives will take any woman
without a husband. Anyone found at home
after receipt of this order will be hanged.
Decree of the Lion of Judah, Emperor Haile Selassie,
as the Italians invaded Ethiopia, 1935

"Ethiopia was fairly jerked out of the middle ages by the Italians," mused the British-educated Somali guy I was drinking tea with at the Oryantal Hotel in downtown Hargeisa. He was chubby and pleasant, and ran his father's hotel. Talking with him in his office at the hotel was like talking to an old friend. He told me of the good old days when he worked as a clerk for a company in the oil centers of Kuwait and the United Arab Emirates and the wild weekends they had in Bangkok, Thailand, flying there on Friday after work, drinking and screwing their brains out with the friendly Thai women, and flying back on Sunday night to be at work on Monday morning. At twenty-eight he was still single, but his parents were trying desperately to get him married to some nice Somali girl.

"Why, when the Italians invaded Ethiopia from their coastal colony of Eritrea, the Ethiopians tried to stop them with spears and shields. Reality was a hard, cold machine gun!" he said, leaning back from his desk.

"It was like when the Polish cavalry rode out to meet Hitler's motorized divisions," I put in.

"Really! Look at the sophistication of the weaponry developed by western culture. They have taken the art of death and destruction to its very heights, and are still refining it! Military technology is the number one industry in the world today, the basis for our whole world economy!"

He sighed, and took a sip of sweet, milky tea. "The Ethiopians have sure come a long way," he went on. "A few years ago they were fighting off the Italians with spears; now they are run by a military council of generals who spend all their money on Soviet machine guns, grenades, rockets, and fighter planes. My country is run by a military council as well, but they want their arms from the Americans now! Why," he exclaimed, slamming down his empty cup on the desk, his mouth contorted with rage at the foolishness of wars, "these countries can't pay for these arms! They can't even feed themselves! The Soviets, Americans, British and French practically give them these weapons of death, but make them sign over their souls first. It is evil!"

"You're right," was all I could say. The horrible reality of the Third World arms race was hitting me for the first time. It made me feel ashamed to be an American. I blew my nose on the corner of my dirty shirt; I was going to wash it that day, anyway.

"Even though people around the world have real problems that have to be solved, like the problems between Somalia and Ethiopia, most of these wars are being fostered by the capitalist and communist powers to enslave the Third World economically. These countries spend the bulk of their foreign aid and national budget on military hardware to arm themselves against their neighbors. These mini-armsraces are going on all over the world and serve the industrial world's purposes very well. It keeps the arms factories going and the huge exports of sophisticated weapons on an ever-upward swing. But like all products, they are meant to be used; they are consumables that will eventually be consumed in a war of some fashion. Did you know," he asked, "that almost every African country is fighting some war or another, many of them secret wars you never read about in the newspapers?"

"No," I said.

"That is true. In a world at war, there will be a never-ending demand for instruments of destruction. Ethiopia is a good example. It is currently fighting three or four wars against sessionist provinces in the north and south. It was fighting Somali nomads over the control of the Ogaden Desert, why, just near Hargeisa! Using Russian MiG jet fighters, they fought back at the Somali guerrillas who are indigenous to the area. But the Ethiopians aren't trained and educated well enough to operate all this sophisticated equipment, so East German and Cuban technicians fly these planes for them, while thousands of Soviet advisors instruct them on how to use millions of dollars worth of military hardware. The Russians want Ethiopia to stay at war! The Americans do the same thing. All these underdeveloped countries are so eager to build up their military strength, making a headlong rush into the 'everything must go Armageddon sale!' The Third World is destined to destroy itself, thanks to the industrial world."

"The industrial world will probably destroy itself at the same time. It's a sorry situation, all right," I said.

"I'm just happy to sit here at my father's hotel and drink tea," he said. "At least, Hargeisa isn't likely to get hit with an atom bomb." Unfortunately for my friend, Government tanks and troops from Mogadiscio would destroy his city—it wasn't going to take an atom bomb for that.

I hung out in Hargeisa for a couple of days, resting up after my trip from Djibouti, sitting in the shady restaurants around the market. On my last night I had a nice dinner with owner of the Oryantal Hotel in a 'small worker's restaurant just near the market—beans and tomato sauce with small loaves of whole wheat bread. Back at the hotel I read the sign on the inside of my door:

Patrons are requested not to spit on the floors and for safety's sake to put out their cigarettes in the ashtrays provided.

I left Hargeisa the next day and took the local bus down to Berbera, the main port of the north. It's not very far, an hour and a half or so. Berbera was like a dream, hot and illusory. Hargeisa was in the mountains, and the temperature difference was incredible. Berbera was so hot and humid my clothes were constantly stuck to me. I walked around in a daze for an hour. It was only noon, and I could still hitch back into the mountains to Burao. I walked out of town onto a flat sandy plain. To the west, the direction I was walking, a range of mountains shot directly up from the plain. My Michelin map of Northwest Africa indicated that this was an especially scenic road, and by the look of the mountains—their ruggedness and steepness—I thought it was right. I caught a ride in a big truck on its way back to Burao to get straw, sitting in the front with a pleasant, balding, English-speaking Somali. We talked of Somalia and America as the truck switch-backed its way up the mountains. Looking back toward the coast, the view was incredible. The shore was visible in the distance, white waves lapping the sand, which stretched back barren and perfectly flat to the mountains that rose out of it suddenly.

We got to Burao just after dark. We had stopped in a small town called Sheik for a late afternoon tea, and then drove the rest of the way without stopping. Burao was a nice little town, small and easy to get a handle on. I took a room at a very nice little hotel just near the central bus court. The hotel didn't have a name, but it was a fairly new cement building, one story and for two dollars I had my own room, with a table and chair. This was luxury! There was also a nice terrace restaurant adjoining the hotel, and after a meal of shish kebabs and spaghetti, I walked through the market. Old Somali men were sitting around the many small shops, little stalls with their goods inside so that the buyer browsed form the street. Everyone had a pile of qat too. Qat is grown just north of Burao in Ethiopia, and it is at its most plentiful in Somalia in Burao. I enjoyed sitting on the street drinking tea and talking with the old men, most of whom spoke English. Burao had a special mellow feel to it, probably because most of the men were into qat, and didn't really give a hoot about anything else.

Burao impressed me as a particularly nice town, but it was really very small, and I had seen the whole place by mid-morning the next day.

I started hitching out of Burao at noon that same day, and caught a ride on a truck to a small little town that wasn't even on my map, about sixty miles from Burao. There wasn't much there, but there was a spaghetti stand by the road, a cement and brick building with three rooms, like the dozen or so other buildings in town. It was here, while sitting in the little spaghetti stand in the Ogaden Desert, that I met Hassan, who was to become my faithful companion on my trip through the Ogaden. Hassan was a Somali kid from Djibouti. He was fourteen, with a fresh, brown face, medium-length tight wiry hair, and a cheap permanent press suit. He carried a small suitcase and was traveling by himself.

"Bon jour," he said, coming up and sitting next to me. He had been waiting all day for a truck, but there hadn't been anything so far. He was on his way from Djibouti to Mogadiscio to visit some relatives. The bus to Mogadiscio from Hargeisa showed up after a little bit. It was packed to the brim with chickens, goats, people and luggage. I could have sat on a goat or vice versa for a couple of hundred miles through the desert, but decided against it. I wasn't in that big a hurry. I'd just spend the night here if I needed to.

I practiced my French with Hassan for a while. He told me he would look out for me. Just what I needed, I thought, a fourteen-year-old kid to keep me out of trouble. We ate spaghetti and talked, looked at my map, and eventually a rusty old Landcruiser pulled up. It was the Shell Oil man who ran the filling station in another town a couple of hours south. He said he would give us a ride, and so Hassan and I jumped into the back, since it was a pick-up. A few other Somalis climbed in back with us; transportation in the Ogaden is rare and best taken advantage of.

"We're off!" said the Shell Oil man in perfect English, yelling out the window. He was bald and cheery, with a smile that showed off his white teeth. I held on to the spare tire, while Hassan looked at the young nomad bride who had also gotten in. A few other Somalis were carrying a couple of large cans of camel's milk. We sped off through the desert, cutting through the chill late afternoon air like a dagger. We passed camel caravans making their way slowly through the sparse, low brush. The countryside was dotted by huge monolithic termite mounds, the tallest soaring up into the sky like a minaret for twenty feet.

This is what traveling is all about; those precious moments of adventure, motion and companionship in the exotic corners of the world. Chills shot down my spine and a smile lit my face. Hassan was smiling too. Our smiles were highly contagious, it turned out, and soon everyone in the back of the Landcruiser was grinning away like there was no tomorrow. For me it was the joy of living and of sharing these moments with strangers. Life was really a gas, I told Hassan, and he nodded his agreement, keeping his eye on the attractive nomad bride, dressed in a black, single-piece dress with a beaded belt around her waist and lots of colorful plastic bracelets. She was a teenager herself.

Our Shell Oil friend let us all off just after dark in Dusa Mareb, a little town in the Ogaden where he had his gas station. Aside from the station, a couple of spaghetti joints and a hotel that Hassan found us, there wasn't much else to this town. The hotel was a couple of rooms, none with doors, each with some rope-strung, wooden frame beds, typical of the beds in the Horn of Africa and Arabia. I threw my sleeping bag down on one, Hassan placed his suitcase on another.

We went out to eat. A glass of camel's milk and a bowl of 30-cent spaghetti was dinner. The cook at the clay stove and small petrol burners that were the kitchen in this one-room, three-rickety-table cafe told Hassan in Somali that this small town, Dusa Mareb, had been bombed by the Ethiopians a few months before. "Fortunately," said the cook, "the pilots were Ethiopians instead of Cubans, and most of the bombs missed the town." Somehow that made me feel better. Hassan was soon busy chatting to a pretty Somali maiden. Life was strange, but it was a gas.

§§§

The next morning, Hassan and I were sitting out in the street waiting for a ride to come by. We had tea and bread in a restaurant by the road, one of three in the town, which was the Somali version of a truck stop. The owner of the little restaurant came up to me and said in broken English, "Somalis drink tea like Americans drink whiskey."

After gagging a bit on my tea, I said, "What!"

"We drink a lot of tea in Somalia," he said.

"Well, I think you have the wrong idea. We don't drink whiskey so much," I returned.

"Sure you do," he said, wiping his hands on a dirty apron, his brown eyes widening in disbelief. "I've seen the movies. Americans always drink whiskey instead of tea or water. We Somalis know that."

"That's just not true," I laughed, wanting to set him straight. "We drink tea, water, milk, fruit juice, and many things. We don't drink whiskey all the time. Many people do not drink whiskey at all. I don't!"

"You don't!" he gasped. I had shattered his image of Americans, poor guy. It was the James Bond syndrome again. From seeing certain movies, James Bond films being the most popular, people think most westerners are secret agents of one sort or another, drink whiskey like water, and that women will jump into your bed when you look cross-eyed at them.

Hassan and I sat there for the morning. We found a truck, paid for it, got in, and then found out it wasn't going anywhere really, got out, and got our money back...

We went back to our roadside tea stall and Hassan handed me a toothbrush stick, a twig cut from the *peelu* bush that was very fibrous, good for chewing on and cleaning your teeth. All through northeast Africa, the Middle East and even India, it is what people use to brush their teeth. Nearby, a donkey was chewing on a cardboard box, occasionally breaking into a braying fit. Dust clouds searched through town, looking for a pile of papers to scatter. I wiped the sweat from my forehead. It was hot, but not as hot as Djibouti or Berbera. I pulled out my Michelin map and attracted a big crowd of people, all curious about this large, colorful piece of paper.

I was getting thirsty for a lift. We had been waiting about eight hours now for a ride, for just anything to come along, but there was not even one truck. I didn't mind waiting so much. It was good to be back in Africa, life took on a whole new time reference, and I knew from my travels in Sudan that waiting for one or two days for a ride was nothing. Just being able to sit there at that cafe for a day was enjoyable enough. I would savor the experience, and quench my thirst with camel's milk instead, thick and creamy, a meal in itself.

Eventually Hassan and I got a ride in a Toyota Landcruiser. We all rode inside. These guys were qat carriers, or at least that's how I translated it from Somali. They were coming from Togo Wichale on the Ethiopian border where another car from Harar in Ethiopia met them with a load of qat. Now they were bombing down to Mogadiscio with their load of fresh qat. They had about two thousand dollars in qat in the truck and we were making good time, rolling through the desert. We all sat up front and chewed qat—such a speedy drug— and drove like crazy. I talked with the driver in a mixture of English, Arabic, Italian, and Somali. He told me about the Somali custom of circumcising women.

"Women, when they are reaching puberty," he said, biting off a bit of qat and chewing it slowly, "have their clitoris and labia cut off with a knife and then a medicine woman, usually an old lady, sews up the vagina."

With a mighty, horrible gulp, I accidently swallowed my wad of qat, something you shouldn't do, as the wad is supposed to be spat out. "That's horrible!" I said.

"It's our custom,' he told me. "The old women, especially the grandmothers, insist on it. It's tradition. Then, when a woman gets married, the husband has no doubt that his wife is a virgin. Usually, a Somali man will have to open his wife up by slipping a dagger in her on the wedding night, because her opening is scarred shut.

"That's terrible," I winced. "I just can't imagine that!" Shivers ran up my spine. How savage! Female circumcision is actually common in much of northeast Africa, but the dagger stuff....

"Of course, a real Somali nomad would never use a dagger," he went on, "they have a great desire on their wedding night, being in their late twenties and never having been with a woman. They don't need a dagger...but today, in Mogadiscio, the educated people don't use these daggers either, they will take their brides to doctors who will open them very carefully. This is the best way, I think."

Suitably shaken by our conversation, Hassan and I were let off in Galciao, the district center just near the "wedge" of the Ogaden, where Ethiopia juts into Somalia. Galcaio was bigger than most of the other towns, had some extra shops, and Hassan, faithful companion that he was, even found us a movie theater to spend the evening in.

The qat carriers had given me a little qat. It had been a while since I had chewed it in Yemen, so I decided, what the hell. I bought a bottle of Vimto, the prune juice pop from Britain that is very popular in Somalia, and sat watching the movie and slowly chewing my small bundle of qat. Hassan was too young for qat, and fortunately wasn't interested. He was absolutely mesmerized by the film, an absolutely atrocious Indian movie about a witch and a rather foolish truck driver somewhere in Podunk, India. It was horrible. The Somalis loved it. It didn't matter that they weren't showing the reels in order. They started with the last reel first, then showed the first reel, and lastly the middle reel. It didn't make much difference, I reflected and if anything, added a little depth to the film. The narcotic effect of the qat was starting to affect me and I found that I was actually enjoying the film as well.

"Great movie," Hassan said in French as the last reel started. He had a Vimto too and took a drink. As the hero of the movie beat up six or seven thugs single-handedly, I put my feet up on the empty chair in front of me and thought about our sojourn through the Ogaden so far.

The next day, I got a lift that let me off in Belet Huen in the mid afternoon, and Hassan went on with them. I waved goodbye; it had been a good trip. Hassan, I thought, was a thoroughly cheerful and resourceful person. Next day, I caught the daily bus into Mogadiscio from Belet Huen, a six-hour trip. I found a cheap hotel: the Hotel Vittoria. Not feeling the least bit tired after the bus trip, I tossed my pack onto one of the beds. It was a small room, cement floor and walls, just big enough to fit two single beds, a desk, and a chair. It was sparse, but by Somali standards, pretty luxurious. There was a balcony at one end. I swung the louvered doors open to get a nice evening view of the market from my second story balcony.

I was out on the street after a quick shower. The market was alive with the evening proprietors, hawking everything from lizardskin daggers to Italian ice cream. Mogadiscio has that Italian feel. People are often heard speaking in Italian and the food is definitely Italiano!

I grabbed the ever-present bowl of spaghetti and tomato sauce for dinner at a street cafe, a vendor whose kitchen was a bucket of charcoal and a pot. I grabbed one of the few stools that the vendor had around for his customers, and squatted on it watching the fascinating night walk by. Old men, their hair receding from the skull caps that Muslims wear and exposing their smooth, chocolate-brown skin, were browsing along the streets. The men would be wearing Somali *ma'ooses*, which were long sorts of skirts, often in blue, brown, or red checkered pattern, which they wrapped around their waists like bath towels. The women often wore black loose robes that were similar to nuns' habits, I thought. Large black hoods that could cover their faces, should they feel the need, were attached to the backs. Most women did not cover their faces, however; after all, this was supposed to be the swinging, modern capital city!

As I slurped up that last bit of spaghetti, I sighed at the incredible beauty of the Somali people. Tall and stately, they had fine facial features, a distinct blend of Negroid and Semitic racial characteristics—high cheekbones and thin faces with full lips, broad noses and tight wiry hair.

Someone else wanted my stool, so I gave the spaghetti merchant a couple of shillings and tripped on into the night. I could hear some music coming from a few blocks away and headed for it. It turned out that the next day was National Unity Day, Somalia's main holiday, and there was a big celebration in the central square near the capitol building.

As I neared the square, I could see a band up on a stage, playing music. All around me people were jumping up and down in one big open-air party. On the stage was a bunch of fuzzy-haired Somali men, young, in their late teens and early twenties, playing a wide assortment of instruments. There were guitars, congas, bongos, steel drums, xylophones and more. The music, a combination of African rhythm and reggae, rolled out of the loudspeakers and onto the street. It bounced, jolted, hopped across the square. The people, too. Through the waves of energy and sound, I could hear the deep sensual voice penetrating the music:

> *I'm talking about Africa,*
> *I'm talking about Zion,*
> *I'm talking about Rasta,*
> *I'm talking about Zion,*
> *I'm talking about Haile Selassie...*

The continuous thump, da thump, thump of music...the bright ringing of the steel drums...

> *I'm talking about Jah...ho...va...h*
> *I'm talking about...*

A hand tapped me on the shoulder. I turned to my left to look at a tall, well-dressed European. "Do you know what this guy is talking about?" he asked.

"Sure," I laughed, "He's talking about Rasta." The man introduced himself. His name was Rolf, and he was a Norwegian diplomat based in Tanzania. He was stopping in Mogadiscio for a few days on his way back to Norway. I was the first other "tourist" he had seen in two days.

We decided to go back toward the beach for a cup of tea and talk. As we left, the band broke into another song:

> *The whole world is Africa,*
> *But it's the violent continent, yeah.*

Rolf and I walked down to the Uruba Hotel and decided to have a beer in the disco rather than drink tea. There at the bar a young lady, a few months pregnant, asked Rolf to dance. She latched onto him, which Rolf didn't seem to mind too much. Her name was Alia; she was short and cute. She spoke good English and told us she was from the north.

Alia introduced me to her friend, Fushia. Fushia was six feet two and built like a Olympic decathloner. With her Afro that making her look seven feet tall, beautiful dark brown skin, and big wide eyes, I fell in love with this Amazon.

We danced and held hands. I tried to talk with her, but she didn't speak English. We communicated as best we could in a combination of Italian, Arabic and Somali, while Alia did a lot of translating for us. Eventually, Alia suggested we go for a taxi ride to a place she knew, and Rolf went and got us a taxi. Alia took charge and directed the driver. Rolf made a joke about liking his women the way he liked his coffee, strong and black. A Norwegian proverb, I guessed.

We drove out of town to what was the Somali version of a drive-in, a bunch of huts you could rent by the hour. The four of us groped around for a while on a camel hair blanket. Alia seemed pretty aggressive with Rolf. Fushia was more reserved.

Fushia announced that she wanted to take me back to her sister's house, and we all got back into the taxi that was waiting for us. Rolf and Alia let us off in a nice area of Mogadiscio where there were many villa-style homes with walls around them. Fushia was the daughter of a Somali Supreme Court Judge and an Eritrean mother, she told me. Her father had remarried an Italian lady, and was quite a well-known person in Somalia, I gathered.

In the house, a three-bedroom cement home with a kitchen and bathroom, were her three sisters. We all chatted for a while and one of her sisters made up a bed for us in one of the bedrooms, where it was announced that we would spend the night. Suddenly someone came to the door.

"Who can it be at three o'clock in the morning?" cried Fushia. She and her sisters quickly hid me in a closet. The rest of the night was like an Italian sex comedy, hiding under beds and in closets from one room to another. I never did find out who came to the door; one or two different people in and out for about an hour. As I stood nervously in a closet, waiting to be discovered by an angry father or brother, I wondered if how it was that I got into these messes. Perhaps my sense of adventure was getting out of hand.

Several days later I was preparing to leave Mogadiscio. I had had what was left of my plane ticket rerouted, and I was flying out of Mogadiscio to Nairobi that day, as the border between Somalia and Kenya was closed at that time. It had been a great trip through Somalia. It was good to be back in Africa, even if it was the violent continent...yeah.

Abba Libanos,
Rock-Cut Church,
Lalibela, Ethiopia

The granite obelisk at Axum, 22 meters high of solid granite. An awesome relic of Axum's glorious past, it is estimated to weigh more than 200 tons.

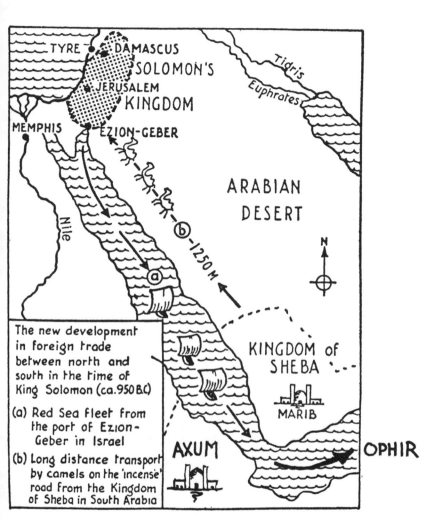

TYRE
DAMASCUS
SOLOMON'S
JERUSALEM
KINGDOM
MEMPHIS
EZION-GEBER
Tigris
Euphrates

ARABIAN
DESERT

Nile

ⓑ - 1250 M

ⓐ

N

KINGDOM of
SHEBA

MARIB

AXUM

OPHIR

The new development
in foreign trade
between north and
south in the time of
King Solomon (ca. 950 B.C)

(a) Red Sea fleet from
the port of Ezion-
Geber in Israel

(b) Long distance transport
by camels on the 'incense'
road from the Kingdom
of Sheba in South Arabia

Sabean inscription in Tigre
province, near Axum. King
Solomon's Phoenician ships
sailed past Sheba's empire
into the Indian Ocean and
beyondon a three year voyage
to Ophir.

Emperor Menelik II
(reigned 1889–1913)

A 1964 stamp issue of famous Ethiopian queens: Sheba, 990; Helen, 1500; Seble Wongel, 1530; Mentiwab, 1730; and Taitu, 1890, the consort of Menelik II

One of the monolithic rock-cut churches of Lalibela in Ethiopia. This is the House of St. George, said to have been cut in the 13th century A.D., but the actual age is unknown.

ANCIENT TRADE ROUTES

LEGEND: ANCIENT SITE—o
SAILING OR CARAVAN ROUTES—➤

Chapter 11

KENYA :
THE LOST CITIES OF SINBAD

...Of the gladdest moments in human life,
methinks is the departure upon a distant journey to unknown lands.
Shaking off with one mighty effort the fetters of Habit,
and leaden weight of Routine, the cloak of many Cares,
and the slavery of Home, man feels once more happy.
The blood flows with the fast circulation of childhood...
afresh dawns the morning of life...
–Sir Richard Burton
Journal Entry 2 December, 1856

One of my associates in Khartoum who worked for the oil companies, a friend of mine with whom I had worked in Sudan, was now in Nairobi. I had written him that I was coming, and he met me at the airport. We went out to dinner and then he dropped me off at the Nairobi Youth Hostel.

The friend picking me up was Rick, a farm boy from Kansas, who was traveling through Africa like me and got a job with our crazy catering company. After working at the oil camps in central Sudan for a while, he was transferred to the Nairobi office.

Nairobi, the capital of Kenya, is one of the most modern and cosmopolitan cities in Africa. With a population in excess of 800,000, it has everything to offer, from super-fashionable shops to appalling crime. The center of Nairobi is best defined as the Hilton Hotel, a tall, circular building right in the middle of the downtown. A block away is the Thorn Tree Hotel, which has a terrace cafe that is known all over Africa. As far away as Algeria or Capetown you will hear people tell each other that they will leave them a message on the bulletin board at the Thorn Tree. A popular spot to sit on a hot afternoon, drinking a cold Tusker Lager on the terrace, you will often meet people who you thought had died of dysentery in Juba two months before or were still in jail in Zaire for being mercenaries. The Thorn Tree is the meeting place for all of Africa, and sooner or later everyone is going to show up there.

After two days of shopping and wandering around downtown Nairobi, Rick and I went out for the evening. Rick had a beat-up old VW bug, so he picked me up at the youth hostel. After dinner at one of the nicer Indian restaurants, we ended up at the New Florida Disco, a big flying-saucer-shaped night club in the central district. It cost a couple of dollars to get in. We were promised live music, a show and a good time. Almost as soon as we sat down, Rick was talking with a short, cute little Kenyan woman with a big Afro and a low-cut dress. I was watching some girls do the limbo in tiger-skin bikinis on the dance floor. Later a Kenyan rock group began playing plenty of loud, western-oriented pop music, the latest disco hits, tinged with reggae and African rhythms. I met a secretary from one of the local offices who was in her mid-twenties and pretty lively, who wanted to dance, and we all, Rick and his new friend too, danced happily for hours.

Glancing around the room, I noticed that the ratio of women to men in the place was about twenty to one. The place was utterly packed with tall, sexy Kenyan women, all wanting to dance, among other things. It certainly wasn't necessary to bring a date to this place.

As it got late, I wanted to go. "I'm ready to go, Rick," I said, turning to him at our table. "It's pretty late."

"You're right," he said and turned to the lady he'd been dancing with all night. "Want to go to my place?" he asked.

She was busy talking with another girl who had joined her at our table. "Not now," she said.

"Later?" he asked hopefully.

"Maybe," she said coolly, turning to her girlfriend. She certainly was in no hurry to leave. This place stayed open till dawn.

"Rick, shall we go?" I asked.

"Yeah, okay. What the hell..." And we all got up and started out the door. Suddenly, with a wild scream, two women at the next table started fighting. Kicking and scratching, they began to tear off each other's clothes. After a few tables, some other women got into the action and soon it was an all-out fight between a dozen or more drunk women. Bottles and chairs flew, other bottles were broken on the tables. We headed for the door, keeping our eyes out for flying objects. Someone grabbed Rick's shirt.

"I'm ready to go now," she said.

"I thought you might be," he smiled, ducking a beer bottle.

§§§

The next day Rick and I were siting on the terrace of the Thorn Tree having a beer.

"Kenya is an economy based on sex," said Rick. We were sitting on one of the sidewalk tables, watching Nairobi pass by. It was a pleasant afternoon, especially after a few Tusker beers. We watched slender Kikuyu women, of the main tribe around Nairobi, walk by, dressed as fashionably and with as much sophistication as women anywhere in the world. Brown-skinned and turbaned Indians, well-tanned whites, sunburned tourists just off the plane, and the globe-trotting black New African businessmen of industrial Africa in their conservative suits were all passing by us on the street.

"Kenya's tourist industry rivals its major export, coffee," Rick went on. "Kenya is by far the most visited of any country in black Africa, with more than three million tourists a year, mostly on package tours. Kenya has everything to offer, miles of tropical beaches, game parks with nearly every African animal in them, snow-capped peaks for mountain climbers, unexplored deserts to the north for modern-day explorers, and best of all, sex!" He took a long, refreshing drink of his beer. "Yeah, the nightclubs of Mombasa and Nairobi are just crawling with women. They're all over the place. You can hardly walk down the street without some girl coming up to you wanting to go out dancing or something. Kenya is well known in Europe, especially Germany, as the ideal spot in Africa for a sex holiday."

"You mean they don't just come here for the elephants and lions?" I asked.

"No, man," he said, his pimply face turned toward a sleek, beautiful, sexy black woman walking by. "It's because of the tigers, man! These girls are tigers in bed!'

"Oh, come on," I laughed.

"Really, there are special hotels along the coast where they cater to orgies for package tours from Germany. The tour supplies Kenyan men and women for a wild time. In all the hotels for tourists there is a certain element of sex, but these hotels are something else. The night club we were at the other night was typical. Places like that are all over."

He was right, I found out later, as I spent more time in Kenya. After all, you do tend to attract into your environment what you think about. And if it's sex, it can't be far off in Kenya.

Kenya started developing its tourism in 1963 when it gained independence from Britain under the presidency of Jomo Kenyatta, the "George Washington of Kenya." He led a rather bloody revolt, the Mau Mau Rebellion, against the colonists prior to independence. Kenya steered a pro-western course that helped its economy greatly after independence. Kenyatta died in 1978, and it was discovered that he had an entire warehouse of ivory in Mombasa that he had been illegally smuggling out of the country for years. He is still revered, though his reputation was tarnished slightly.

Kenya has a large population of whites and Asians, some 200,000 of them. In the early twentieth century, the Kenyan highlands to the west of Nairobi became a popular spot for European immigration, being a cool and healthy climate, and sparsely populated. Originally there was competition between white settlers and Asian immigrants over the choice farm land, but the Europeans gained the upper hand by having 16,700 square miles of the highlands reserved for whites as opposed to Asians. This area became known as the "white highlands" and many successful farms began producing most of Kenya's exports.

Most Kenyans would be surprised to learn that Kenya has its share of lost cities. The idea that some ancient civilization that built cities, roads and canals throughout East Africa more than three thousand years would startle most Kenyans. However, in an anonymous article in *Nature* in 1932, it is stated, "Capt. G. E. H. Wilson discusses in *Man* for November the evidence for the existence of a forgotten civilization in the Rift Valley, East Africa. The existence of ancient works, terracing, grading roads (the so-called elephant tracks) and irrigation works—canals and drainage—is now established not only in Tanganyika, but also in Abyssinia, Uganda, Kenya and Northern Rhodesia. The terraces, averaging in width at the top about one foot, but probably originally about three feet, follow the contours of the hills. The depth between terraces is about three feet. The roads, clearly not elephant tracks, point to a high state of civilization. They are difficult to locate, though in places they are part of roads in use to-day. The points at present located suggest a system of communication running north and south on the eastern side of the Great Lakes, pointing to outlets by way of the Nile in the north and by Rhapta in the south, with possibly an intermediate route via Mombasa, the origin of which may prove very much more ancient than is thought. There are traces of an extensive system of irrigation at Uhehe, and in low-lying districts, such as the Mgeta River near Kisaki, there are river diversions which may be artificial. As to the authors of this civilization, there are legends of an alien race dominating local peoples in both north and south Tanganyika. At present there is a great diversity of language and culture where these ancient works are found; but at some time the people may have been more homogeneous. If there has been an alien immigration, it is possible that it may have taken place so early as 1500 B.C., and that by the time of Solomon (970 B.C.) a flourishing trade already existed and the Sheban port of Rhapta had been established. It is suggested that this ancient civilization may have originated in the north, spread through the Rift Valley over the highlands of the Great Lakes and have reached Zimbabwe."[33]

1932 was a good year for mysterious ancient roads in Kenya, largely because European farmers of good education were starting to create the productive agricultural farms that are now the backbone of Kenya's economy. What they discovered was that someone else had been changing the land before them. In an article in *Antiquity* in 1932 entitled "Azania Civilization in Kenya," G.W.B. Huntingford says, "Under the term Linear Earthworks are included (A) artificial works which are beyond doubt roads; (B) works which appear to be ditches rather than roads. Undoubted roads, which in some places are graded, and in others pass through hillsides in cuttings not unlike railway cuttings, and cross swampy ground over carefully made embankments, occur in Kenya and Tanganyika. Such a road, with cuttings and embankments, crosses the east side of the Uasin Gishu plateau. Sometimes a ditch may be really a sunk road, as in the case of a series which encloses on two sides the farm where I live. Here in 1922, before the land was ploughed, two lengths of ditch were visible for a distance of over 600 yards, running between low banks from one river to another, and reappearing across the river, which must have been crossed by a wooden bridge, as it is not fordable there. On Tilolwa Ridge in Nandi occurs a ditch which is plainly not a road, and equally plainly not of natural origin; a length of some 250 yards (no more is recorded) has a higher bank on the upper side than on the lower."[33]

Who were these ancient engineers building roads, bridges and canals all over East Africa? Whoever they were, they must have lived somewhere, so where are their cities? The first article suggests that they may have been part of the Sabean Empire of three and a half thousand years ago, an interesting theory that would have Saba or Sheba extending through all of Ethiopia, much of East Africa and part of Southern Yemen. Such an Indian Ocean Empire would have naturally had strong links with India, South-East Asia, the Persian Gulf and Egypt. One wonders what relationship they may have had with Madagascar and the Malaysians living there.

The second article ascribes the works to the Azania culture, something of which we know only because of the strange and ancient earthworks that abound throughout East Africa. Were the Azania the ancestors of the Bantu of East Africa today, or did they precede them? A 1915 article by M.W.H. Beech, in the British anthropological journal *Man,* is entitled "Pre-Bantu Occupants of East Africa" and discusses the legends of the Kikuyu around Nairobi. "During a conversation with some A-Kikuyu elders I was informed that in the land they now occupy in the Dagoneta district, which was until quite recently covered with dense forests but is now cleared and cultivated, if they dug down low enough (which they seldom do) they not infrequently came across pieces of ancient pottery of a workmanship entirely different from their own.

"Although I left the district before succeeding in obtaining a piece, all the elders agreed that this pottery was the work of the *Gumba,* a people who inhabited the Kikuyu country after displacing a race of <u>cannibal dwarfs</u> called *Maithoachiana,* and that further information could doubtless be obtained from the elders of the Fort Hall district, whence the present occupiers of Kikuyu had come less (probably) than 100 years ago."[33]

Beech then quotes a Mr. Northcote, the District Commissioner of Fort Hall who had collected some of the stories of these "cannibal dwarfs" and the *Gumba,* "The Maithoachiana appear to be a variety of earth-gnomes with many of the usual attributes: they are Rick, fierce, very touchy, e.g., if you meet one and ask him who his father is he will spear you...

"Like earth-gnomes in most folklore, they are skilled in the art of iron-working. They originally lived round this part (i.e. south of Mount Kenya), but they were driven out by another legendary people called the *Gumba,* who dwelt in caves dug in the earth, and who disappeared one night after teaching the Kikuyu the art of smelting. Another account says that they lived in the earth themselves. It is a Kikuyu insult to say 'You are the son of a Maithoachiana.'"[33]

Beech goes on to draw a few conclusions from the tales of the Kikuyu, "The references might well be to Bushmen, Pygmies, or both, and it is, perhaps, not unreasonable to suppose that the Maithoachiana were an indigenous pygmy or bushman race of the who made and used the many stone implements which are to be found everywhere in the Kikuyu district. In this case Mr. Northcote's informant may have erroneously attributed to them the skill in iron work which was in reality only possessed by their successors, the Gumba.

"The Gumba are said to have made pottery and to have 'taught the Kikuyu the art of smelting,' which is equivalent to the A-Kikuyu admitting that they did not bring a knowledge of iron with them or it out for themselves.

"Assuming the tradition to by substantially true, and unless the Gumba who, be it noted, are not described as dwarfs, were pre-Bantu Harnite invaders (a supposition for which, as far as I know, there is absolutely no warrant) the legend would appear to be in favor of the first discovery of iron having been in Africa."[33]

Were the Gumba also the builders of the roads throughout East Africa? It is certainly interesting to relate some of this information to the Bushmen and Hottentots of Southern Africa, a mysterious group, who are classified by anthropologists as "whites". Had they once inhabited East Africa and been forced into the Kalahari by invading Bantus? The mystery of civilizations in southern Africa and their connections with the rest of world is undoubtedly connected with Zimbabwe, the roads in Kenya, the ancient kingdom of Sheba, and Phoenician voyages in search of gold, ivory and other goods to trade.

At the Youth Hostel, I met a young lady from Chile named Marie. She worked as a receptionist for an airline, and had gotten a free ticket to Kenya. Marie and I decided to rent a car and drive out to some of the game parks around Nairobi for the week. I had my tent, and we both had sleeping bags. We drove to Masai Mara Game Park, to the southwest of Nairobi, first. It was almost dark when we got to the outskirts of the park, and were looking for a campground to spend the night in, when we met an Englishman in a Landrover who was the engineer of a black road crew just outside the park. After talking to him for a bit, he invited us to spend the night with him. He was in his forties and had a wife and teenage sons in England. Economics had forced him to come to Kenya to work. He had a nice little two-bedroom trailer home that was his office and the headquarters of the road crew. There were twenty or so Kenyans that lived in tents and a quonset hut who also worked with him. We had a nice dinner and talked about Kenya and England. He missed home, but liked Africa.

"Come on, I'll show you the savanna at night," he said suddenly. We got into his landrover and drove out into the flat, grassy bush. It was incredible. We ran into herds of zebra and wildebeest, who were momentarily frozen by our headlights.

"Yee-ha!" he shouted, chasing them through the low bushes and sparse thorn trees. We were barreling along among a hundred zebras who were running in all directions. The Englishman cracked open a beer, his fourth, and guzzled it. He was really letting himself go; I took it that he didn't do this often. With more yelling and shouting, he ran into a pack of hyenas, and Marie cried out in disgust at their ugly canine heads, big teeth and oversized haunches. The seven or eight hyenas stared at the car for a moment and then broke out running in different directions.

We went on driving through the dark moonless night, not a road to be found anywhere. We came on another herd of zebra and with a whoop were chasing them off through the bush again. This time the zebras got back at us. As we chased a cluster, there was a sudden whomp and Marie and I hit the roof. We had driven into a ditch.

"No problem," said our inebriated zebra rustler, "these Landrovers can do anything!" With some spinning of wheels and shifting of gears, he managed to get us out. Marie wanted to go back to the camp. We all slept well, and he waved goodbye to us from the gate after breakfast. It had been an exciting stay.

We drove into the park and then watched a water hole from the Keekorok Lodge. Just at sunset, we drove out into the savanna a short mile, and then built a campfire. I pitched the tent and we cooked up a vegetable soup.

Masai Mara is just opposite Tanzania's Serengeti Park, the best game park in Africa. Masai Mara has Kenya's greatest abundance of animals, and every year in May or June wildebeest and zebra migrate to the permanent waterholes in Kenya form the plains of Serengeti in Tanzania. The sight of over 50,000 wildebeest and zebra migrating across the plain is overwhelming.

Marie and I then drove down to Ambosselli Game Park where you can camp near majestic Mount Kilimanjaro, the highest mountain in Africa and a spectacular sight. Kilimanjaro is actually in Tanzania, and Amboselli is right on the border. After a week of camping out in the game parks we headed back to Nairobi for a taste of civilization. A friend of mine in Sudan has written to tell me I could use his Yamaha 125 motorcycle for a couple of months while he was up in Sudan at a camp. I had it tuned up and told Marie I'd meet her in Mombasa. I left the youth hostel in the late afternoon, bombing down the road toward the coast. Marie had gotten a free plane ticket to Mombasa as an airline employee and was flying down two days later.

As I wound around the green hills outside of Nairobi and into the lower grasslands, I glanced into my rearview mirror to see a couple of giraffes, silhouetted against another African sunset that was burning down the hills behind me with its deep reds and murky oranges. I rode on through the night until I came to a truck stop, a place along the road with three bars and a hotel. It wasn't even on the map. I got a room for a dollar and a half in the hotel and was going to chain my motorcycle to a pole outside, but the manageress, a friendly but streetwise old African lady, made me keep in inside my room.

"Won't be there in the morning if you leave it in the street," she said. "There's a lot of crime in Kenya!" At least it was good to have people looking out after me, I thought.

I was up early the next morning and zooming down the highway again. It felt great to have the wind in my hair. I kept my backpack with the shoulder straps on but the belt off. It sat right on the seat behind me and took the weight off me, but was held firm so it wouldn't shift around. I couldn't take any riders, though.

As I passed a particularly parched and dry section of the savanna, I saw a large baboon squatting on a road post, just beneath an elephant-crossing sign. With his chin in his hand, elbow resting on his knee, he looked like he was in such deep, contemplative thought, watching me roar past on the bike, I felt that I just had to stop and take a picture.

Baboons can be dangerous, so I parked the bike a safe distance away, a hundred feet or so, hoping I'd be able to hop on and start it before the baboon attacked, if he had a mind to. I didn't know what this guy was thinking about, but he seemed to be taking it very seriously.

It took me a while to get out my camera, which was well inside my pack, put on a telephoto lens and then get a light reading with my hand-held light meter. Meanwhile, the baboon got impatient, lost his concentration, and decided to split. He hobbled off his post and meandered into the bushes. He not only wasn't going to attack me, he was just plain bored with me.

I arrived in Mombasa in mid-afternoon. Kenya's major port and second largest city at 350,000, it was named by the early Arab sailors "the island of war." In the fifteenth century Mombasa took over from Kilwa in Tanzania and Mogadiscio as the most powerful city on the East African Coast. Significantly, it was here at Mombasa that Vasco da Gama came to find a pilot to guide him to India in 1498. Mombasa had become the center of a flourishing trade between Arabia, Persia, India, Indonesia and China that had been going on for thousands of years. Now with the arrival of Europeans, the established trade patterns were disrupted and some of the Swahili towns were reduced to subsistence agriculture.

Mombasa did not succumb to Portuguese rule until 1589 when a Portuguese attack coincided with turmoil inland caused by the fierce Zimba tribe who had stormed up from the Zambezi region to face their final defeat by the Segeju near Mombasa. The Portuguese built their Fort Jesus here in 1593 to secure the area from the Turks and Arabs who were raiding the coast. The city's well known sleazy night life reflects its long history as a major sea port. The old city with its narrow winding streets of cobblestones and the Portuguese port are the main attractions, plus the miles of sandy beach to the north and south of the city.

Mombasa is fun shopping. You can walk the old city for hours looking at African kangas, printed cotton material with colorful local designs that women wear as wrap-around skirts and that make good beach wraps and sheets. The male version of the kanga is the kikoi, a smaller woven fabric that is also sometimes printed, and is worn around the waist.

After five days of shopping and going to the beach in Mombasa, Marie and I headed for Malindi, three hours north up the coast. Marie took the bus and I rode the motorcycle up. We stayed in the Lucky Lodge, recommended to me by a friend because he liked their yogurt. It was nice and clean too, with mosquito nets on the beds.

We stopped for a few days at Malindi's beaches and I investigated the vast acres of ruins of the mysterious lost city of Gedi. Near to the Kenyan town of Watamu is the Gedi National Park where the extremely well-preserved ruin of Gedi can be found. The city is believed to have been deserted in the thirteenth century, several hundred years before the Portuguese showed up on the coast. The first archae-ological excavation was only started in 1948, yet it is a mystery why the city, virtually still intact today, was deserted. This is similar to the deserted cities of the Anazazi in the American Southwest, who theoretically abandoned their dwellings at about the same time.

Excavations at Gedi have unearthed blue and white porcelain bowls, brown and black flasks and beads from China, glass flasks and beads from China, glass flasks from Persia, and carnelians from India, besides local cooking pots, eating bowls and utensils, all of which are on display at the local museum. Many of the buildings are massive, megalithic carved walls and arches, something similar to the Nabatean carved cities of Petra or Mada'in Salih, or the rock-cut churches at Lalibela in Ethiopia. Some of the structures are said to be tombs, mosques or private residences. Yet, there is no real evidence that the Arabs or Swahilis built the city, only that it was probably occupied by them for at least a while.

It is unknown who built the city, when it was built, and for what reason it was abandoned. Evidence suggests that while it was abandoned by at least the thirteenth century A.D., it may have been abandoned even before that, and if the Arabs did occupy the city (and later abandon it) it is quite possible that they did not build the city, but merely occupied the ruins of a much older civilization. Certainly the architecture and construction are quite different from the Swahili towns like Lamu or Zanzibar. Furthermore, it is apparent that Gedi was once a port city, yet, like the Sumerian city of Ur in present day Iraq, it is now inland, without any access to the sea. This might well be the reason for its abandonment, as a port city is quite useless if it is not near water. As David Round-Turner says in *Project Kenya,* "It's a mysterious place, full of colorful birds and shy, small animals. How a flourishing seaport became stranded so far from the sea remains a mystery, even to the archaeologists."[99]

Indeed, tourist literature, not wanting to admit that very little is known about Gedi, generally says that it was an Islamic city built in the thirteenth century and that is all. This is definitely false. The city was nothing but ruins in the fourteenth century, and it seems unlikely that it would be built and abandoned within a hundred years. It would be more correct to say only that the city was abandoned at this time. Still, who would build a port city that was not on the coast? Has the East African coastline changed so much in just a few hundred years? Other ancient port cities such as Lamu do not seem to have been affected. Gedi is probably an ancient port, dating back at least to Sabean times, about 1000 B.C. and possibly earlier. The last major shift of the earth's crust which changed water levels and geography probably occurred about 3,000 B.C. This would put Gedi in about the same time-frame as Ur in Sumer or Lothal in Gujerat, both similar port cities now miles from the coast.

As Marie and I walked away from Gedi and I started up the motorcycle she said, "Well, I guess I can choose my version of history, the guidebook's 600 year old city, or your 5,000 year old city." She paused for a moment, "I'll take your version," she said.

"Thanks," I replied, putting the bike in gear. "I need *someone* on my side!"

§§§

We continued farther north to Lamu. Lamu is the legendary home of that most famous of Indian Ocean sailors, Sinbad. For thousands of years sailors have been visiting Lamu and the east coast of Africa. In the first millennium BC, Egyptian, Phoenician, Arabian, Indian, and even Chinese ships sailed down the east coast. King Solomon's ships may have stopped on the East African coast on his way to the gold mines at Ophir, though it is more likely that they headed for Southeast Asia directly from the Horn of Africa. The Egyptian Pharaoh Necho II had Phoenician/Libyan sailors circumnavigate Africa for him about 600 B.C., 400 years or so after Solomon's and Hyram of Tyre's ships had stopped trading across the Indian Ocean. The geographer Ptolemy reported Necho's claimed expedition, but denied that it could be done, because, he said, Africa was linked to Asia by an "Austral" continent that locked in the Indian Ocean. Time, of course, was to prove the ancient Egyptians right, and prove as well that the ancients had a much better knowledge of geography than did the classical scholars who succeeded them.

One problem in recreating the history of East Africa is that it was under Arab/Islamic domination for a thousand years, and the Arabs have a penchant, not unlike other cultures, and especially conquerors, of wanting history to start with them. In Islam particularly, discussion or investigation of pre-Islamic civilizations is very definitely discouraged. Cities along the coast and inland rose and died out. The Arabs and Persians eventually established permanent towns of stone, mud and wood that still exist today; Lamu is one of them. They developed the trading language for east Africa, KiSwahili; Arabic for "coast." They lived in comparative luxury with silks, porcelains and spices from the East. Rarely venturing inland, they depended on the interior natives to bring the goods to them, on their island cities. Even those traders who did venture inland were often forced to wait on the borders of tribal kingdom while the trading goods were brought to them, such as with the Kikuyu of the Kenyan highlands. Generally, they traded metals for ivory, ostrich feathers and animal skins. Also, unfortunately, there was the slave trade.

From these ancient sailors of Arabia we get our tales of the Arabian Nights and the adventures of Sinbad the Sailor. Residents of Lamu will proudly tell you that Sinbad lived in Lamu in those glorious days of yore. From the safe island city-nation, he and his crew ventured out into the Indian Ocean to mysterious, unknown lands. Certainly Sinbad must have sailed regularly to Ceylon and India, as well as back to Oman with the monsoon. As for the Arabian legend of the roc, a gigantic bird that carried one of Sinbad's sailors away, such birds were known to have existed at Sinbad's time. The largest bird known, the aepyornis, a bird ten feet tall, weighing almost a thousand pounds and laying an egg six times the size of an ostrich egg, was still extant in parts of Madagascar until the nineteenth century. Marco Polo was told by the Chinese that the roc came from islands to the south of Madagascar and Herodotus was told by Egyptian priests about a race of gigantic birds "beyond the sources of the Nile" which were strong enough to carry off a man.

The ancient sailors of Arabia sailed to Madagascar, where they would probably have seen these huge flightless birds from their bases on the coast. It seems likely that these sailors ventured as far as Indonesia and Australia to the east, to Zimbabwe and even around the African Cape.

It was a full day's trip north to Lamu from Malindi. There are daily buses to Lamu form Malindi and Marie was on one with all our luggage while I rode up on the motorcycle. Lamu is an island, just off the mainland, and even today there are no roads or cars there. As a result, Lamu retains much of its traditional flavor as an ancient trading town. I had to park the cycle in a government parking lot with a watchman on the mainland, and take a ferry over to the city.

Lamu was the perfect place to wind up our romance in Kenya. Marie had to fly back to Europe soon, but we would have a week to spend in the medieval atmosphere of Lamu. I was fortunate enough to rent a room on the top floor of a private house, with a fantastic view looking out over the three- and four-story stone and cement homes that rose up the hill from the port into the palm trees. Lamu is a small place of around 10,000, with narrow stone streets and a thriving market, scores of little tea shops and a population of people whose quiet, unhurried lives have yet to be touched by the industrial world outside. Lamu lives in its own peaceful little world of palm trees, sunsets, beaches, and Sinbad the Sailor. There are few places like it in all of Africa.

Marie and I spent our week together shopping and walking along the port, watching the fishermen with their catches and the ships coming in from other ports, small Arab dhows with two sails and a cargo hold in the hull, just like the ones Sinbad used to sail in. The beach is two miles away from town to the southern end of the island, a pleasant walk, and is perfectly deserted, with sand dunes behind it. Although it is against the law in Kenya, some people swim nude here.

After a swim at the beach we would walk back into town and stop at Petley's Inn on the seafront, a small charming tourist-class hotel with a nice cafe out front for a cup of tea of a late afternoon beer.

We kissed each other goodbye on the wharf at Lamu. Marie would take the ferry back to the mainland and the bus back to Malindi and Mombasa. She would fly to Nairobi and then to Amsterdam and back to South America. I would miss her. She got the last bus back to Malindi before a torrential rain completely washed out sections of the road to Lamu, and an important ferry across the Tana River at Garsen. The rest of the travelers in Lamu were stranded for three weeks and no more could get in unless they came by dhow.

I settled into an easy routine, drawing sketches in my penthouse suite of sorts, going down to one of the tea shops in the morning for sweet milky tea and donuts, and walking out to the beach. Often I would stop at a friend's house along the way, and before I knew it the day would be gone and I had never gotten to the beach.

One day I got a letter from Catharina in Sweden, saying that she was coming out to Kenya. We had kept in touch, and now she wanted to come to Africa for a while. I promised to meet her at the airport in Nairobi, and we would go to Tanzania together. My motorcycle was still on the other side of the strait on the mainland and there was no way I could ride it back now. I only hoped that in a few weeks the roads would be passable again.

One afternoon I was sitting with some friends on the steps of the Castle Lodge in the heart of Lamu. An English traveler named Doug and a fellow from Quebec and I were chewing marungi, a Kenyan form of qat, and playing music. Doug was playing his guitar, Jean was blowing a mean harmonica, and I beat a piece of wood with a stick in my best rhythm.

We were grooving that afternoon on those steps, feeling good from the marungi, chewing the tender red stems and spitting out the fiber, but swallowing the juice mixed with our saliva. Soon it seemed like every kid in Lamu was sitting with us. Their smiling faces and eager looks kept us going. An old Somali man, grizzled and unshaven, sat next to me and began clapping his hands. He wore a kikoi wrapped around his waist, his brown thin chest was bare, and on his gray tight curls was a white Muslim cap.

He also had a small bundle of marungi, offered me a stem, and began chewing his own. We talked of Lamu, Somali bandits known as "shiftas" and the current wars in Uganda, Ethiopia, and Zimbabwe.

"Since the beginning of history," he said in excellent English, "mankind has tried to solve its problems by killing each other. We'll never learn that killing doesn't solve problems, it just creates more."

Doug and Jean stopped playing. There were more than fifty kids from fourteen to four sitting around us. The old man offered to show us all a palm wine place on the outskirts of town. We waded through the sea of kids and walked through the narrow streets to the south and then out into the fringes of the palm forest.

The palm wine place was a shack with some benches around it, and forty or fifty people standing and sitting around with old wine bottles filled with fresh palm wine. Palm wine is made daily by a bucket and tap system, similar to the method of getting syrup from maple trees. It is allowed to ferment overnight and is slightly alcoholic, about the strength of beer, but costs only a few cents a bottle. We each bought a bottle and I took a good swig. The locals didn't have much of a problem drinking it, but the stuff tasted a bit like vomit to me. It was a challenge getting a bottle down. Still, each mouthful got easier...

The old Somali man went on about violence. "Killing, no matter how justifiable the cause, unless you are protecting your very own life or your family's, just begets more killing," he said.

Jean, the French Canadian, said, "Isn't that the law of karma, that what you do will return to you?"

"That's true," said Doug.

"I've never heard of karma," said the old man, "but I do believe that you will reap as you have sown. The prophets have always taught this. But mankind does not listen. Those in power cause nations to go to battle. It is a sad situation."

By now, we'd each had one bottle of palm wine and were starting into our second bottles. Empty bottles lay all around us from the other drinkers. It was getting pretty dark, and those people who could still walk were slowly puttering off, the men lifting their kikois to let the wind blow up between their legs.

Jean, Doug and I bid the old man good night and walked back toward town, slightly inebriated. We wandered the twisting back streets, a winding dark maze that turned and ended in a myriad of alleyways and black shadowy doorways. It was wonderful. Lamu felt like such a timeless, mysterious place; I thought that anything could happen here. A certain mystical feeling hung in the air.

The timeless words of the ancient teachers were quietly echoed down the flagstone streets: Love your neighbor as you do yourself.

§§§

I finally made a break from Lamu and tried to ride the motorcycle back to Malindi. I had gotten a late start from Lamu, so I spent the night at a road crew station just outside of Witu. The next day I rode as far as I could, often through several feet of water, until I came to a spot where water flowed across the road for as far as I could see, at least a mile. There was no way I could ride through that on a motorcycle. Just when I thought I would have to go back to Lamu, a Landrover came along, filled with Africans, and driven by a black game warden. For a dollar he agreed to put the motorcycle up on top of the four-wheel-drive vehicle and take me over the worst part. We drove through the swamp, which was about a foot and a half to two and a half feet deep all the way to the ferry. At one point a virtual river was flowing across the road. He disconnected the fan belt and we rolled up the windows while driving through four feet of water.

At the ferry at Garsen, which was still washed out, I paid a guy in a dugout canoe to take the motorcycle and myself back over the river. It was shaky, the motorcycle balanced precariously over either edge of the canoe. One slight tip and the bike would be dumped overboard. It was a long ten minutes making the crossing, and to keep from worrying about the possible loss of the bike, I thought about some of the mysteries of northern Kenya.

In 1927, an article about mysterious rock-cut wells and cairns in the Northern Frontier Province of Kenya was run in the British journal *Man*. The author, C.B.G. Watson, says that wells bored through 16 to 40 feet in limestone rock exist all through the northern desert area of Kenya. Underground passages lead between some of them, and a large number of cairns or barrows are scattered throughout the area. Watson relates that some believe the wells and barrows to be of natural origin. The other view is that they are the remains of some "ancient civilization of people who have left behind them no trace save their feats as water engineers and cairns that yield no bones, implements or records of any kind. This tribe has been called *The Medenli.* " [33]

Watson believes that the area, now scorching desert, was once fertile at the time of occupation. Says he, "The present inhabitants—Boran to he west and south, Ogaden Somalis to the east—are bone-idle and only work sufficient wells for their needs, not attempting to repair them or to dig out those that go dry. Figures will show that the Medenli must have been: (a) immensely numerous and wealthy; (b) clever water diviners; (c) cunning engineers; and that their country was fertile, though now a barren desert."

Watson then goes on to describe the incredible wells throughout the territory, "At El Wak (the wells of the Wak tree) are some 50 wells spread over an area of about 20 square miles. Only a dozen are now used by a mixed Galla-somali tribe called Gurre. They consist of (a) a deep pit averaging 60 feet deep and 15 broad at the top, (b) underground passages varying in number and length. Wanderers have been known to have been lost forever in these passages."[33]

Watson continues by expounding on implications of these wells, "At Wajir one can give figures which give an idea of the size of the Medenli tribe and its stock. In an area of about 2 square miles there are known to be more than 400 wells of this type, all of which could be used if required today. Now in the Government *boma* a few whites and about 300 natives get all the water they require from two wells. Water can be pulled up tribesmen for a day continuously before a few hours' rest is needed for the water to filter through the sandy subsoil to restore the level. Who then could these people have been who required this immense water supply obtained by such laborious feats of engineering? At Arbo the wells number about sixty. At Buna we see wells made on the same lines but about 40 feet in depth. To my mind they explode the 'volcanic' (natural) theory."[33]

I was just wondering whether the so-called Medenli "tribe" had anything to do with the ruins at Gedi, the mysterious Gumba or legends, or the strange roads that crisscross ancient Kenya, when the canoe almost tipped over. I screamed in fear, but the canoe suddenly righted itself and we made it safely to the other side. It was clear riding from there.

I rode to Malindi that day and spent another night at the Lucky Lodge. Then on to Mombasa the next day, and up to Nairobi the day after that, just in time to meet Catharina, the Swedish kindergarten teacher who was flying in from Copenhagen.

We spent a few nights with Rick at his place and inquired about getting into Tanzania. The border was still closed between Kenya and Tanzania, because of the collapse of the East African Economic Community a few years ago. Tanzania hadn't been too pleased abut the way Kenya seized most of the planes that belonged to the joint-owned East African Airways, and at how the tourists would stay at the nicer Kenyan game lodges and just drive over to the better Tanzanian game parks for the day. So now it was impossible for anyone to drive or even fly between Kenya and Tanzania. Since Catharina and I wanted to go to Tanzania, our only choice, we were told, was to go through Uganda, which was currently at war: Idi Amin and his troops in the north and the invading Tanzanians in the south.

Catharina decided to try it. She had a lot of spunk and wasn't deterred too easily. Besides, if I thought it was safe... We said goodbye to Rick and began hitchhiking up toward Lake Naivasha and Nakuru. We made a quick detour to Mount Kenya on the way, the second highest mountain in Africa, at 17,058 feet, and a majestic sight. Then Catharina and I hitchhiked up to Lake Nakuru, a great bird sanctuary with some game. A friend showed us an abandoned, half-built house just past the lodge on Lion's Hill. It has a great view of the lake, which is often absolutely pink with thousands of flamingos.

In front of us the sun was setting. It was good to see her again. Our parting more a year before in Israel had not been very good. Our adventure through East Africa was just beginning. War-torn Uganda lay ahead of us. What dangers awaited us there? I am thankful to this day that we did not know at the time!

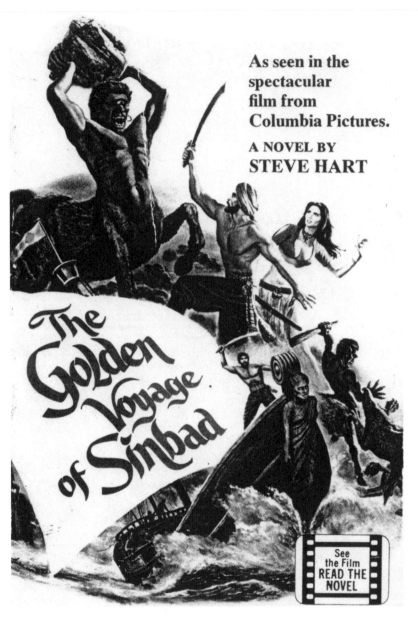

Movie poster for a film of the adventures of Sinbad.
His real adventures may have been just as exciting.

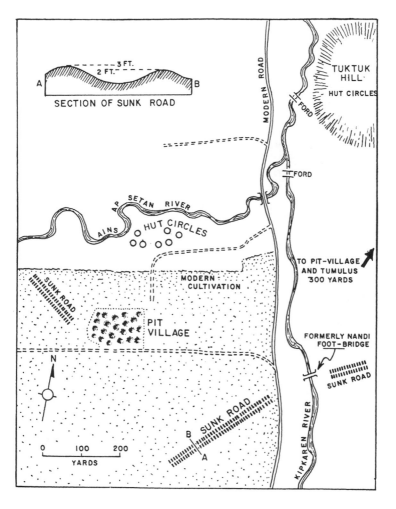

Kenya is crisscrossed by an series of ancient roads.
Generally, they are attributed to the Azania Civilization,
thought to be conected with ancient Sabean Ethiopia.

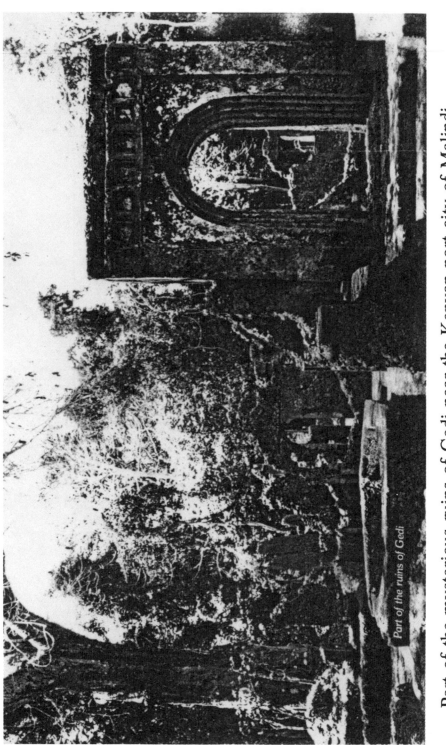

Part of the ruins of Gedi

Part of the mysterious ruins of Gedi near the Kenyan port city of Malindi. Strangely, Gedi is a port city but is several miles from the ocean!

Chapter 12

UGANDA & ZAIRE:
DINOSAUR HUNTERS &
THE MOUNTAINS OF THE MOON

*Sometimes people mistake the way I talk
for what I am thinking.*
—Idi Amin

With a wild screech and clattering of nuts and bolts, the rusty bus ground to a halt in some muddy village near the Ugandan border. Catharina and I had hitched up from Nakuru in a day and were about to enter Uganda. I had been riding up on top of the bus, perched on the dusty luggage rack with the bed rolls, trunks, suitcases and bicycles. Swarms of insects were gathering above the trees and the colorful African birds were making more noise than usual. It was sunset once again.

I handed our packs down to Catharina, who had been riding inside the bus. She looked so cute in her pink T-shirt and safari shorts, long dark hair tied in a pony-tail behind her, falling down over her small curvaceous body. She was lovely and radiant, and the only thing that made me wonder about her was that she was willing to follow me into Uganda....

Grabbing our packs, we headed down the dirt road to a muddy crossroads. I decided to ask directions of a suitably gnarled, bent and intelligent-looking man, who was leaning on a walking stick nearly as old and twisted as he.

"Excuse me, sir," I inquired politely, "but can you tell us the way to Uganda?"

He didn't move at all for several long quiet moments, then his tight, closely-clipped gray beard twitched against his wrinkled, black face. "Uganda!" he winced. "What do you want to go there for?"

"We're interested to see it," I said.

"We're trying to get to Tanzania," said Catharina.

The old man leveled his dark brown eyes at us. "Don't go," he said. His voice was even and relaxed.

"No, it's OK, really. We'll be fine. We've already made up our minds." This was all I could think of to say to him.

The old man let off a loud spontaneous snort that cut through the gathering evening mists of the Kenyan highlands. Bits of phlegm clung to his sleeve and my shirt. His face was concerned as he put a gnarled, wrinkled hand on my shoulder and pointed down a road to the southwest.

"There! There!" he cried. "There is the Montes Lunas! The Mountains of Madness! And that is what you find there, death and madness!" He squeezed my shoulder ever so gently. "Be careful," he said.

Just then a flock of flamingos went flying over us, heading east, away from the setting sun. The noises of the evening insects became noticeably louder. Thanking the old man for his advice and directions, Catharina and I held hands and strode forth down the road to the Mountains of Madness, the Montes Lunas, the Mountains of the Moon.

Called the Jewel of Africa by Winston Churchill, Uganda was once considered to be the potential paradise of the continent. These days, just the mention of the word Uganda can strike terror into the hearts of men. From the lush tropical shores of Lake Victoria to the snowcapped Ruwenzori Mountains, the Montes Lunas, Uganda is as varied and beautiful as a country can be. Wild life abounds, herds of elephants roam the country, great flocks of birds fly overhead, and even that shyest of primates, the great mountain gorilla, resides in this landlocked central African state.

The mists of Ugandan history go back to the Buganda tribe's legends of a people named the Bacwezi who are the ancestors of the ruling house of Buganda. The Bacwezi were said to be a race of giants who could accomplish miraculous feats! Margret Shinnie, the author of *Ancient African Kingdoms*,[112] believes the Bacwezi came from the north, but does not state why. Possibly because that is the direction from which the Nile flows.

Massive mounds, ditches and other earthworks abound all through Uganda and they are universally attributed to the Bacwezi. Many trenches are cut into the earth and are part of a fortified series of ditches and circlular earthworks over a mile across, oval in shape and with a large mound in the center. The largest of the earthworks is at Bigo, along the Katonga River. At Bigo is what Margret Shinnie calls the most impressive monument, a ditch cut out of solid bedrock 12 feet deep. What the purpose of this impressive feature? She did not know, though she excavated it herself. Was it part of an ancient canal? Pick marks could be seen and many iron and quartz implements were discovered.[112]

In the fifteenth century Uganda was a series of small kingdoms, fighting and allaying with each other. By the eighteenth century, the two main rivals were the kingdoms of Buganda and Bunyoro-Kitara. Swahili-speaking tribes and Zanzabari traders began penetrating central Africa. In their wake came the European explorers of the nineteenth century; the most famous was John Speke, who discovered, in a sense, the source of the Nile, Lake Victoria, near Jinja in 1858. By 1894, the British established a protectorate over Buganda and intense missionary activities of all kinds ensued, eventually dominated by the Anglican Church. This protectorate spread over the rest of Uganda, with the Buganda chiefs dominant over the entire country, but dependent on the actual British rulers. The stubborn independence of the Bugandan rulers lasted well into the twentieth century, which hampered the development of national unity. Still, Uganda prospered more than most African countries under colonialism. Education was good and Uganda had a high literacy rate, while the standard of living continued to increase dramatically as the British showered attention on their African prize.

In 1962, Uganda became an independent state and Milton Obote became the first prime minister. Meanwhile, a Muslim sergeant, Idi Amin, who had served in the British African Rifles, was rising fast in the Ugandan army. Smuggling is sometimes seen as the hobby of African politicians, and Milton Obote was busy smuggling coffee out of Uganda for his own personal wealth. He needed an accomplice, and thought he had found it in Amin.

Thinking Amin to be dumb and controllable, but at the same time unscrupulous enough to help with his smuggling schemes, Obote raised him to the rank of major general, the second most powerful position in the country at the time. Idi Amin turned out to be more ambitious than anyone thought. In 1971, while Obote was in Singapore for a Nonaligned Nations Conference, Amin seized control of the country by merely proclaiming himself the new president for life and telling Obote he would be killed if he came back to Uganda. Obote went into exile in Tanzania, one of the few countries that refused to recognize Amin's government. For the next eight years, Amin's troops, mostly ruthless hired mercenaries from southern Sudan who owed their dominant social position to Amin, terrorized the country.

Finally, after Amin had killed a large number of his own soldiers in southern Uganda in an attempt to suppress a coup, he made his big mistake by claiming that Tanzania had attacked Uganda. He then counter-attacked Tanzania, which had never attacked Uganda in the first place, and claimed a small area of extreme northern Tanzania as a part of Uganda.

Tanzania could never put up with Amin anyway. He was the laughingstock of the entire world with his insane political antics and murderous internal policies. So they responded by attacking Uganda. The entire Tanzanian army was used and this became the first time in African history that one African country actually invaded another. Catharina and I were now approaching the border about five months after the beginning of this war. Idi Amin's troops were still fighting in the northern part of Uganda, while the Tanzanians, so far the victors, had "liberated" the southern half of Uganda, including the capital, Kampala. Amin had retreated to his own homeland and powerbase in the north, an area that would take a long time to clean up, as they say in army jargon. Catharina and I were about to enter the newly liberated Tanzanian part of Uganda, the first travelers to do so since the war.

"Welcome to Uganda!" said the immigrations officer, nodding his head and taking a bony hand out of his torn, faded blue uniform to shake our hands. Cheery and friendly, this old Ugandan was pleased as punch to see a couple of tourists coming to beautiful war-torn Uganda. The border had opened a few days ago, and nobody really know what was going on inside. I held Kay's hand and gave her a gentle squeeze of reassurance.

We had to declare all our money and weapons, having none of the latter, except a Sudanese lizardskin dagger and a tear gas pen that looked like a big black felt tip marking pen. I declared neither.

I had gotten a visa in Nairobi for Uganda, but had been told that Scandinavians didn't need one. I had wanted them to give Catharina one anyway at the time, but they had refused. The immigration officer at the border was awfully nice but said that Catharina would have to get a visa. "You'll have to go back to Nairobi," said the officer.

"That's OK," said Catharina, who wasn't all that crazy about going to Uganda anyway.

"Are you sure?" I asked.

"Well, I guess I could give her a visa," he said, winking. "I'll have to charge you 200 shillings, though." I figured that up in my head real quick. It was about thirty dollars, quite a bit of money, I thought. But then, since he couldn't find any receipt books because they had all been looted in the war, and because we were the first travelers into Uganda, he decided just to stamp us in.

"With a giant, sincere smile, he tipped his faded official black hat and said, "Enjoy your holiday in Uganda!" as if Uganda had the reputation of being Africa's playground.

We laughed and thanked him. Somehow, I suspected our holiday in Uganda would hold a few surprises for us. Why not enjoy it?

We found a ride with an American Baptist missionary, the first missionary to enter Uganda since the war, he said. He was taking some medical supplies to Kampala, driving in the dark. He had a big two-ton truck and was pulling a small

French auto. Going from the border to Tororo, the first town inside the border, we gave some Tanzanian soldiers in a VW a tow, saw quite a few machine gun nests and a white Peugeot that was completely blasted apart. The missionary, who had lived in Uganda before, said it was one of Amin's cars.

Things were pretty expensive in Uganda, I reflected, or at least some things. A plate of rice and vegetables at the border had cost nine dollars at the official bank rate. On the other hand, a Soviet AK-47 assault rifle could be had for a carton of cigarettes. "It's all in what you value," said Catharina.

We let off the Tanzanians and their VW in Tororo, and then drove through the forest between Tororo and Bugiri with an armed guard as some of Amin's soldiers were said to be lurking in this area still. We passed a number of schools and hotels which were all just empty shells of buildings, burned and blasted out by bazookas. We also passed a small town where Amin's troops had slaughtered 37 schoolboys only two weeks before. It was like driving through a ghost- country inhabited by countless confused ghosts.

The missionary let us off in Jinja and continued on to Kampala, another hour's drive to the west. We checked into the Crane's Head Hotel, the only hotel in Jinja, and found we were the first Europeans to stay there since the beginning of the war. In our hotel room we found some binoculars and a hand grenade in the dresser. I wondered if they were standard issue in Ugandan hotels, like soap and towels.

We walked down to the lake the next morning. This is where the Nile starts its long journey northward: here is a plaque indicating when and where Speke made his historic discovery. On our way back through town to the hotel, a crowd was gathered on the main street. One of Amin's soldiers had been apprehended in civilian's clothes; a Tanzanian soldier kept the crowd from killing him then and there....

We hitched into Kampala that day, getting some rides in the most beat-up, dilapidated vehicles I had ever seen anywhere. One car was an old fifties sedan, the insides completely gone, so we sat on wooden boxes, as did the driver. We ended up going the last twenty miles or so in a bus. A drunken Tanzanian soldier sat next to us and talked loudly in Swahili about "*Muzungus* ", which is Swahili for foreigners. He threw a dead fish onto Kay's lap. We laughed and thanked him in Swahili for the fish. When he finally got off the bus, the people around us apologized for him.

We stayed at the Makere University Guest House while we were in Kampala. Makere University is situated on a hill above Kampala, looking out east over the city. The guest house was very nice and inexpensive. As we were going to bed that night, we heard gunfire, but during the day everything seemed pretty safe. The Tanzanians were certainly in control of the city, and had set up their headquarters at the Nile Hotel. Kampala is a major city of 380,000 people, and wasn't so destroyed by the war as were the smaller towns. Most shops and businesses had been looted, however, in the anarchy that followed Amin's retreat from Kampala.

One day, while sitting at the sidewalk cafe at the Speke Hotel, to our surprise we met a young British couple. They had been teaching English in Uganda and were on holiday, like us. They would have left the country and gone to Kenya, but they had only Ugandan shillings to spend so they had to stay inside the country. I remembered the smiling immigration officer at the border, "Enjoy your stay in Uganda!"

We had coffee and sweet-potato french fries at the restaurant there while we chatted. "Want to see Idi Amin's house?" asked Rose, the wife of the English guy, whose name was Don. She was tall and blonde with glasses, pleasant and fun. She was raised in Kenya and spoke Swahili. Don was thirty, with long brown hair down to his shoulders, though practically bald on top. With thick glasses and an average build, he seemed like the schoolteacher type. "Idi Amin's Kampala residence is just a few blocks away."

Why not? We all trucked off up the street past the modern homes of this upper class neighborhood. Idi's house was a large two-story home, typical of suburban America, but luxurious by African standards. We just walked right up the walk and into the house.

Inside, it looked as if a hurricane had hit it! Papers, photography and movie films were everywhere. You could hardly step without walking on some TOP SECRET or CONFIDENTIAL document. All the furniture and valuable household items had been taken, leaving just the papers. I imagined that Idi Amin's house was one of the first in Kampala to be looted. For some time we looked around and read documents. Suddenly Catharina screamed, "God, look at this!"

We ran downstairs to the garage where Catharina was. Inside was, literally, a mountain of T-shirts. They were different sizes, but all had the same thing printed on them. "Idi Amin Dada; King of Africa, Conqueror of the British Empire." There was a picture of Idi Amin being carried on a palladium by some British businessmen, based a well-known incident in Idi's rule.

These were the only things in Idi's house that no one would take during the looting; pardon the pun, but no one would be caught dead in one of these. I picked one and held it up. It was a Hanes T-shirt, made in America, the label said.

"Go ahead and take one," said a Tanzanian soldier who was suddenly standing behind us. Tall, bulky and friendly, he was dressed in camouflage gear, an officer of some sort. "Go on, take as many as you like. I've got quite a few myself." He spoke English like someone very well educated and was knowledgeable about the war. We all grabbed some T-shirts and he showed us around the grounds.

"Idi Amin is feared by the people here," he went on. "Many actually believe him to be the devil himself. He was certainly evil enough, and he had some sort of mysterious, charmed life."

"Why do you say that?" said Don.

"Well, there were many attempts to assassinate Amin. One time a hand grenade was thrown into the back seat of Idi's limousine where he was sitting. It bounced off his fat stomach and exploded harmlessly outside the car. Another time an assassin was waiting to shoot Amin as he was driven into Kampala from his Entebbe residence, but on that day, Amin decided to drive, and his chauffeur was sitting in the backseat. The chauffeur, instead of Amin, was shot through the head!"

"Wow! He was a pretty lucky guy," said Rose.

"That is nothing," the Tanzanian soldier went on. "Another time, just recently, Amin was driving a Landrover pickup with some troops in the back while our troops were advancing. He was heading back from Entebbe when a Tanzanian mortar landed right in the back of the vehicle Amin was driving. Everyone was killed except Amin! Every time he missed death by a hair! Amin has survived at least fourteen assassination attempts!"

"What about reports that Amin was a cannibal?" asked Rose.

"We believe that Idi Amin was using black magic," said the Tanzanian. "I don't know what you English believe, but in Africa we believe in spirits, demons, and magic, which can be used for good or bad. Idi Amin practiced many black magic rites, including cannibalism. We have reason to believe that he was used by evil witch doctors, most of them only on the spirit plane. There are many interesting parallels between Amin and Hitler, who was also involved in black magic. Hitler also escaped certain death many times, almost as if by miracles. I read a book once on how Hitler used black magic to gain his power in Germany. Like Idi Amin, he was the tool of evil witch doctors!"

We thanked the Tanzanian officer for his talk and went back to the Speke Hotel, our minds blown by what he had said. It seemed incredible, yet possible at the same time.

Catharina and I went back to the university and made dinner on the stove in the kitchen, then went to bed early. The next morning, we met Don and Rose at the Speke Hotel by arrangement, and this time they brought another friend they had met, an Australian journalist who was writing a book on Uganda for the United Nations. He was on his way to the State Research Bureau, near Amin's house, and invited us to go along. The State Research Bureau was heavily guarded by Tanzanian troops, but we were able to get in because the journalist had a note of permission from Jules Nyere, the President of Tanzania himself.

An officer met us at the gate and then took us inside. It had the unassuming look of a large, three-story house in the Kampala suburbs. At one time that's probably all it was. Now it was a captured jail and torture chamber, the former headquarters of Amin's secret police.

The first floor was an armory. The floor was covered with bullets of all shapes and size; crates of grenades, bazooka shells, dynamite and all sorts of arms were everywhere. A carelessly-tossed cigarette butt could have blown us all to smithereens, remarked Don. Nobody laughed.

The first floor could have kept Idi's army cooking for a long time. The second floor was full of files and offices. Papers were strung out all over the floor as in Idi's house. One room was full of old magazines, *Time, Newsweek,* even *Grit.*

One magazine caught my eyes. It was the July, 1977, issue of the Italian periodical *Grand Motel.* On the cover was a painting of Amin casing a lustful gaze at a distraught Jackie Onassis. The headline translated as "Jackie's Latest Problem: Amin wants to Seduce Her!" Hmmm.

The third floor was a couple of jail cells and not much else. We weren't allowed to go down into the basement, where the "personal research " went on, namely torture. All the better, really. The Kenyan newspapers had been full of the most horrifying stories of what went on down there, most of them true, no doubt. I tried to still my imagination.

It was good to be back out in the warm sunlight. I really wondered about the people who saw daylight for the last time as they were dragged into that deceptively plain building for questioning. I remembered a quote from a book, *The Ultimate Frontier,* by Eklal Kueshana: "They seek absolute power over every activity of every person—the kind of limitless power to make another suffer privation or die at their whim. They exult in their terrifying grip on throats of men...Slavery, torture, horrifying savagery and extermination are the devices of sheer evil."

Idi Amin is not the kind of guy to stay low forever. After the war, he escaped to Libya and then went on to Jeddah in Saudi Arabia, where he lived for many years, pious Muslim that he is. On January 3, 1989, ten years after he had been deposed, Idi was arrested in Kinshasa, Zaire as he slipped into the country on a false passport. He was held under arrest for two weeks in Kinshasa while Ugandan and Tanzanian authorities attempted to have him extradited to either Kinshasa or Dar Es Salaam. However, Idi was quietly put on board a Zairean military transport plane on January 20, 1989, and returned to Saudi Arabia. Perhaps even this will not be last we hear of Idi Amin.

Back at Makere University it was dark. As Catharina and I lay in bed, I could once again hear gunshots down the hill in the city. A student had given me this poem that evening:

The sound of rapid gunfire
Wakes me from my sleep
I think of those in torment,
And those lost in the deep.

§§§

It was getting dark. We had just passed through a bombed-out town, its buildings mostly leveled by the shelling of the advancing Tanzanian Army a few weeks before. I had to swerve the little yellow Honda Civic from side to side on the road to avoid shell craters and spots where tanks had torn up the asphalt. Catharina was sitting next to me, the English couple was in the back seat. We were trying to make it to Kabale, a town in southwest Uganda where we thought there might be a still-operating hotel, and eventually to the Montes Lunas.

It was crazy, I thought, and a huge grin lit my face. How insane, how wild and stupid! What am I doing here and how did I get to be driving this car? I asked myself.

The mundane answers to these questions were really very simple. The English couple had rented this car and procured a large ration of gas to make their planed holiday trip to the Volcanic Park in the southwest where they wanted to look for mountain gorillas. There was just one catch; neither of them knew how to drive.

"Oh, well, I know how to drive," I said one morning at the Speke Hotel Cafe.

"Great! Let's all go tomorrow. We've got the car rented and everything, we just have to pick it up."

It seemed so simple...

With a wild screeching of brakes and a hard lurch of the car we came to a stop within an inch of a makeshift road barrier across the two-lane highway. It was quite dark now, and there were no lights or reflectors to indicate a roadblock, just a long pole across the road. Maybe I should have just run it...

Several men came running out from the small guard hut to the road. Two of them had old rifles and they were all dressed as poor farmers; they were not Tanzanian soldiers.

"Get out!" said one of the men in Swahili.

"Don't get out!" said Don from the back seat.

I felt it was better to alleviate the tension, and swiftly popping the lever, jumped out of the car. "Jambo!" I greeted them with the Swahili greeting, my hands outstretched to show that they were empty. The Englishman got out of the car then too, and Catharina got into the back seat. The two rifles were leveled at us.

"We are English teachers on our way to Kabale," Don told them in Swahili.

"We're just coming today from Kampala," I said.

By the smell of their breath, they had been drinking. We gave them our passports and told the girls to stay in the car. They shined their flashlights in the car, looked briefly at our passports, and told us we could go.

"You really should put a light on your roadblock," I said as we got in, "It's quite dangerous like this, you know." One of the guards nodded and gave a grunt. They waved goodbye and we drove on toward Kabale.

About half an hour later, just as we were coming into Kabale, a soldier in camouflage uniform leaped out from the bushes right in front of us and aimed a machine gun straight at the car. It was his way of telling us to stop. He was one roadblock that I had no intention of running. Crouched and ready to fire, he didn't move at all as the car screeched to a halt in front of him. This night driving was getting hazardous, all right.

Don and I got out with our hands up. He ordered everyone out and Rose explained to him in her fluent Swahili just what we were doing there and where we were going. He searched all our luggage and the car very thoroughly, which took him a good half hour, and then he let us go. We were impressed by his efficient, no-nonsense manner. He was friendly and courteous under the circumstances. What did we expect, anyway? Here we were, four loony tourists off looking for mountain gorillas in a country at war. It was kind of hard to believe. I had trouble believing it myself sometimes.

He let us go, and we asked him directions to the White Horse Inn in Kabale, a pleasant little town in the beautiful hill country of southwest Uganda. We checked in at ten o'clock. It was a relief to be in a hotel room. Catharina looked at me and sighed as we got into bed. I put my arms around her. All this danger and excitement made us both pretty romantic. "

"Night driving is always hazardous," she said.

We left for Kisoro around mid-morning the next day. Kisoro is only thirty or so miles from Kabale, but it takes several hours to drive there as the road winds quite a bit through the hills. Arriving at Kisoro in the early afternoon, we checked into the local hotel. The first thing we did was to arrange for a guide to meet us and take us up one of the tall volcanoes nearby to look for gorillas the next day.

Walking around Kisoro that afternoon, we found that it was a very small place. There were a few shops and a small vegetable market. The view of the volcanoes was spectacular. The main feature of the town seemed to be The Traveler's Inn, a small but nice little lodge with an African-style restaurant serving mutokee, a big glob of cooked green bananas served with gravy and Rwandan beer.

At some time, Kisoro was probably quite a tourist spot. There are seven lakes in the vicinity and the valleys are filled with terraced farms and green hills; it was like a mountainous garden of Eden. To the south were a series of tall volcanoes, five or six of them, all inactive, the highest about 15,000 feet. These volcanoes, which straddle the Ugandan-Rwandan border, have a thick, green forest on their slopes, so dense that there are few trails, and only pygmies can live on the hillsides. It is here that the rare mountain gorilla lives. As tall as six and a half feet and weighing as much as 700 pounds, this rather fearsome primate was discovered only in 1901, when a French hunter brought a skin back from Rwanda, proving the "outrageous" legends that zoologists had been pooh-poohing for decades.

We hoped to get close enough to meet one of these beasts. We met our guide early the next morning and he took us to his home near the base of the volcano. He was a large, hearty mountain man, who climbed these volcanoes barefooted. He wore an old tweed sports coat and brown denim pants and carried a long walking staff. He used to have an old rifle, he told us, but the Tanzanians took it away when they arrived. His broad, square face betrayed a certain simple honesty. He seemed a straight-talking fellow who could be trusted. His name was Zacariah and he brought with him a young zoology student from Kampala. We parked our car as close to the base of the volcano as we could and began our hike. Straight up the side of the volcano we went. It was strenuous walking through the terraced fields on the lower slopes that grew bananas, potatoes, carrots and other vegetables. It was hard going and we were soon out of breath. Rose had to stop, and said she couldn't go on any more. She would wait there until we got back.

We rested for a while at the edge of the jungle and then we plunged in. Incredibly thick and overgrown, it all but choked out a slim, faint trail up to the edge of the gorge. Suddenly Zacariah plunged into a bamboo thicket. Following him, we found ourselves hunched over in a tunnel of bamboo that cut vertically across the steep slopes of the volcano. After several hundred yards of bamboo thicket, we arrived at a small clearing where the bushes and small trees had been pushed down. Zacariah set his rusty machete down and reached for a little something on the packed underbrush. Lifting it to his nose, he sniffed it and passed it silently to Don, who took a big whiff.

Coughing and gagging, Don cried, "Phew, what is that stuff?"

"Gorilla shit, and fresh, too," said Zacariah.

"Fresh! I'll say!" said Don, still choking, his face gone blue.

"Here, smell," said Zacariah, passing it over to me.

I took a step backward into the bamboo and fell down. "No thanks, Zacariah," I said, "I believe you." Don was still trying to clear his throat.

"This way," said Zacariah, and we dove into the jungle, our guide hacking his way through the vines, bushes, and thick trees. The forest was constantly in a dim twilight of its own. Sunlight could not penetrate the upper foliage canopy. We never saw the ground, but walked on a dense mat of growth and fallen logs. Sometimes my foot would go down through a hole in the undergrowth and I would sink up to my knee.

We seemed to be following some kind of vague path through the jungle, marked occasionally by a broken branch or slightly compressed spot, but only Zacariah could see it. Suddenly, without any warning, a five-hundred-pound male gorilla leaped out of a tree and landed about thirty feet in front of us. Our hearts stopped and there was a general loosening of the bowels in our party. The gorilla pounded his chest for twenty seconds and then stopped to stare at us. This was gorilla talk for, "You've come far enough!" His clan of babies and mothers must have been behind him somewhere and he had come out to scare us off. It worked.

We all looked at our five-hundred-pound friend for one long silent moment and then started backing up. We headed down the mountain and picked up Rose on the way, all still a little high from the adrenalin rush we had gotten from our meeting in the jungle.

It was late afternoon when the four of us got back to the hotel. We sat in the center courtyard to rest at the lawn chairs and tables, exhausted from our adventure. As we were having our afternoon tea, we watched the volcanoes, which we could see very clearly. A perpetual cloud hung at the top, with a white tail streaming off form the constant wind at the high altitude. We were writing in our diaries, looking at maps and quietly sipping our tea.

What we didn't know was that all day something had been brewing in the hotel. This area of Uganda had never seen any fighting. A small unit of Tanzanians had been sent to it to secure the area and keep the peace. Everything had been peaceful until a week ago when there had been mysterious killings. Some Tanzanian guards at road blocks had been found dead, their throats cut. Therefore, the Tanzanians were naturally a bit on edge and rather suspicious.

Suddenly, four Europeans cruise into town and say they are looking for gorillas! This was an unlikely story, many of the Tanzanians felt, and there had been rumors of Palestinian Liberation Army (PLO) guerrillas fighting with Idi Amin's troops, as well as Rhodesian soldiers. It was a fact that Libya had sent some troops to fight for Amin and the Tanzanians had captured quite a few.

So, two Tanzanian soldiers had been sitting in the restaurant all day drinking beer. They were discussing the many different theories of whether we were actually schoolteachers looking for gorillas, or trained killers that were knocking off Tanzanian soldiers one by one. As the afternoon went on and on, they got more and more drunk. And as they got more and more drunk, they got more and more positive that we were the culprits and would have to be arrested.

Now, we didn't know any of this. And suddenly one of the two men, a sergeant, came into the courtyard and walked up to our table. He said he wanted us to go with him to the barracks, began looking at our maps of Uganda and Africa, decided to seize them, and took Catharina's Swedish diary as well. "Let's go," he commanded in Swahili, and then fell over backwards onto the ground.

This was our first indication that he was very drunk. I got up and went to Catherina's and my room, got the keys and the tear gas pen, which I put in my front pocket, suspecting that this fellow could be dangerous.

Another soldier came out, a young private named John, who spoke pretty good English. He was also drunk, but seemed to be in a better mood than the sergeant. The sergeant, a husky, wide-shouldered guy with very short hair, gave orders to John, a taller, thinner fellow, who translated for us, even though Rose could speak Swahili perfectly well.

"He wants to see your passports," said John. We gave them to the sergeant, who looked at them for a moment and then told Don and me to with him to the parking lot. We walked through the restaurant to the parking lot and the hotel manager followed us. He was a very nice man, concerned and sincere, who watched the whole thing rather nervously. He sent a waiter out to get some military police in case things got out of hand with these drunk soldiers.

The girls sat in the courtyard finishing their tea, while Don and I stood out in the parking lot with the two drunk soldiers and the hotel manager.

"Let's go to the barracks and talk to your commanding officers," I suggested.

"You're under arrest," said the sergeant. "Give me the keys to the car, I'll drive."

Not realizing the circumstances that had led up to all this, and being something of a rebel against authority, I didn't feel I should give him the keys. He was much too drunk to drive the car.

"No," I said, "I'll drive."

It is fortunate that the sergeant did not have a pistol on him, as I grossly misjudged his reaction to my refusing to give him the keys. Unfortunately, he did have a very large and sharp military knife with an eight-inch blade that was definitely designed for killing people.

With a growl, he drew his knife and attacked Don and me with it! He moved toward us with the knife raised over his head, his eyes bulging, saliva foaming at his mouth and a look of murderous frenzy on his face. We were all shocked by his reaction. It was certainly uncalled for. At that point, I had to decide whether to run or to stand and fight him, spraying him with tear gas, dodging his knife and wrestling him to the ground. We were all frightened, including the hotel manager, and it was a very dangerous situation.

I decided to run. Don did too.

We both ran around the Civic with the crazed, knife-wielding sergeant after us! After we ran around the car three times, John, the private, dashed in and grabbed the sergeant. With John gripping the wrist of the sergeant's hand that held the knife, I quickly took the keys out of my pocket and threw them on the ground at the sergeant's feet. The hotel manager was totally aghast. Both Don and I were shaking.

The sergeant took the keys and tried to unlock the door, but he was too drunk. He and John then went back to the restaurant just as two other soldiers came along. The hotel manager stopped them and told them quickly what had happened. They went into the restaurant and the manager told us to go around to the side, grab the girls and lock ourselves in one of the rooms.

We grabbed the girls, who had no idea what was going on, and took them to Don and Rose's room. We played cards until the district commissioner came and told us that everything was all right. We had to meet with the local Tanzanian High Command who apologized to us, and who wanted to know who we actually were and what we were doing in Kisoro anyway.

§§§

Aye! There's places in Africa where you get visions of primeval force... In Africa the Past has hardly stopped breathing.
—Trader Horn

The next day, we headed north in the Honda to Fort Portal, the gateway to the Mountains of the Moon, the Ruwenzori Mountains. On the north side of the mountains is the Ituri Forest, where pygmies live in their ancient traditional lifestyle. Mount Ruwenzori is the third highest mountain in Africa at 16,763 feet. It is covered by clouds 360 days out of the year and has one of the highest rainfalls in the world. The very top of Mt. Ruwenzori can often be seen sticking out of the clouds. The hot springs of Bundibugyo and the Toro Game Reserve are also good sights. The women of this region are famous all over Uganda for their beauty and dignity.

In the afternoon, we sat having tea on the veranda having tea and looking at the Mountains of the Moon, shrouded with clouds to our west. They were an impressive site.

"Out there are the vast swamps and rivers of the Congo," said Don. "And in those swamps are all kinds of nameless animals, huge, man-eating animals!"

"Now Don, what kind of animals are you talking about?" said Rose, taking a sip of tea.

"Dinosaurs! That's what I'm talking about! Swamps full of dinosaurs, brontosauruses and tyrannosauruses and critters like that! That's what I mean!"

"Don, be quiet!" said Rose. "You know perfectly well that there aren't dinosaurs out there. No one has ever seen a dinosaur. They've been extinct for 65 million years. It's scientific fact!"

"I'm serious, Rose! Don't tell me to shut up!" said Don. "Here, let's ask David, he's a rational fellow. I ask you, old chap, is it scientific bloody fact that dinosaurs have been extinct for 65 million bloody years?" With that Don poured himself another cup of tea.

"I'm sorry, Rose, I'm afraid Don's right, it is not a scientific fact that dinosaurs became extinct 65 millions of years ago," I said.

Don smiled as Rose snorted and said, "How can you say that? Why, everyone is taught that dinosaurs are extinct. Isn't that true, Catharina?"

Catharina just looked at me, a bit bewildered. Calmly I answered Rose, with a bit of a smile. This was a fascinating subject for me, and it's always fun to find someone with whom one can discuss dinosaurs intelligently. "It is a theory that dinosaurs became extinct 65 million years ago, Rose, not scientific fact. There is a great difference. Most scientists do believe that dinosaurs are extinct, and in the strange dating of uniformitarian geology, they give that extinction date as an arbitrary 65 million years ago. There is nothing factual about it. It is just one of many theories, though I do admit that, like many scientific theories, it is frequently taught as fact, and is treated like dogma in many cases. What is a fact, however, is that dinosaurs, literal living dinosaurs, are reported around the world every year, and in some cases, what we would call living dinosaurs have in fact been captured."

"Right!" said Don, "why just look at the capture of a coelacanth in 1938 off South Africa. They were supposed to be extinct for 70 million years, yet they're still around."

"Well, coelacanth's aren't actually dinosaurs," I said, "but they do prove our point. The dating and even the mechanism for creating fossils is really not known. There is no way to carbon date a fossil. The age, like so-many-million- years for such-and-such a fossil is pretty arbitrary. Those dates are taken from geological strata and assumes that geological sedimentation and change happens very slowly, over millions of years. When paleontologists find a fossil and it has ten feet of gravel over it, with a few layers of clay and what not, the paleontologists, all uniformitarianists, say to themselves, 'it would take so-many millions of years for that ten feet of gravel to accumulate over this fossil. Therefore, it is so-many millions of years old.' However, that ten feet of gravel could have been dumped onto the bones of some dinosaur in a few days or hours in a cataclysmic geological change. What to the uniformitarian geologist is millions of years of built-up sediment, might really be the sudden eruption of a volcano or the dumping of sediment from a tidal wave to the cataclysmic geologist.

"In fact, the very cause of fossilization has to do with cataclysms. When they die, animals do not normally become fossils. Fossils are anomalies. When hundreds of thousands of buffalo were slaughtered on the great plains of America in the early part of the last century, and just left there to decay, not one of them became a fossil! That is because under normal circumstances a dead animal will decay and turn to dust. Only an animal that suddenly dies and is instantly covered by volcanic ash or debris from a tidal wave or landslide will become a fossil. It is even worse for marine fossils. An ocean bed must suddenly rise from the ocean and then the marine animals, left high and dry, must then be covered in volcanic ash or other debris, before they can decay. The slow draining of prehistoric seas would not create a fossil! Therefore you can see that the same mechanism that creates fossils, cataclysms, is probably the mechanism causing the disappearance of these animals. Witness woolly mammoths flash-frozen in the Arctic with buttercups in their stomachs. They were apparently flash-frozen in a sliding of the earth's crust. The huge piles of mammoth bones and ivory found in huge piles in Alaska and Siberia seem to be the result of the same pole shift and a resulting tidal wave that pulverized the animals and left them in huge piles on the tundra. Why shouldn't dinosaurs have survived in remote swamps in Africa, South America and other places?"

"Exactly!" said Don. "People have been searching for dinosaurs for years in the Congo. In 1919 Captain Leicester Stevens held a press conference at Waterloo station in London and announced that he was setting off, with his half-wolf companion, to hunt dinosaurs in Africa. Even the great animal collector Carl Hagenbeck, probably the greatest animal collector of all time, believed that dinosaurs still lived in parts of Africa. He claimed that he had two different eyewitness accounts of brontosaurus-type dinosaurs in parts of Northern Rhodesia and the Congo."

"Even here in Uganda people have reported such animals," I offered. "In 1902 Sir Harry Johnston, the man who discovered another large unknown animal, the okapi, drew attention to a large unknown creature living in Lake Victoria known as the *Lukwata.* Other natives around the lake told of a similar giant creature called the *Dingonek* that was fifteen feet long with scales, a long broad tale and huge footprints like a hippos but with claws.[94]

"But the most famous of all the dinosaurs is the Chipekwe and the Mokele Mbembe," said Don. "It was a German expedition to the southern Cameroons in 1913 that first brought back tales of the Mokele Mbembe. At that time, the Cameroons included what is now the northern Congo region. From different native guides who could not have known each other, the expedition came up with this description of the Mokele Mbembe, 'brownish-gray color with a smooth skin, its size approximating that of an elephant; at least that of a hippopotamus. It is said to have a long and very flexible neck and only one tooth but a very long one; some say it is a horn. A few spoke about a long muscular tail like that of an alligator. Canoes coming near it are said to be doomed; the animal is said to attack the vessels at once and to kill the crews, but without eating the bodies. The creature is said to live in the caves that have been washed out by the river in the clay of its shores at sharp bends. It is said to climb the shore even at daytime in search of food; its diet is said to be entirely vegetable.'"[94,96,98]

"I didn't realize you were such an expert on dinosaurs," laughed Rose.

"I'm a latent dinosaur hunter me-self, lassie," Don laughed and gave her a kiss on the cheek.

"What about the Chipekwe?" asked Catharina. "If the Mokele Mbembe lives in the swamps of Zaire, the Congo and Cameroon, where does the Chipekwe live?"

"The Chipekwe is said to live in swamps around Zambia, Angola and southern Zaire," said Don, "and there have even been reports of the natives killing a few. It is a huge animal, with a smooth body and a head with a single ivory horn."

"Authors like Ivan T. Sanderson and Bernard Heuvelmans like to compare the Chipekwe with the dragon on the famous Ishtar gate at Babylon," I chipped in, using a bit of my obscure ancient history culled from my various travels. "This animal, called a Sirrush by the Babylonians was supposedly kept in a dark cavern in the temple and was worshiped by the Babylonians. In the book of *Bel and the Dragon* in the Apocrypha of the Old Testament (the Apocrypha is an appendix to the Old Testament, *Bel and the Dragon* is the name given to Daniel chapter 14), Daniel is in Babylon and is shown the 'great dragon which they of Babylon worshiped. And the king said unto Daniel, wilt thou also say that this is of brass? Lo, he liveth, he eateth and drinketh. Thou canst not say that he is no living god: therefore worship him.' Daniel tells the king that he shall slay the dragon, and takes pitch, fat and hair, sews it up and feeds it to the dragon who then dies of indigestion. Daniel is then thrown to the lions, but is protected by god."

"You mean to tell me that the ancient Babylonians worshiped a brontosaurus!" exclaimed Rose.

"That's what Heuvelmans, Sanderson, Willy Ley and Roy Mackal all believe," I said. "The relief on the Ishtar gate seems to be of a real animal with a single horn, scales, feet like a bird or lizard and a snake's head on a long neck. A forked tongue darts from its mouth. Depicted on the gate is the Sirrush alternated with the giant extinct ox called an *auroch*. There are also lions on part of the gate. The famous Swiss zoologist Bernard Heuvelmans says that the Babylonian Sirrush is not an accurate picture of an Iguanadon or a Ceratosurus, but its a good stylized likeness. 'The Sirrush could have lived in Central Africa, where it has been proved that the Chaldeans went, and where they could have seen a giant lizard (and captured it). When Hans Schomburgh came back to Europe with native tales about the Chipekwe, he also brought back several glazed bricks of the same type as those on the Ishtar Gate!'"[94]

"Wow," said Don, "there's fuel for your lost cities material! A lost Babylonian city in the jungles of Central Africa from 3,000 B.C. Perhaps it was a combination mining city and zoological outpost, gathering exotic animals, perhaps slaves, and even dinosaurs! Now, that's my kind of history!"

"Well, that's not my kind of history," exclaimed Rose. "I refuse to have any part of the history you two expound. As far as I'm concerned, dinosaurs are extinct, and that's all there is to it. When I see a dinosaur, then maybe I'll believe that they are still alive!" With that Rose got up to go to her room. Don followed her, and Catharina decided to take a shower before dinner. I sat on the veranda watching the sunset and musing over dinosaurs at the Mountains of the Moon. It was one of my favorite subjects. Perhaps I had just seen too many King Kong movies.

The dinosaur hunters of Babylon started a craze that continues to this day. University of Chicago zoologist Dr. Roy Mackal has been leading expeditions into the People's Democratic Republic of the Congo since February of 1980 in search of Mokele Mbembe. His latest book on the subject, *A Living Dinosaur,* [96] chronicles his various expeditions and researches in quest of this legendary creature. Mackal's book is a wealth of personal adventure and interviews with natives and their descriptions of unusual animals in the swamps of the Congo. According to Mackal, there is not just one dinosaur out there in the swamps, he says that the natives chronicle at least six different types of unknown animals including the *Mokele Mbembe,* a brontosaurus-type animal; the *Emela-ntouka,* the "Killer of Elephants", a triceratops-type animal; the *Mbielu-mbielu-mbiellu,* the "animal with planks growing out of its back", a stegasaurus-type animal; the *Nguma-monene,* "the snake dinosaur"; *Ndendeki,* a giant turtle; and *Mahamba,* a giant crocodile.

Mackal and his expeditions in search of Mokele Mbembe have been the subject of at least one recent movie, *Baby,* by Walt Disney films, in which a baby brontosaurus is captured and brought back to the U.S. As the last rays of the sun shown on the ever present clouds of the Ruwenzori Mountains, I wondered if some pygmy tribe was right now moving their entire village to the scene of a recently slaughtered brontosaurus in some remote Congolese swamp. As Trader Horn had said, "In Africa, the past has hardly stopped breathing."

We said good bye to Rose and Don the next morning. They drove back to Kampala and we were going to head for Tanzania. After two days in Fort Portal, we headed back south to Kasese, a couple of hours from Fort Portal near the Zaire border. Catharina and I spent the night at the Margerity Hotel, the first tourists to stay there in six years!

Catharina and I hitched out of Kasese toward Mbarara in the south, getting a long ride in a big five-ton truck full of Ugandans. In Mbarara, we stayed one night, and then finally we were on the road out of Uganda on our way to Tanzania. We sat in the empty back of a Tanzanian army truck, huddling together in the cool evening while, up front, two Tanzanian soldiers on leave were happy as anything to be heading home. There wasn't even a border post on this main road to Tanzania so we never got stamped out of Uganda, or into Tanzania either. No matter. It had been an exciting holiday in Uganda, that was for sure. Catharina put her head on my shoulder and fell asleep; the stars were bright and white. It was fine to be alive.

Fortean Times

SUE NO. 34 | The Journal of Strange Phenomena. | PRICE: 95p. $2·50

IN SEARCH OF DINOSAURS

+ Phantom Hitch-hikers; Mystery Cats; Beached Whales; Ice Falls; Synchronicity.

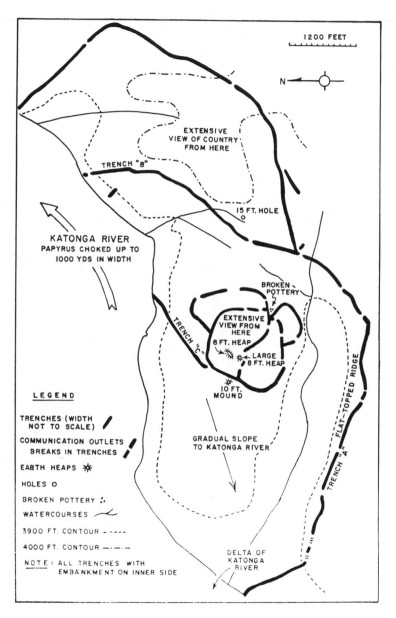

Extensive earthworks at Bigo, Uganda. Attributed by Bugandan historians to a race of giants, the complex includes rock-cut canals.

It may seem fantastic to suggest that there could be apelike creatures eight feet tall and weighing more than a thousand pounds, but actually a big male gorilla like this one would fill the bill if he had legs as long in proportion to the rest of his body as a man's are.

MAGAZINE SECTION.

THE NEW YORK HERALD.

PART I. PAGES 1 TO 8.

NEW YORK, SUNDAY, FEBRUARY 13, 1920.— BY THE SUPPLEMENT, THE NEW YORK HERALD COMPANY.

PRICE FIVE CENTS.

Is a BRONTOSAURUS
ROAMING AFRICA'S WILDS?

HERR CARL HAGENBECK Deeply Interested in Stories to the Effect That a Creature Which Is Supposed to Have Disappeared Millions of Years Ago Has Been Found Alive in the Dim Recesses of the Dark Continent—Dr. W. D. Matthew, of the American Museum, Thinks a Native Has Seen a Gigantic Sea Cow in an Inland Lake.

THE PLESIOSAURUS
Copyright by the American
Museum of Natural History

Mysterious creature, resembling the Brontosaurus, drawn from a description by two natives of Northern Rhodesia.

(CONTINUED ON PAGE TWO.)

New York Herald for February 13, 1920.

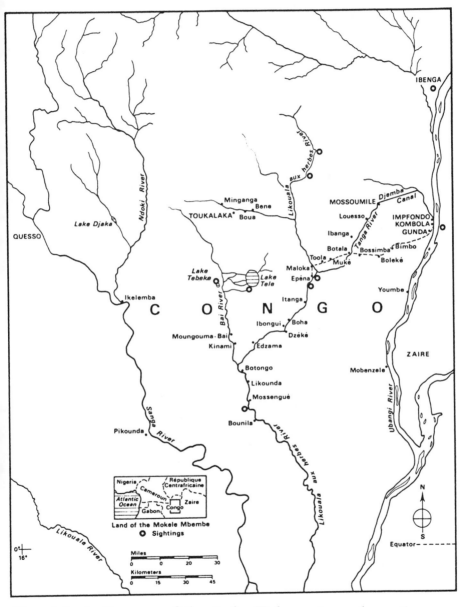

Roy Mackal's map of the Lake Tele region where the Mokele Mbembe is reported to live.

The *Mahamba* based on descriptions of native African observers.

Mackal's illustration of one of the unusual animals reported in the Lake Tele area. From his book, *A Living Dinosaur?*

Carved dragon from the Ishtar Gate in Babylon. Did the Babylonians go to Central Africa to capture this strange beast?

A tyrannosaurus attacks a brontosaurus in this Charles Knight drawing.

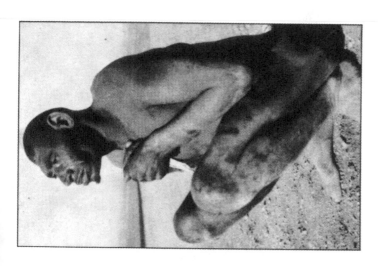

Photos of "Neanderthal" man taken at the Skura Oasis in Morocco by explorer Willy Fasso.

Chapter 13

TANGANIKA & ZANZIBAR: TALES FROM KILIMANJARO

We carry with us the wonders
we seek without us;
there is all Africa
and her prodigies in us...
—*Sir Thomas Browne*

I had just arrived in Tanzania and was sitting on my pack, the tall savanna grass all around me, the plain dotted with thorn trees looking like gigantic surrealistic mushrooms. In the distance, to the south, I could see a couple of giraffes grazing on the tender top leaves of the trees, their long necks stretched up and towering above the plain like polka-dotted construction cranes.

It felt good to be on the edge of Serengeti Park, lounging the day away on a lonely crossroads waiting for a lift. It could easily take all day to get a ride—maybe even two or three days. This is because there aren't any cars or public transportation of any kind—no buses or taxis of anything. If I were lucky, a Landrover full of tourists might come along or maybe a truck full of Africans. Hell, I wasn't fussy. And I wasn't in a big hurry, either. This was a good opportunity to work on patience, a virtue that Western people often lack, at least by African and Asian standards.

Catharina had met some Danish Overseas Volunteers, sort of like our Peace Corps workers, at a small village near Mwanza and we had mutually decided that she should remain with them for a while. So I went on by myself to hitch through Serengeti and climb Kilimanjaro. I had mixed feelings about leaving Catharina; I was somewhat relieved to be on my own again, but also sad and lonely that we weren't traveling together. She felt I was too reckless and adventuresome. She was mostly interested in finding a nice beach and soaking up the sun. Since we couldn't find a compromise, mostly due to my stubbornness, we agreed to separate and meet again on the coast in a couple of weeks.

Two days of hitching had brought me here. At times I was elated to be back living for the moment, with no immediate cares or concerns. At others I was sorry I hadn't made more effort to make both of us happy, and wished we were still traveling together. Staring out into the yellowish-brown tall grass, I was lost in thought about my life, my relationships, my future—certainly I wasn't living in the immediate present.

A low droning sound brought me out of my contemplation and I stood up to peer over the grass. It was a truck! Oh boy! I had only been waiting about eight hours for a ride, and here came a big, rusty, dusty red cattle truck, right my way! It looked beautiful.

Smiling foolishly out of sheer joy, I began jumping up and down and waving my arms wildly in true African hitchhiking style. There was no way in the world they weren't going to stop. I made sure of that.

With a grinding of gears and a squealing of well-worn brakes, the ancient cattle flatbed truck ground to a halt just in front of me. It was an old British Bedford lorry, circa the 1950's I guessed, judging by its style and front grille. In the back, standing up in back of the cab, were half a dozen Africans in various stages of dress and undress, all covered with gray dust, all smiling big white grins as they gazed at the crazy young hitchhiker standing in the grass at the edge of the game park.

I grabbed my pack and slung it over my shoulder, striding happily up to the driver and cab.

"Where are you going?" called a short, sort of purple fellow from up in the cattle cage.

"Seronera!" I answered, speaking of the main village in the center of Serengeti Park where there was a youth hostel, a game lodge and a small cooperative village called an Ujama.

The English-speaking guy in the back said a few words to the driver and they had a quick exchange.

"Come on," he called finally, and the driver smiled. I handed my backpack up to the fellows in the back and climbed up myself. Before I was inside, the truck was again hopping down the dusty wheel tracks that pass for roads through the African bush.

"It's great to have a ride," I said, holding onto the front railing and moving next to the man I had spoken with, who was standing just behind the cab.

Everyone was all smiles. Most of the men in the back were wearing traditional Masai tribal garb, a faded red cotton loincloth with a strap over one shoulder. They wore sandals made from old tire treads and strips of inner tube on their hardened, cracked feet. They were carrying spears and their hair was long and braided, tied back behind their heads with string and an occasional glass bead.

With broad grins of delight and amazement, they all stared at me, and I stared at them.

"Jambo!" we all said in greeting together, and I laughed. They laughed too. The African standing next to me put his arm on my shoulder and asked, "Where are you from, brother?"

He was dressed differently, in a blue flowery polyester shirt, red polyester slacks and machine-made plastic shoes; he had a certain city look about him. Spoke good English, too.

"The United States of America!" I said. "Where are you from?"

The Republic of Tanzania," he said, white gritty teeth showing. The wind from the truck blew in our faces.

"Where in Tanzania?"

"Arusha," he said, "I'm just here to visit some relatives in this part of Tanzania, and now I'm heading back to Arusha."

Suddenly we had to duck a low bunch of branches as the truck plowed through the bush. I told him about my trip to Uganda and my breakup with Catharina.

"Ah! Well, you'll just have to change your attitudes if you're sad," he said.

"I'm not sad," I told him.

"You look sad," he said. His thin face was young but full of character. He was obviously well educated for a Tanzanian, and he seemed to have that good native African savvy that western educated Africans in the cities often didn't have. "Your thinking is important to what happens to you in life."

"Are you sure?" I said. "I mean, aren't we really just victims of fate? How can we be responsible for our situation in life?"

"I don't believe in chance," said my friend. "Our thoughts and desires usually cause the things that are happening in our lives. Like your break-up with your girl-friend—you must have been wanting that on some level."

"I suppose you're right," I said, slightly distracted by a small herd of zebra to the left of the truck. The sun was getting low on the horizon; the great, flat grassy plain was blending into the dusk. There were giraffes, wildebeest, zebra, and gazelle to be seen all around, in the distance and next to the truck as we passed by them.

We came into a small town just as the sun was setting, one part of the sky a deep yellow and the other orange-red. I jumped off the truck to take a photograph, and as I stood there, the English-speaking man told me we would have to spend the night here, because the truck didn't have any headlights and couldn't drive at night.

"OK," I said, "by the way, my name is Dave."

"Mine is Samuel," he said. "Here, follow me, I'll find us a place to stay."

We walked into the center of the village, which was quite small, maybe twenty or so grass huts, and a few other small, square cinderblock buildings. The last glow of the sunset was fading away and I was amazed to see the that the entire village was gathered around a fire, and everyone was drunk! They had been drinking the local African millet-corn beer, easy to make and plentiful, and were all fairly soused.

Samuel said a few things in the local dialect and suddenly a great fight started between two women who began shouting and arguing and pushing each other. Somehow I knew it was about me, and became very uncomfortable, wondering what I had done wrong, and what kind of trouble I was in now.

"They're arguing about you," said Samuel.

"I know," I said, "What did I do wrong?"

"Nothing," he replied, "They're arguing about whose guest you are going to be tonight. The one lady says she saw you first—do you remember taking a picture of her at sunset? And the other lady says that she is the wife of the chief and that you will stay in their hut!"

"Oh, is that all!" I was greatly relieved and accepted a bowl of the brown milled sludge that was their beer. Like the palm wine in Kenya, this has a sour fermented taste, reminding me of stomach fluids. I could drink the stuff down, though, having gotten used to it before at similar native drinking sessions in Sudan.

Samuel and I sat by the fire and drank the beer. I made a pot of soup out of some food I kept in my pack while the chief's wife settled the argument with the other lady. It was decided that I (and Samuel too) would stay in the chief's hut.

As I got into my sleeping bag that night, in the doorway of the hut appeared two nubile young teenage girls who took a particular interest in me and sat next to me. One offered me more food, and when I told her I was full, began stroking my stomach. I realized they were drunk (like everyone else in the village) and was a bit afraid that one or both were going to attempt to crawl into my sleeping bag.

They didn't, and as I drifted off to sleep I recalled a sign I had seen at the entrance of the game park when we had crossed it late that afternoon: HITCHHIKERS FORBIDDEN. IF FOUND INSIDE THE PARK THEY WILL BE PROSECUTED.

I was already on the run.

§§§

Tanzania is a nice country, I thought, as I stood at the crossroads of Seronera trying to get a ride south toward Oldavai Gorge and Ngorongoro Crater. I had been waiting all day for some kind of ride, but so far had not had any luck. Only one car had passed, containing a couple of Japanese tourists who looked right through me. I tossed a pebble at a sign that said NGORONGORO: 92 km. I wouldn't get there today, I thought.

Tanzania is the only country in the world to set aside nearly one quarter of its land as national parks, game preserves, and forest reserves. It is a beautiful land, rich in wildlife and scenery. It is perhaps the ideal African country from the tourists' point of view.

It is the largest country in east Africa, cut on its western border by the Rift Valley, forming Lakes Malawi, Tanganyika, and Victoria. The southern central area is a high, arid plateau, while the northern area is known as the Masai Steppe, a semi-arid plateau with occasional hills and mountains, including Mt. Kilimanjaro, the highest peak in Africa and still an active volcano (a fact recently discovered). Then there is the narrow coastal plain, a hot, humid strip which varies in width from 10 to 30 miles.

The United Republic of Tanzania was formed in 1964 when Tanganyika, the mainland which achieved independence in 1961, and the island of Zanzibar agreed to unite, forming a republic. Julius Nyere, a politically active and idealistic schoolteacher, was elected the first prime minister. He remains the prime minister today, having been reelected no less than five times, indicating his popularity among his people.

Nyere was instrumental in setting forward a form of African socialism known as ujama, which means "familyhood" in Swahili. It is an agricultural commune system based on "love, sharing, and work." It is similar to the better-known kibbutz system of Israel. On a broader scale, ujama means a joint effort by which all Tanzanians can theoretically gain an equal share of the nation's resources. In 1973, the ruling party, the Tanganyika African National Union (TANU) decided that all peasants had to join ujama villages, making the movement no longer voluntary.

Nyere's government has progressively nationalized private businesses, banks, and privately owned homes. They have sought aid from all countries, but, because of its strict policy of neutrality, its major financiers and aid donors are the Scandinavians and the Chinese, who built the Tazara Railway between Dar Es Salaam and Zambia.

Back in Mwanza, the fourth largest town in Tanzania, we stayed with the driver of the truck from Uganda, who shared an apartment with a friend. A huge, friendly, wonderful guy, he had us all go down to a photography shop and have our photo taken together before we left Mwanza. I finally got a lift in the late afternoon with a Landrover on its way to a safari camp outside the southern border of the camp. It was a couple of hours from Seronera to the southern entrance of the gate where I was left off. The driver, half Asian-Indian and half African, gave me a terse goodbye. Maybe he'd read the sign about no hitchhiking in the park. His helper, a cheerful African, gave me a bright toothy smile and a grand wave. "Take care!" he called as they drove off, "and good luck!"

"Bye," I called, "and thanks for the ride!"

There were still a few hours before sunset, and I could see that I wasn't going to get a ride any farther that day. A tall articulate game warden at the small fortress-like ranger station asked me what I was doing, where I was going, and where I was going to stay.

"I'm on my way to Ngorongoro Crater," I told him, and said that I would camp out by the park entrance if that was OK with him.

"There are many wild animals," he said, "it's too dangerous."

"I have a tent," I told him, which wasn't strictly true, as I didn't have one with me at the time.

He agreed to let me camp out, and I walked up a hill by the ranger station where I got a great view of the Serengeti Plain at dusk, wide and grassy, stretching out in a sophisticated blend of browns, greens, yellows and blacks in every direction. Thorn trees dotted the savanna like chess players on a faded mahogany board. There were herds of zebra, wildebeest, gazelles, and giraffes in the distance, and even a small pride of lions lounging around the carcass of a wildebeest in the shade of a tree at the bottom of the hill. They looked satiated, so

I wasn't too scared that they would mistake me for an afternoon snack. Even so, I found myself wondering, as I strolled back to the ranger station, whether tear gas was effective against a charging lion.

Back at the ranger station, I played a game of frisbee on the lawn with a young boy in the growing twilight. There were quite a few people here, I noticed, at least one family, plus a half-dozen workers, a ranger and a cook.

Later I tried to light a fire behind the back of the ranger station. As it was too windy to get a good blaze going, I went into the station and asked the cook if I could borrow some hot coals to start my fire. Instead he suggested that I cook my food inside on his little charcoal grill, and the chief ranger said I could stay in his room since his roommate was gone for a week and therefore his bed was empty.

As my soup simmered on the fire in the soot-black billy can I carried with me, I chatted with the chief ranger about Tanzania.

He talked about being a ranger, and how Somali poachers would hunt in the park. "There will usually be four or five of them," he said, "each armed with a machine gun." He got out his ancient carbine rifle and showed it to me. It looked like a relict from the first World War.

"All that we rangers have is a bolt action like this! My best friend was killed by poachers! They machine-gunned a whole herd of elephants and my friend and two other rangers crept up on them to arrest them and they were all killed."

"What happened to the poachers?" I asked.

"They escaped. Always do," he lamented.

I asked him what he thought about the government, Nyere, and the economy. Nyere is often called affectionately Mwalimu, or teacher.

"Mwalimu is wonderful, but he is too idealistic and not practical enough. He does not know what is happening in the country. He thinks everyone is as idealistic as him. When ujama was started, everyone loved it, even the privileged class. They thought ujama was for the poor people in the bush, not for them. But now the privileged must also be part of ujama, they must move to the communal villages and live without their cars and other luxuries.

"Here in the bush, there do seem to be a lot of shortages of consumer goods," I said.

"The whole country is corrupt, except Mwalimu," said the ranger bitterly. "Everyone is trying to make an extra hundred shillings on the side to get ahead. The government controls the prices of everything, but the demand exceeds the supply. People will buy out a stock and then sell it again on the black market to their friends, doubling or tripling their money. This is so common, it makes me sick!"

We sat in silence for a while. The glow of the hot coals caused red shadows to climb and dance on his face. His eyes were wide and bright; sweat dripped off his broad, black nose onto his upper lip. After finishing a cup of tea he had made for us, we retired to our rooms.

When I awoke in the morning, he was busy making crepes. He handed me a cup of sweet milky tea as I sat down next to him by the stove. Such a refined, loving person, I thought. Not what I had expected to find out here in the wilds of Tanzania. He was like the Boston lawyer who decided to take up fur trapping in the Yukon.

He gave me a warm, firm handshake when I bid him goodbye. Swinging my pack on my back, I began walking east down the road away from the park.

It wasn't long before a Landrover came by and picked me up. It was five park employees on their way to Ngorongoro, only about fifty miles away. I sat in the back on an old tire, crouched over and uncomfortable but glad for the lift.

We stopped briefly at Oldavai Gorge, one of the most important archaeological sites in the world. Because of finds at Olduvai and other areas in East Africa, it has been theorized (and virtually accepted as dogma) that the "cradle of mankind" was an area east of a line drawn from Khartoum to Capetown. In the remote past, the area had been a lake and early homonids and other animals clustered around the shores. It is an area of heavy volcanic and earthquake activity and frequently volcanic ash would cover unfortunate dwellers by the lake, creating fossils. Eventually the lake dried up and and a grassy plain (the Serengeti) concealed the thick stream and lake deposits below, until a massive earthquake uplifted part of the plain. As a result, a new stream cut a gorge about 75 meters (about 200 feet) deep through the old deposits, revealing a wealth of fossils.

The area was first discovered by a German entomologist named Kattwinkel while he was chasing a butterfly just before World War I. He took some fossils back to Germany and as a result the German geologist Hans Reck visited the gorge and revealed that it was an important source of fossils from the Pleistocene epoch (the Pleistocene is a vague, though important "geological period" starting about 1,600,000 years ago and ending only 8,000 years ago). Because of World War I, Reck was not able to return to Olduvai until 1931.[18]

It was in 1931 that Reck first introduced the young Kenyan archaeologist Louis Leakey to the site, but Reck told him that there was no evidence of human culture, namely artifacts, in the gorge deposits; there were only fossil bones. Leakey was to prove Reck wrong. Stone tools, skeletons of men and other animals, including giant baboons, have been found in these deposits. In 1959, Dr. and Mrs. Leakey unearthed a skull which they dated at over one and three quarter million years old. This skull was said to belong to "Nutcracker Man," or *Zinjanthropus,* because its enormous jaws and huge molar teeth. Zinj is the ancient Arabic name for East Africa, and the Indian Ocean was known as the Sea of Zinj, and so the Leakeys named their new find, *Zinjanthropus* or "East Africa man".

For creationists who believed that the world had been created by God in 4004 B.C. (a Tuesday, think many ministers) the idea of some primitive proto-man being around 1.7 million years ago rocked their boat a little, as one can imagine. The dating is tricky, at best, but it is certain that the fossils discovered are very old, yet some of the conclusions are questionable. Nothing over about 70,000 years old can be carbon-dated. There are two methods used to date *Zinjanthropus.* One is potassium-argon dating and the other is palaeomagnetism, sometimes called archaeomagnetism. Palaeomagnetism is quite questionable, and both forms of dating require an assumption of uniformitarian geology, that is, the geological changes in the earth are slow, gradual, and occur gradually over millions of years. Palaeomagnetism as a technique for dating depends on knowing that the Earth's magnetic field has subject to changes in direction and intensity over time and makes use of the magnetism in fossil material. When rocks or clay are heated to a certain temperature the magnetism is destroyed, but returns when the object cools down again, taking on the direction and strength of the magnetic field that the object is lying in. It is therefore theoretically possible to date an object by fitting it in to *the known curve of the earth's magnetic field* at any given epoch of time.

The main problem with this is that we really don't know the curve of the earth's magnetic field thousands or million of years ago, except for the fact that it was not the same as today. In cataclysmic geology, with the earth's crust slipping and rotating on the inner core of the earth, the earth's magnetic field is changing constantly, every ten thousand years or less. This possibility is not considered in the field of palaeomagnetism, and subsequently makes any dating questionable. It does however at least offer the clue that the object being so dated was before the last shift of the earth's magnetic field (and/or crust) which at least helps establish some antiquity.

Potassium-argon dating is somewhat more accurate and scientific. It is based on the decay of a radioactive material into a non-radioactive material at a known rate. Potassium is present in most rocks and minerals and has a single radioactive isotope, ^{40}K, which decays in Calcium 40 and Argon 40. The eleven percent that decays into the gas argon can be measured, together with the potassium left in the sample, to give a date. At Olduvai, different lava flows above and below the fossil remains were dated by Potassium-argon decay. The layer below the fossils was dated at 1.9 million years old, and the all-important layer above the fossils was dated at 1.7 million years old. Other skulls, found in Northern Kenya, Ethiopia and South Africa, have been dated at 1.5 to 2.5 million years old, while one skull of an *Australopithicus afarensis* from the Lake Turkana area has been dated at 5.5 million years old, though these dates come from strata and the assumption in uniformitarian geology that each strata takes many hundreds of thousands or millions of years to be created. For instance, according to the Potassium-argon dating of the layers surrounding *Zinjanthropus*, these volcanic eruptions took place 200 thousand years apart! 200 thousand years is a long time (especially for a cataclysmist) for the geologically unstable area in the Rift Valley not to have any activity. Even uniformitarian geologists would probably find this a bit far-fetched.

At any rate, a great age for these fossils of what appear to be primitive ancestors of man has been established. The fact that many layers of sediment, volcanic ash and lava flows have covered the remains is testimony to at least a certain amount of antiquity. Were primitive homonids with stone tools lurking around East Africa hunting, gathering and fishing? Certainly! What is more incredible is that there is a certain amount of evidence that these same "proto-humans" are still around today doing exactly the same thing!

Strange, hairy "men" still apparently roam the world long after their supposed extinction, and are reported all over the globe every year. Tales of big foot, yetis, yowies, almas and other strange "wild men" who are often very tall, though in other cases of a more normal stature, are taken seriously by many scientists and photos do indeed exist. The noted British anthropologist Myra Shackley has proposed that yetis and bigfoot are the remnant of Neanderthal men. More incredibly, in the 1950's in Morocco, the French anthropologist at a remote oasis known as Skura on the edge of the Sahara, east of the Atlas, was able to photograph the last descendant of a "Neanderthal tribe" whose name was apparently "Azzo". Azzo was the spitting image traditional pictures of Neanderthal man, with a sloping forehead and large brow ridge; naked, and using primitive tools.[100]

Even Neanderthal men may have undeserved reputations. In his book, *Shanidar, The First Flower People*,[113] Ralph Solecki discusses the important archaeological excavation of the Shanidar cave in the Kurdistan area of Iraq. According to Solecki, the inhabitants of the cave were Neanderthal men and women who lived there forty thousand years ago. Most importantly, thought Solecki, these "primitive" Neanderthal men buried their dead in a huge pile of flowers! Solecki's book paints a picture of Neanderthal men as sensitive, art and nature-loving people who cared for the sick and aged, used tools and fire and buried their dead in elaborate flower ceremonies.

Considering both the way we date things geologically, and the possibility thatour assumptions about early man's extinction may be completely erroneous, the Olduvai dates may be way off and the presumption that East Africa is the home of mankind may also be incorrect. More interesting is the question, typically skirted by orthodox archaeologists, of why, if proto-men were making tools and living together around some lake in East Africa 1.7 million years ago, they did not advance one iota further in terms of civilization during the next 1.69 million years? When the so-called Pleistocene period ended a mere 10,000 years ago, these same "experts" tell us that mankind was essentially at the same state of "cultural evolution": living in caves, using stone tools, and occasionally grunting. Furthermore, the development of the different races and the strange distribution of those races around the world cannot be solved if East Africa is the cradle of civilization. Many scientists have sought to place the origin of man in the

Americas, and in Central Asia, and the mystics find it in a lost continent in the Pacific. Indeed, most anthropologists place the distribution of mankind from an area just east of the Black Sea in Russia. Cataclysmic pole shifts, lost continents, or extraterrestrial visitors, are of course never given any consideration in their theories. And from their point of view, why should they be? As with the Olduvai Gorge and other important archaeological sites of the world, it seems that the more we discover, the more complicated the solution has to be!

After our brief stop at Olduvai, we continued on to Ngorongoro Crater, the road being rough gravel with lots of potholes. We began climbing uphill after Olduvai and then came to a small, green grassy crater that was absolutely teeming with zebras and wildebeest. After another 12 kilometers, we came to Ngorongoro Crater itself, the largest intact crater in the world. Actually Ngorongoro is a caldera; the cone collapsed and slid back into the volcano leaving a crater 17 kilometers in diameter. With a total floor surface of 160 square kilometers, the rim of the crater where the hotels are located is 2,286 meters above sea level. It's 610 meters from the rim down to the crater floor. It's so steep that only four-wheel drive vehicles can make the trip down the narrow switchback road into the crater.

It was barely noon as I sat on the terrace of the Ngorongoro Wildlife Lodge, drinking cool Kilimanjaro Lager Beer.

"Excuse me," said an African gentleman, pulling up a chair and sitting at the table next to me. "Mind if I sit down next to you?"

"Not at all," I said, looking at him carefully. There was no one else on the veranda. He was dressed in a light green safari suit, wore corduroy shoes and carried a dark mahogany cane. He looked friendly and honest; in his mid-thirties, I guessed, his hair a bit more fluffy and longer than most city-bred Africans.

"I'm a game warden here at the park," he said. "This afternoon's been a bit slow."

"There aren't many tourists here at the moment, are there?" I replied.

"No, it's slow this season." He signaled to a waiter and ordered a beer. "Must be the war in Uganda. Most Europeans aren't that familiar with African geography and they know that Tanzania is involved in the conflict. Where are you from, by the way?"

"Montana," I said, "in the northwestern United States."

"A nice place, I've heard. Myself, I'm from a small village near Shinyanga, southwest of Ngorongoro." He gave a small smile and added, "My father was a witch doctor."

"Really?" I sat up from my slouch and finished by beer. "Wow! A witch doctor!"

"Right. It's no big deal. He was familiar with local herbal remedies and such cures for maladies of the tribal people. Western medicine is replacing most of this nowadays. My father passed on much of his knowledge to me as a child, hoping I would be a witch doctor too, but I became a game officer instead. I suppose," he said, "you'd like to go down into the crater."

"Of course," I responded, "I'm just looking for a lift down. Perhaps I can share the rent on a Landrover for a day."

"I've got my Landrover just outside," he said, "I'd be happy to give you a ride. You can pay for gas. I love going down to the crater myself."

"That sounds great! I'd love to."

"My Christian name is Thomas," said the Ranger, taking some brownish-red bark out of a worn small leather pouch and putting it in his mouth.

"My name is David," I said. "What is that stuff you're eating?"

He washed it down with a bit of beer and replied, "I don't know the English name; it's bark from a rare tree that grows in some dry parts of Tanzania. I like to take some before I go into the crater."

"May I try a little?" I asked.

"Sure, just a bit, it's kind of strong," and he took out a pinch and put it in my open palm. Popping it in my mouth, I washed it down with the last of Thomas' beer.

"Let's go," he said. I grabbed my camera and we headed out the door. Within a few minutes we were bouncing and jouncing down the road to the edge of the crater where the road heads down to the bottom.

Over 30,000 animals live in the crater, the most dense concentration of wild animals anywhere in the world. Looking down into the crater I could see a large lake at the southern end, and there was an off-pink foam all around the edge of it. A funny feeling of tickling was happening at the salivary gland on each side of my mouth, sort of forcing my mouth open in a grin. I began feeling a little light-headed about then, twenty minutes or so after I had taken the bark. I recall asking abut the odd pink foam.

"Those are flamingos!" bellowed Thomas, laughing heartily. Indeed they were. Thousands, millions of them, feeding off the algae that grows exclusively in these soda lakes of the Great Rift Valley.

At the bottom of the crater we headed for the edge of the lake, and then stopped and got out. I was dizzy with the wild vibrations of herds of zebras, wildebeest, and gazelles, plus the pink foam of flamingos that totally covered the shore of the lake.

I felt good and was grinning ridiculously. The crater rim seemed to flow and move. The entire crater was vibrating and alive with wildness, I could feel it everywhere, surging through my bones and out into the lake. The grass, flowers, and lake were a symphony of green, yellow, red, and blue, playing its special time to the tune of nature.

"I feel funny, Thomas," I said. "What was that bark?"

"Don't worry," he laughed. Soon my anxiety was gone and I relaxed. We jumped back into the Landrover and chased a rhino along the lake and then barreled past some elephants wallowing in a prehistoric swamp. As we passed herds of zebras and wildebeest, I watched fields of yellow flowers whizz by the car in psychedelic splendor, to be suddenly interrupted by a small spotted cerval cat which pounced upon a snake in the road. We stopped and watched it for some time and then continued on our search for a pride of lions.

It was incredible. Every blade of grass, every flower, every bird leaped up at me in vivid vibrating color. I had the feeling that I had gone back in time a hundred million years where the ecology was intricately more wild, balanced, alive.

We stopped again on our way out of the crater. The sun had already gone down over the rim and it was getting dark. For one last time, Thomas and I felt the vibrations of Ngorongoro. There were the voices of the evening, clouds of insects buzzing around, the chirping of crickets, the occasional snort of an animal or the call of a bird. Silently a covey of quail crossed the road in front of the Landrover.

At the top of a crater a full moon was rising in the east and the sun was setting in the west. Unable to speak anymore, I watched the sky while Thomas drove back to the lodge.

After thanking him, warmly shaking his hand and giving him gas money, I went to the terrace of the lodge alone where I had been sitting some six hours before. God, how the time had flown! And how my mind was blown! I drank beer the rest of the night, but never got drunk, just staring out over the crater, watching a lightning storm that hung above the crater but did not affect the rim; still mind-boggled by such an intense experience with the wild.

It was late when I left the lodge to walk the mile back up the lonely road between the lodge and the village where I was spending the night. After a quarter of a mile, at a dark spot by a tree, I heard a noise of rustling in the bushes. There was some animal there. I paused cautiously for a moment, knowing that leopards, lions, rhinos, and water buffalo were everywhere. I could be charged if I continued, but I had to go on up the road to my village.

"Courage," I told myself, and decided to continue on up the road. One hundred yards further up there was a sudden loud crashing in the bushes beside me as if some large animal was coming down to the road. I turned and fled in terror, yelling at the top of my lungs and running faster than I have ever run before in my life.

I didn't stop until I was back at the lodge. Luckily, a Landrover came slowly bouncing along just then. Standing in the middle of the road, I managed to stop it, and the two Africans inside gave me a ride to the village. Safely in my bed for the night, I lay awake for the longest time thinking about this incredible day.

§§§

I had been given a 24-hour pass token when I entered Ngorongoro Park, and was three hours over this when I left the park the next day. I got a lift in a Landrover pickup with half a dozen spear-carrying, half-naked Masai warriors and a policeman. It was about thirty miles to Lake Manyara National Park, a game park famous for its tree-climbing lions and giant herds of buffalo. On those roads it took about four hours to drive there. It's 128 kilometers from Lake Manyara on to Arusha, the central town of the northern highlands. Arusha is also probably the main tourist center in Tanzania, situated between Kilimanjaro and the game parks in the west.

Just near Arusha, in the Rift Valley southwest of Lake Natron, a "large ruined city" was discovered by a Tanganyika district officer in 1935. Called Engaruka after its location, it lay on a steep escarpment which was littered with rocks and slithering scree, and spiked with thorn. There were large ruins and no one had apparently reported them before. Dr. L. S. B. Leakey then visited the site and found more than the familiar scattering of stone fragments, solitary graves and ancient terracing, although all these were present. What he found were the ruins of an immense city. "I estimate," he wrote, "that there are about six thousand three hundred houses in the main city of the scree slopes....and that there are about five hundred houses in the valley ruins, where burials are far more commoner than houses." The population figure, he thought, "was probably between thirty and forty thousand, and this may be an under-estimate."[106]

Leakey went on, "The houses of the main city are all built upon very well made stone walls. The terraces include pathway terraces and house terraces... [and] there is a vast mesh of stone walls and terraces in the valley ruins which I take to be connected in some way with cultivation and irrigation, but this in not proved." Leakey found no inscriptions, though he did find "cup marks" (circular carvings in rocks) and oddly, no burial remains. Therefore, there was nothing to carbon date. Incredibly, the date that Leakey placed on this city was not more than three hundred years old![106]

This, to most knowledgeable researchers, would seem impossible; at the least, grossly underestimated. Leakey was claiming that this huge city of thirty or forty thousand people was built only in 1630, yet, there were no bones left from the graves that "are far more commoner than houses"? In 1630 the Portuguese were already well in control of the East African Coast! Where did these people go? Did they build their huge city and then leave a few years later? Not even the local natives were aware of the lost city of Engaruka. Also, what of the ancient terracing, roads, ditches, etc. that are found all through East Africa? Does Leakey see no connection? Even Basil Davidson[106] could not accept Leakey's conclusions. It is interesting that a scientist like Leakey, who finds the skulls of "ancient man" around East Africa and dates them as hundreds of thousands of years old, and even millions of years old in some cases, cannot come to admit that there might be cities in East Africa that more than a few hundred years old. We are back again to mankind living in ignorance in his cave for a million years and then suddenly starting civilization seven thousand years ago in Sumeria. It was 6,700 years later that civilization finally made it to the hills of East Africa. This is the logic of the "experts".

If I may be allowed to differ with Dr. Leakey, I would venture that Engaruka is the remains of a city many thousands of years old, and goes back to the ancient and mysterious Azania civilization that covered most of East Africa in the past. They created vast roads, terraced hillsides, huge cities of 30 to 40 thousand people, mined many metals, traded across the Sea of Zinj (or Zanj, the Indian Ocean) and probably were in direct contact with the rulers of Axum, if not even part of the Empire of Sheba, at one point. The "cup marks" mentioned are very interesting, and such marks are found in Ireland, Scotland, Malta, and even ancient New England. This would give evidence of ancient Celtic explorers, and even, perhaps, of the mysterious Atlantean League. The very fact that there is no skeletal material to be found at Engaruka indicates the city's great age. Probably, the original inhabitants of Engaruka, possibly Bantus, or even Bushmen or Celts, were forced out by invading Masai coming from the Horn of Africa. At this point, very little is known about Engaruka, and it remains a lost city of mystery.

In Arusha I met a couple from New Zealand. Melvin was tall, dark and athletic; Sonja was of medium build, and pretty with long auburn hair; it turned out that we were all staying in the same hotel. We talked at dinner that night and Mel told me, "If I came to Africa and didn't climb Kilimanjaro I'd kill myself!" Since I felt the same way, we decided we'd climb the mountain together. We set off the next day for Moshi, and within two days we were prepared to make the climb.

The climb, which was not too difficult and proved well worth the effort, was basically five easy days and four nights spent in comfortable large Norwegian mountain huts. We carried food in the form of dried soups and milk, dried fruits and meats, candy, and nuts, crackers, cans of luncheon meat, beans and fish, plus a few oranges and bananas.

The park entrance fees were pretty steep: it cost about $50 dollars in park fees (it is more today, I understand) just to make the climb. The first day was easily done in four or five hours from the park entrance to Mandara Hut. The second day was about a five-hour walk through the jungles of the lower slopes, sunlight filtering through the moss-covered trees dripping with lichen, and ferns crowding the undergrowth. Monkeys chattered and played in the trees overhead. Before we reached the second hut, we came out onto a kind of treeless alpine moor, open and covered with grass and moss. The second hut was Horombo Hut: it was actually a group of ten small huts and a large A-frame lodge.

The third day was probably the most difficult; six hours of pretty steep climbing, though at a slow pace to Kibo Hut, the last hut before the summit. It was important to take all three days to reach the last hut even though it could be done in one day by a healthy adult, because of the acclimatization necessary to avoid altitude sickness. Even so, many people notice slight headaches and insomnia at Kibo Hut. Melvin, Sonja, and I had brought firewood and water with us to Kibo Hut, and we made some soup. We were all tired from the day and retired early because we knew we would be getting up around one o'clock to make the final push for the summit. We all lay in bed for hours, unable to sleep.

Insomnia is one of the first signs of altitude sickness, along with loss of appetite and shortness of breath. People become more irritable at high altitudes as well. There is really no danger as long as you don't ascend to a much higher altitude and stay there for a long time. The best prevention for altitude sickness is to ascend slowly to acclimatize your body and to breathe deeply to give extra oxygen to the brain. If the sickness becomes too uncomfortable to handle, the cure is to descend to a lower altitude.

I hardly slept at all that night. I finally dozed a little, when suddenly our guide woke us. We were already packed, so we merely dressed, had a cup of tea to warm us, and started, at about one o'clock in the morning, out on the rocky, winding trail that zig-zagged up to the final steep face.

We had left the hut a bit early, apparently. Our guide didn't realize that we could do the last three thousand feet so fast. We arrived near the summit by four-thirty and sunrise wasn't until six o'clock. The freezing temperature combined with windchill at over 18,000 feet posed the serious problem of how we were going to keep warm. We were all wearing down jackets, sweaters, and wool hats, but at the subzero heights of Kilimanjaro we were literally freezing to death. My experiences mountaineering in Colorado, Montana, Washington and the Himalayas told me one thing—keep moving to keep warm!

But we couldn't continue on to the summit; we would arrive at the top an hour before sunrise, and there we would be most exposed to the wind and elements. So we huddled in a small cave, taking what little shelter from the wind we could, and beating our legs, chests and arms to keep up the circulation and stay warm. After twenty minutes or so, it was just too cold to keep this up and we had to continue slowly on up to the top. We stopped again at another small shelter near the summit for another twenty minutes or so, beating ourselves and stamping our feet to keep warm. Eventually we slogged our way to the glaciated summit, our boots crunching on the hard snow. Seracs, huge blocks of ice as big as freight cars, were strewn around the crater rim of this famous volcano. The center was a huge, white icy bowl dotted with jagged blocks of ice sticking up in the sky.

We were all breathing hard. With a final burst of energy, I took the last few steps to the summit and collapsed on the snow. Pant, pant. "God, what a view." Gasp, gasp. I swung my pack off my back and lay on a rock that was sticking up out of the snow.

So this was the top of Mt. Kilimanjaro! Fantastic! It was still dark, only ten minutes before sunrise. To the east, a faint lightening of the sky was just visible. It felt great to be on top of the highest mountain in Africa. 19,340 feet was quite an altitude, and it was exhilarating to say the least. My breathing was becoming more controlled. I closed my eyes and rested for a minute. It seemed like ages ago that I had been standing on those lonely crossroads on a hot dusty afternoon waiting for a ride.

We did it! My New Zealand friends and I, with our trusty guide, had climbed to the top of Kilimanjaro, about five and a half hours from Kibo Hut. It was certainly worth the trip, I thought, and got up to take a photo from the icy summit as the sun rose over the plains of Kenya to the east. The sight was spectacularly beautiful, and I appreciated it even more when the shutter froze on my camera as I tried to take a picture. As the sun warmed the air slightly, we spent a good long time enjoying the view.

Our fourth day was to descend from the summit back to Kibo Hut, get our packs and descend on down to Horombo Hut where we spent our last night on Kilimanjaro. The last day was a long one from Horombo back to the park entrance and Moshi.

> We, the people of Tanganika, would like to light a candle
> and put it on top of Mount Kilimanjaro
> which would shine beyond our borders giving
> Hope where there was Despair, Love where there was Hate,
> and Dignity where before there was only Humiliation.
> —*Julius K Nyere, plaque on the summit of Mt. Kilimanjaro*

§§§

There was something about the forced-steam sound of train engines and the general confusion of the Moshi train station that excited me. I was taking an overnight train to Dar Es Salaam, and it was going to be quite crowded, that was for sure. Fortunately I'd met a conductor who bought me a third-class ticket so I wouldn't have to wait in line for three hours. He told me to try and get a seat in the Buffet car, as the train would be jammed.

"Veddy, veddy crowded," he said.

"I'll bet," I answered and thanked him for getting me the ticket. My mind was already flashing back to my train trip in Sudan—I knew what these trains could be like!

As I walked around the train yard, I saw a young boy selling peanuts and roasted lentils in a little flat basket. I came up behind, meaning to buy a snack from him. He suddenly turned, saw me reaching toward him, and gave a horrified shriek, backing away.

"Shillingi, nagapi?" (How much are these?) I asked, reaching toward him. He shrank away, absolutely horrified, his face full of terror of this white vagabond pursuing him. Did I look so bad, I wondered? I had showered and shaved this very afternoon. Some older boys next to him started laughing and one put a reassuring hand on the small boy's shoulder and obviously explained to him that I just wanted to buy some peanuts, not eat him. Just then the train came, and with a mad dash, showing all my expertise from living in India and Africa for three years, I dove through a window of the Buffet car and procured a seat. I could have gone second class and had a reserved seat, but that would have been missing half the fun.

It was rather a restless night on the train, but my mid-morning I was in Tanzania's main port, Dar Es Salaam, the "Haven of Peace." One of the prettiest ports in the world, Dar is a fairly modern, but rather dirty and run-down, city. There are several ways to reach the island Zanzibar from the mainland. I climbed aboard a rusty old tramper one night and slept on top of the third deck with the permission of the captain, who was a Norwegian working for the United Nations. By morning we were steaming into Zanzibar harbor, all green and tropical in the early morning sun. You get stamped into Zanzibar just as if it were a separate country, and you have to have a visa, too.

Zanzibar is one of the truly idyllic, relaxing places in Africa. Full of charm and history, it was not always the peaceful place it is today. The last open slave market in the world was closed here in 1873 by the British who had sought for years to destroy the East African slave trade. To combat the British anti-slave raiders, the Arabs told the captured slaves that the British were cannibals and they would be eaten if the British captured them. Therefore, when the British tried to capture a slave ship, the Arabs would set their slaves loose and let them do the fighting, as the slaves thought they were truly fighting for their lives.

Zanzibar gained independence in 1961, and the last Sultan of Zanzibar fled the country in 1964 when the Afro-Arabian conflict came to a head and 5,000 Arabs were killed in the peasant uprising. The land which the Arabs had controlled for so long was redistributed among the peasants in three-acre plots.

Today, Zanzibar has the highest per capita income in black Africa, and keeps its currency reserves separate from the mainland. The economy is largely dependent on cloves, most of which are actually grown on the neighboring island of Pemba.

I had been resting in Zanzibar for two weeks, lounging around on beaches, wandering the wonderful twisted streets and market places and drinking tea in the small tea shops. One day there was a knock on the door of my room: it was Catharina! How wonderful to see her again! We embraced and held each other for a long time. We were in a sense, meeting each other at our destination.

I sat with Catharina down by the harbor; young kids were selling sweet roasted casava root and peanuts. Another vendor sold ice cream cones. The sun was slowly lowering; the people were milling around the harbor park to watch the evening glow. What a wonderful place, Zanzibar! I thought of Livingston preparing for his last expedition to the mainland. Catharina and I was preparing for ours too; I would take the train to Zambia, she would fly back to Sweden. I squeezed her hand gently; we had had a memorable time on the beaches of Zanzibar together over the past two weeks, but now it was time to move on.

Like Livingston, my steamer was leaving for the mainland shortly. Africa the unknown continent called once again. On our last day we made a final tour of the nearby beaches on rented bicycles. Catharina turned to kiss me goodbye.

"Dr. Livingston, I presume?" she whispered.

§§§

Within a few short days I was on the famous Chinese railway on its thousand mile journey to Kpiri Mposhi in Zambia. This is a two-day ride straight through the heart of Tanzania and touching the northern tip of the Selous Game Reserve, the largest and wildest of Tanzania's many national parks. As we neared the Zambian border, I found myself talking with three Chinese technicians who were in Tanzania to inspect the railway their country had built, and to observe how the Africans were maintaining it. Not too well, from what I gathered, although it is by far the most efficient and modern train in Tanzania. Suddenly into our compartment burst a Zambian health official. "Let's see your vaccination certificate," he demanded. I showed him my yellow booklet with my cholera, small pox and yellow fever vaccination shots.

"Where is your plague shot?" he asked.

"I don't have a plague shot," I answered.

"Then you'll have to go back to Dar Es Salaam," he said. "You can't enter Zambia."

"Wha-what?" I stuttered. "You must be joking! I didn't know I needed a plague shot! Are you sure?"

"Positive! You can't enter Zambia without a plague shot." He then showed me a United Nations brochure which stated that the plague still existed in parts of Kenya, Tanzania and Madagascar.

"I can't go back to Dar Es Salaam," I pleaded, "it's 900 miles and two day's ride back there. This is crazy. I'll get one in Lusaka (the capital of Zambia)."

"No way," he stated, and as he left, he said, "I'll talk to you in your own compartment privately."

I went to my own compartment, a first-class compartment. I had it all to myself, because a Swiss couple who had purchased all the tickets for this compartment had gotten off at the last stop. I had had my usual third class ticket, but they had generously invited me to share their compartment.

I sat alone wondering what to do. How frustrating! I figured I would just get off the train at the nest stop and then sneak in Zambia. No way was I going to go back to Dar after coming all this way.

Suddenly the officer burst into my compartment again and said, "OK, let's change money!" He wanted to change fifty American dollars at the bank rate for Zambian kwacha, and I realized this was a scam. He would scare travelers like me and then force them into changing money with him at a terrible rate so he could make a killing on the black market. I figured him out fast, but played along with him and changed what few Tanzanian shillings I had into Zambian kwacha, but no dollars.

"OK," he said, and before he turned to go, he asked, "By the way, do you have any English magazines or paperback books—for my kids, you know."

"No, sorry," I told him, "not a one," which was true. For a con man, he seemed a reasonably likable fellow, even though I'd been a nervous wreck for half an hour. Later I talked with a Dutch girl on the train to whom he had told the same story. She had changed a hundred dollars with him at the bank rate.

"That's African bureaucracy for you," she said.

I just smiled and peeled a mango. "Dr. Livingston," I thought, "never had to deal with customs or health officials." I wondered what he would think of the white man's civilization in Africa today.

David Livingston and his guides enter Lake Nysasa
(Malawi) in 1859 through a swamp of towering reeds.

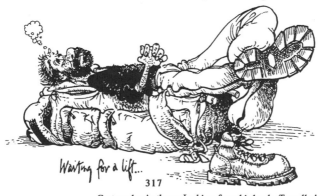

Waiting for a lift...

Cartoon by Anthony Jenkins, from his book, *Traveller's Tales.*

Ancient terracing in East Africa. The civilization that created the roads, cities and terracing throughout Africa is still a mystery

Chapter 14

ZAMBIA AND MALAWI: THE BATMEN OF AFRICA

> *So Geographers, in Afric-maps,*
> *With savage pictures fill their gaps;*
> *And o'er unhabitable downs*
> *place elephants for want of towns.*
> —Jonathan Swift (1667-1745)

I arrived at Kpiri Mposhi, the end of the Chinese-built Tazara railway, and within the hour I was off on a bus to Lusaka, the capital of Zambia. It was a short ride on a smooth paved road, only 150 kilometers from the train station.

Within two hours of my arrival in Lusaka I was arrested.

In most African countries it is rare to be arrested looking for a hotel, but Zambia was under a lot of strain at the time, and apparently the muscles I'd gained from toting a fifty- or sixty-pound backpack, plus my short cropped hair, T-shirt and beardless face, made me seem a suspicious character.

The trouble started because I didn't know much about accommodations in Lusaka and had decided to try the Sikh Temple, which someone had told me offered clean accommodation at a reasonable cost. I left my pack at a Seventh Day Adventist church and took a bus into a fashionable suburb of Lusaka, where the temple was located. After a bit of walking around a neighborhood of nice, modern western-style houses, I finally found the temple that had closed down a few months before. "Oh, well," I thought, and started to walk back toward the main street to catch a bus back to the center of town.

A security guard caught up with me and began talking to me, very curious about what I was doing there. It was getting late and I was in a hurry to make some sleeping arrangements, so I answered his questions briefly and kept up my fast pace through the quiet neighborhood. The security guard left me and then reappeared with a policeman a minute or so later.

The policeman escorted me to a small police post and called the Lusaka police headquarters, referring to me as the "suspect". It was slowly dawning on me that I had done something wrong and was being apprehended for ques- tioning. As the questioning continued, I began to realize that they thought I might be a Rhodesian soldier, here to bomb a house, assassinate some nationalist African leader, or maybe just pass through on my way back to Rhodesia. (There were tensions between Zambia and Rhodesia because the Zimbabwe-Rhodesian civil war for independence was at its height.)

After about fifteen minutes (it seemed longer) a squad car pulled up. Four black policemen and, to my astonishment, one white woman got out.

"In the car," said one rather burly fellow dressed, like the others, in a green military police uniform. He sandwiched me in the back seat between another large policeman. By now I was thoroughly apprehensive. In the car, the well-muscled officer on my right asked all the questions, but, unfortunately, I had the wrong answers.

319

I didn't sound like an American, he said, looking through my documents. My English sounded as if it were a second language for me; he decided I was a Jew, seeing that my student card was issued in Jerusalem, and therefore I was probably an Israeli spy—I was too young to have so much money on me, a thousand dollars or so in travelers' checks—I was already heading for the firing squad, so he offered me my last cigarette. "I don't smoke, thank you," I told him.

At the police station, I was told to wait in a chair in the desk sergeant's office until Detective Somebody could come.

"Detective who?" I asked. "Who's that?"

"Be quiet and sit down," someone said.

"I want to call the American Embassy," I responded, more calmly than I felt, "and let them know I'm here." But no one answered, and no one seemed to care.

"I'll be all right," I told myself. "After all, I've done nothing wrong."

When I had a chance I explained to several officers my story of how I got there, what I was doing, and where I was going. "Just coming down from Tanzania for a visit," I told them, "In fact, I only arrived today from Tanzania. Look at my passport."

Detective Somebody or other was too busy to see me, so they told two policemen to take me to my luggage and search it. We piled in the car and went downtown to the church, where they let me out.

"I'll show you my luggage," I said. They seemed like nice guys and I was as friendly and helpful as I could be, naturally.

"No, that's OK," they said and drove off. "Goodbye, and enjoy your stay in Zambia!"

"I will!" I called back, waving goodbye. Gee, I thought, I've hardly been in Zambia a day and I'm already making new friends.

I eventually checked into the local camp ground where I pitched my tent. Lusaka is the country's largest city with a population of almost half a million. It is a new, modern city of cement buildings and empty supermarkets. Zambia has had some economic difficulties in past years and often suffers shortages of one type or another. Sometimes their shortages can be inconvenient, like having no bath soap. When I was there for the first time, I could not find a bar of soap anywhere! Out of desperation, I eventually stole a small used bar from a hotel bathroom. There can also be shortages of bread or cornmeal, the staple diet, and you will often see huge lines several blocks long of people waiting to receive a ration of cornmeal. The meal is mixed with boiling water and serves as a sort of gruel with gravy. This is the main staple of eastern and southern Africans.

While standing at the bar having a beer at the Lusaka Hotel one afternoon, I struck up a conversation with a white Zambian businessman who was down from the Copper Belt in the north, where the mines dig out Zambia's major foreign exchange commodity, copper.

"This country's going down the tube!" said the white Zambian plaintively, lifting his beer up to his unshaven face. He wore a brown safari suit and suede safari boots—standard African casual-business dress. "Everybody knows it, especially the young people. We need a change of government, everything's going to hell."

"What do you do here in Zambia?" I asked him.

"Don't get me wrong, I love Zambia," he said. "I was born here, up in Ndola in the Copper Belt. I have a little business manufacturing soap up there. Make a killing. *Make a killing!*"

This was interesting to me as I was still looking for a bar of soap at the time. "Hey, that's great," I said, "Could I buy a bar of soap from you?"

"Oh, sorry, kid," he snorted, "it's all detergents I make, laundry soap and stuff like that. Yeah, I know it's tough to get a bar of soap around here. I don't know what to tell ya. Boy, if I could get a truck load of handsoap into the country, I'd make a killing—a *killing,* I tell ya." He finished his beer and ordered another.

"This government is crazy," he went on. "Take for example a bottle of Coca Cola. The government controls the price of Coca Cola at seventeen cents a bottle. The best hotel in Lusaka has to sell it for seventeen cents. The hotel has to buy it for sixteen cents—sixteen cents! They can only make one cent on a bottle! If a bartender drops one and breaks it, they've got to sell sixteen more bottles of Coke just to break even. It's insane!"

It did seem pretty silly. African socialism could get out of hand occasionally; it often sacrificed practicality, so the economy suffered. However, Coca Cola was cheap, where you could get it. I glanced around the bar full of upperclass blacks in business suits having a cocktail after work. The Lusaka Hotel is the main hotel on Cairo Road, though it's pretty old and run down. Nobody seemed to be paying much attention to our conversation, which was a relief. It is often advisable not to discuss politics too loudly in African countries.

"Yeah, the hotel here doesn't serve Coca Cola," he went on, "couldn't afford it. A couple of months ago the Zambian government sent some ministers down to South Africa with three million kwacha to buy supplies for the stores and supermarkets up here, and you know what they did? They blew the first two million at a gambling casino in Swaziland and spent the last million on booze and wool blankets. God! No soap or cheese or bread in the stores, but lots of whiskey, gin, and wool blankets! It's crazy, but still I love my country. I'm a humanist."

A humanist, according to my dictionary, is someone who studies literature or history, or someone who studies human nature. The president of Zambia, Kenneth Kaunda, was very fond of saying that he was a humanist, and this became sort of the unofficial slogan of Zambia. Anyone, white or black, might say to you, "I'm not a racist, I'm a humanist!" At least, I suppose, it did tend to uplift people's thoughts.

Indeed, Zambia seemed like a pretty well integrated society, where people were judged on their abilities and productivity, whether they were white or black. Blacks could be just as corrupt and intolerant as whites, while whites could be as unmotivated and impractical as blacks. This was human nature; the idea was "we all have to shape up and pull together and build our country." The official motto of Zambia is "One Zambia, one nation," reflecting their aspiration to build a united society undivided by race, color, or tribe.

"I love this country," the man went on, "it has many wonders, not the least of which is the Batmen of the Jiundu Swamp up near the Zaire border. Now, there's a *strange* tale!"

"Batmen!" I exclaimed. "What do you mean by Batmen?"

"The Batmen of Africa, lad!" the man said loudly, ordering us each another beer. "Deadly and quick. Eat you for supper! That's what the natives say anyway. You see, up north in the swamps on the borders of Angola and Zaire the Kaonde tribe has legends of the *kongamato,* or batmen of the swamps. Just to see one means death!"

I took a deep drought of my beer and looked around the room in a daze. I'd heard of the old 30's cliffhanger serial named *The Batmen of Africa* (1933-6) with the famous lion-tamer and actor Clyde Beatty. He saves a woman from a lost city full of batmen in the depths of Africa, but this seemed to be something different. "What is this kongamato that you're talking about?" I was naturally compelled to ask.

The man looked me in the eye and said, "The kongamato is some sort of flying lizard, with wide wings and sharp teeth. It overturns boats and attacks people. The natives up there in the swamps are so afraid of it, they believe they will die if they just see it. This bloke wrote a book about it back in the twenties. Talked all about it. It's true, swear to God!"

Now I knew what he was talking about. However, my beer-swizzling friend was incorrect when he said that a book was written on the subject. One book did, however, briefly discuss the subject. In 1923, a book entitled *In Witch-Bound Africa* [101] was published. It was written by a very respectable British explorer and anthropologist named Frank H. Mellard who was the Magistrate for the Kasempa District of Northern Rhodesia from 1911 to 1922. He was probably the first person to discuss the *Kongamato,* but he wasn't the last. When Mellard himself asked the Kaonde natives of the Jiundu Swamps in north-west Zambia what a *kongamato* ("overwhelmer of boats") was, they told him, "...it isn't a bird really: it is more like a lizard with membranous wings like a bat...the wing-spread was from 4 to 7 feet across...the general color was red...no feathers but only skin on its body."[101]

Mellard goes on to say that the natives believed that the *kongamato* had teeth in its beak, though no native seen the animal close up and lived to tell the tale. Mellard sent for two books which had pictures of pterodactyls, "and every native present immediately and unhesitatingly picked it out and identified it as a *kongamato.* "[101]

Mellard could not get any native to take him into the Jiundu Swamps to show him a kongamato for any price. The swamps are "an ideal home" for the kongamato, says Mellard, and these swamps are a spooky, strange place, at best. According to Mellard, the natives were genuinely terrified by this beast and he said that he himself believed the beast existed, at least until recently. Other explorers backed up Mellard's tale of the kongamato. Colonel C.R.S. Pitman wrote in *A Game Warden Takes Stock* (1942, London): "When in Northern Rhodesia I heard of a mythical beast which intrigued me considerably. It was said to haunt formerly, and perhaps still to haunt, a dense, swampy forest region in the neighborhood of the Angola and Congo borders. To look upon it is death. But the most amazing feature of this mystery beast is its suggested identity with a creature bat-bird-like in form on a gigantic scale strangely reminiscent of the prehistoric pterodactyl. From where does the primitive African derive such a fanciful idea?"[94]

The South African professor J.L.B. Smith writes in *Old Fourlegs: The Story of the Coelocanth* (London, 1956): "The descendants of a missionary who had lived near Mount Kilimanjaro wrote from Germany giving a good deal of information about flying dragons they believed still to live in those parts. The family had repeatedly heard of them from the natives, and one man had actually seen such a creature at night. I did not and do not dispute at least the possibility that some such creature may still exist."[94]

In 1928 a book was published in London called *A Game Warden on Safari* by A. Blayney Percival. The author states, on page 241, "..the Kitui Wakamba tell of a huge flying beast which comes down from Mount Kenya by night; they only see it against the sky, but they have seen its tracks; more, they have shown these to a white man, who told me about them, saying, he could make nothing of the spoor, which betrayed two feet, and an, apparently, heavy tail."[94]

Could a small pterodactyl (seven foot wing-span, according to Mellard) be living in the unexplored swamps of Zambia? A number of reputable scientists and explorers believe so. The famous Swiss zoologist Bernard Heuvelmans suggests that if the creature is not a pterodactyl, it may be a giant bat. Indeed, the late and popular zoologist Ivan T. Sanderson, one of my personal heroes, claims to have been attacked by what he believes to be a giant bat in a river in the Cameroons in 1933. He and his companion Gerald Russell were in the Assumbo Mountains of Cameroon when Sanderson shot a large fruit-eating bat, which fell into the river.

He went into the water to fish it out, but stepped on a turtle which moved and lost his balance. He was floundering in the strong current when Gerald suddenly shouted, "Look out!"

Sanderson turned and looked, and saw to his horror a large black thing the size of an eagle flying directly at him. Sanderson vividly recalled the semi-circle of pointed, white teeth in the gaping mouth. He ducked and the animal flew over him, circled and attacked Gerald Russell who also ducked. Back at the camp, they asked the natives, "What kind of bat is it that has wings like this and is black?" asked Sanderson, spreading his arms out to show the extent of the large animal.

The horrified natives replied this was the *Olitiau,* and on hearing that they'd just seen one at the nearby river, the natives immediately gathered their guns, left their other belongings, and ran out of the camp for the nearest village.[61,94,98,97]

The kongamato may well be a giant bat, or, indeed, a pterodactyl. Giant bats and pterodactyls have been reported in many parts of Africa (see the chapter on Namibia for more tales of pterodactyls), in Sumatra, South America, North America (including Texas and Arizona) and even New Zealand. Incredibly, the *Illustrated London News* for February 9, 1856 (page 166) carried a story of a tunnel being cut to unite the St. Dizier and Nancy railways in France from the *Presse Graylouse* newservice which described how when a rock was blasted open, a creature exactly like a pterodactyl with a wingspread of about ten feet and "livid black" emerged from the rock, walking with the aid of its wings, emerged into the light and expired after uttering a hoarse cry. Naturalists reportedly identified it as a pterodactyl and the rock strata as being millions of years old.

I turned and looked at the soap merchant who was finishing the last of his beer. "Yep, mate," he said, "it's a strange world out there in the bush. Full of strange critters, Batmen, giant snakes, lost cities, even dinosaurs. I know!"

"Yea," I agreed, "a world of pterodactyls, seventeen-cent cokes, and no soap!" We both laughed at that!

§§§

Zambia's namesake, the Zambesi River, spills over the country's main tourist attraction, Victoria Falls, in the southwest corner of the country and flows along its southern border. Definitely one shouldn't miss the falls when in Zambia; Luangwa Game Park in northeast Zambia, near Lake Tanganika, is also worth seeing. This is one of the few parks in Africa where you can walk among the animals in the company of armed guards. Game is prolific and the "big five"—elephants, giraffes, lions, rhinos, and hippos—are all readily seen.

Most of Zambia is a vast, thinly populated, tree-clad plateau in the heart of Africa. Completely landlocked, it depends on its rail links with the surrounding countries. Zambia is the third largest producer of copper in the world, and when copper prices fell drastically in the seventies, Zambia's economy took a plunge. This and the boycott of white-ruled Rhodesia, which served as Zambia's traditional rail link for exporting copper, sent the country on the road to economic ruin. The Tazara Railway on which I had just traveled was meant to open up Zambia's copper to Indian Ocean ports, but proved inadequate.

Meanwhile Zambia was actively supporting one faction of the Zimbabwe-Rhodesian war. (Zimbabwe is the name the rebels used for Rhodesia; during the civil war reporters referred to the country as Zimbabwe-Rhodesia in order to be even-handed; the country is now called Zimbabwe.) The Zambian-based wing of the Patriotic Front was headed by Joshua Nkomo, whose sister just happened to be married to the president of Zambia, Kenneth Kaunda. Nkomo's fighters were Soviet-backed, trained and supplied, while his opposition in the Patriotic Front,

Robert Mugabe (who eventually became prime minister), was based in Tanzania and Mozambique and was backed by the Chinese. Zambia, at the time of my travels there, was full of Zimbabwe-Rhodesian nationalist guerrilla camps. The white Rhodesian government forces would occasionally drive up to these camps and blow them up. Once they even blew up Joshua Nkomo's house in Lusaka. The Zambians were therefore a bit jumpy, and if someone thought you were a Rhodesian soldier, he was apt to shoot first and ask questions later. As I came to understand the political situation better, I realized gratefully that I was lucky the security guard had called the police that first day, rather than shooting me on the spot, for I have no doubt he suspected me of being a Rhodesian mercenary.

Preferring to forget all this political turmoil, I just acted my usual carefree self, and hitched around Zambia as I pleased—and as usual got into trouble. But before that, I met some delightful people.

Stepping into a crowded African beerhall one night in Lusaka, I casually glanced around, then worked my way up to the crowded bar to order a beer.

"One beer, please," I said.

Suddenly, to my left, I heard the strangest American accent yet.

"Wa'y-y'all from? A-merica?"

I turned and saw a big smile stretched across a round black face topped with short curly black hair; gold-rim glasses didn't hide the merriment in his eyes.

"Yeah," I said, "I just got here in Zambia."

"Way'el, Ah'm glad to meet ya'll," he said, coming over to stand next to me. I was all smiles by this time too. His accent seemed to be from the deep south. "How neat," I thought, "Some black American guy traveling around Africa!" Far out, you don't meet them every day—hardly at all, in fact.

"Where are you from in America?" I said, wetting my whistle with a beer.

"Way'el, Ah'm not from America. Ah'm from Liberia."

"Liberia!" I shouted above the hoot and holler in the bar. "Hey, you sound just like an American with a southern accent!"

"My ancestors came from America," he said. "Ah've been there once mahself, t'see relatives in Virginia. Ah live in Monrovia mahself."

I had to put down my beer and stare at my feet, I was grinning so hard. So, I had finally met a Liberian. Great! Liberia was settled in the 1820's by freed American slaves who returned to West Africa to colonize it. Its constitution, institutions and flag are modeled after the United States and its currency is actually American dollars—worn and tatty, I've heard. Liberia's only political party, the rather antiquated True Whig Party, ran the country until a few years age when a military coup placed a sergeant in power.

My beerhall companion was named John. He told me a little about Liberia and how he was in Zambia as a student, his strong Liberian accent in full display. It was so wild hearing this Southern Plantation accent in a Zambian beerhall that I couldn't help grinning the whole time he was talking. Before I left, he gave me a great quote:

> The darkest thing about Africa
> is America's ignorance of it.
> —Rev. James J. Robinson

§§§

It took me one full day to hitchhike from Lusaka south to Livingston, the main town near Victoria Falls. Several short rides got me well into the country, along a two-land paved road that just sort of melted away at the sides into the bush. Eventually I got a ride with a cheerful little black truck driver and his semi-trailer full of beer; he was transporting it to Livingston. The truck ambled along slow but steady—a comfortable ride. Toward late afternoon the driver picked up two young African girls hitching into Livingston. Teenagers, I would have said, and pretty too. They had on nice neat cotton dresses and their hair was pleated into rows that lay flat on their heads. They didn't speak English, but, like all Africans it seemed, they smiled a lot, which was good enough communication for me.

Just before sunset, the driver stopped the truck in the middle of the forest. We all piled out to help him collect wood for his evening fire in Livingston, then he opened up the back of the truck. Climbing over the beer cases, all returnable bottles naturally, he pulled out eight beers, each from a different case.

We piled back in front and roared off toward Livingston, an hour away. I was trying to figure out how to open the beers as the driver passed them out to us. Then one of the cute young girls gently took it from my hand and quickly opened it with her teeth, using her mouth as you would a bottle opener on a pop machine. I winced as she handed me mine and watched her open the rest the same way.

Out the window to my right was a great orange African sun setting through the leafless, dry-season trees. It hung on the horizon like a great ball, and I found I could look directly at it. It seemed to be following the truck as we drove through the barred forest. In that parched, dry land the warm beer tasted as good as any I'd every had.

Livingston was once the capital of Northern Rhodesia (the capital was moved to Lusaka in 1935). The old Government House is now a national monument, and the Zambia National Museum is here, renowned for its collection of Livingston's possessions and "early man" artifacts. The falls are a few miles out of town and are truly spectacular. They are rightly considered one of the great natural wonders of the world. At the height of the floods, from March to May, more than five million liters of water pour over the falls every second, causing a deafening roar and a cloud of vapor to rise over 300 feet that can be seen 50 kilometers away. Hence the falls received its African name, Mosi-o-Tunya, "The Smoke that Thunders."

Around the falls, the recently completed Knife Edge Bridge takes you right through the spray to a rock tower on an island downstream opposite the eastern cataract. The bridge is slippery, but it's an exciting trip through the mist and thunder—looking down into the surging cauldron below is a dizzying experience. From the island you get a good view of the main falls and Rainbow Falls. There is always a rainbow on the east corner of the falls when the sun is out.

On my first day in Livingston I went down to the falls and spent the morning admiring their incredible beauty. Towards noon, I walked out on the main road looking for an ancient baobab tree reported to have a platform high up in its branches. It was known as the Lookout Tree.

By coincidence, I met a beautiful young Zambian maiden named Ann who happened to be on her way into Livingston to visit her sister. We were both walking down the road; I asked her if she knew where the Lookout Tree was. She did, and took me there—a solitary baobab spreading out like an upside-down ginseng root, its thick puffy branches stretching out to the sky.

Instead of visiting her sister, she spent the afternoon with me, talking about life in Livingston. We ended up at the hotel restaurant drinking beer in the hot afternoon and talking of ourselves. She was only eighteen, a dancer at the cultural center, and she had this evening off. She also had a boyfriend, a teacher from Germany who had gone home. "I like Americans," she said. "I'd love to travel through Africa like you. Can you take me with you on your journey?" she asked, giving me a seductive look. "If I can't go with you, maybe you could tell me some stories?"

When tromping around Africa, hot, dusty, unshaven, caked with bits of the local savanna, a traveler of my disposition longed intensely for two things, a tall, ice cold beer, and a warm generous lady, in whose arms one could recount everything one had seen and done, not just in the forgotten deserts of Africa and Arabia, but throughout life. The first is a lot easier to come by, although sometimes it may take a week or two of walking and hitching to find a restaurant with a refrigerator and beer.

The second was a sight more difficult. The African traveler has to rely on his luck and determination, knowing that sooner or later he will fairly collapse in some strange woman's arms and blab out his life story like an idiot.

It was my great good fortune that day to end up back at my bungalow beside the Zambesi with lithe and lissome Ann of the warm and generous heart.

Later that afternoon, as we lay on my bed, and I had finished telling her how the soldier had attacked us at the volcanoes in Uganda, there was a sharp knock on the bungalow door. I opened it a crack, and to my dismay saw a policeman standing there.

"Get your clothes, documents, and luggage and come down to the police station. We want to talk to you. Bring the girl too." He was carrying a rifle and had a very serious look on his face.

I forced a smile and said, "Sure." My whole body tensed in an all-over physical grimace that was the opposite reaction to my smile. I was in trouble, and worse yet, I was afraid Ann might be in trouble as well.

We dressed and the policeman escorted us both to a small police station with one room and a desk near the falls. After searching my luggage very carefully, they found what they considered to be some incriminating evidence: a large Sudanese dagger, an American tear gas pen (given me by an American friend), and some souvenirs from my raid on Idi Amin's house in Uganda. These included some documents I had picked up off the floor of his house with the official seals of Uganda, and CONFIDENTIAL or TOP SECRET marked across the top of them, a Christmas card from Idi Amin to the Libyan Ambassador to Uganda, and of course, my twenty or so "Idi Amin Dada, King of Africa" T-shirts, which had been looted from Amin's garage.

I faced the officer sitting opposite me at his desk with all these items piled in front of him. The other policeman who had brought us in was standing behind me with his rifle. There was Ann sitting in the corner, totally embarrassed and undoubtedly thinking, "How did I get mixed up in all of this?"

In all seriousness and with a certain tone of final gravity in his voice, the officer looked me in the eyes and said, "So you're a spy for Idi Amin! Confess!"

I was so tense and this remark was so absurd that I burst out laughing at the Kafkaesque incongruity of it all. They weren't laughing, however.

I explained to them that I had only been in Uganda at the end of the war and had "looted" Idi Amin's house. These were just some of the souvenirs. "Say no more," said the officer dialing the police headquarters in Livingston. "It will only be worse for you."

"Oh, come on," I said, "what would a spy for Idi Amin be doing with twenty Idi Amin T-shirts? Passing them out to other agents? This Christmas card isn't to me, I've never even met Idi Amin! And everyone in America carries one of these tear gas pens!"

No matter what I said, he was convinced that I was a spy. While we were waiting for a squad car to come and pick up Ann and me to take up to headquarters in Livingston, I made friends with them as best I could. My main concern was for Ann; I didn't want her to get into any trouble.

The officer was still going very carefully through my passport pouch and as he found the small leather pouch which the ranger had given me at Ngorogoro Crater.

"What's this?" he said, picking it up and eying it carefully.

"Oh, no!" I thought. I had completely forgotten that I had it; in side was the strange bark we had taken at the crater. "It's a ..." I said, thinking hard, "some herbs for diarrhea. We tourists sometimes get the runs out here in Africa."

"Oh," he said, "I've been kind of sick the last few days..."

"It's all I have left," I countered and prayed. Remembering the bizarre effects that it had had on me, I knew I would spend the rest of my life in prison in Zambia if he ingested any.

"Oh, all right," he said. "I was going to stop by the pharmacy this evening anyway."

"Let's go over to the restaurant at the Rainbow Lodge and have a beer while we're waiting for the squad car," I suggested. He said OK and sent the other policeman with Ann and me.

We sat by the Zambesi drinking beer. I bought one for the officer, whose rifle was trained on me all the time, and we watched the sunset over the Zambesi. On the other side of the river was Zimbabwe-Rhodesia. The thought occurred to me that I might escape by swimming across. If I could swim across, and the policemen missed me with his rifle, if I wasn't swept over the falls which were only a hundred meters downstream, and if the crocodiles didn't get me, I'd be safe...

"Don't try and swim across," said the policeman, reading my mind. "I'd hate to have to shoot you."

"Never even occurred to me," I said, finishing my beer. "Would you like another one?"

"No thanks. You seem like a nice guy," he said, "too bad you have to go to prison."

"I'm just an American tourist. Really," I said confidently. He raised a knowing eyebrow of disbelief, and steadied his rifle.

Eventually the squad car came to take Ann and me to Livingston. Two policemen dressed in light gray, neatly pressed uniform shorts escorted us into the gray Landrover pick-up. Ann rode in back, and I was put up front between the two policemen. I told them my story, hoping to make friends with them, for they seemed like nice guys.

In the Livingston headquarters, the police, half a dozen of them, spread all the evidence on a table to be viewed.

"Wow! Look at these!" they exclaimed, picking up the T-shirts. "Are you selling these?" they asked hopefully. They browsed through the documents, knife, tear gas pen, etc., and when I told them the story of how I had picked up this stuff, they were fascinated.

Meanwhile, a rather suave, handsome and quite intelligent detective took Ann into a back room to question her. I talked with the police for a while and then wrote in my journal. After more than an hour had gone by, I became worried about Ann. In my mind I imagined the most brutal things that could be happening...she could be being raped, it had happened before...

To my relief Ann came out, and I could see by her composure that she was all right. The handsome detective now beckoned to me and I followed him into a plain room with white walls empty except for a table and two chairs. He had a big lamp to shine in my face, but it was turned away toward the wall. The detective was dressed in plainclothes—a white shirt with the sleeves rolled up, a shoulder holster and pistol, gold watch, black dress pants. He sat across the desk from me and looked at me with a calm intelligent face. We talked for an hour or so. He wanted to know how I met Ann, where I got all these things, what I was doing, where I was going.

After I had told him the whole story, including my involvement with Idi Amin,

I said, "So you see, I'm not a spy for Idi Amin, the Rhodesians, America, or anybody, I'm just an American student coming from Tanzania."

"Yes, I believe you," he said quietly. "But there's one more thing before I let you go..." He paused, trying to frame the words. "Umm, could I have a T-shirt?"

"Of course," I grinned at him, and he smiled back. I gave him a T-shirt and got ready to go, packing up all my stuff. By now it was after ten o'clock. The detective ordered a taxi to take me back to the Rainbow Hotel, but wouldn't allow Ann to go back with me. I'd already caused enough trouble for one day, so I left without arguing. The taxi driver, however, after letting off two other passengers, told me it would cost me about ten dollars to go out to the Rainbow Hotel.

Thoroughly incensed at this, I told him to take me back to the police station, where I blew up; I told them I had been there for seven hours, that I hadn't had any dinner, and the restaurants were now closed, and on top of that they wanted me to pay for the long ride back to my hotel. That was rude, I informed them. They agreed with me and said I could take the next available squad car and just then the two fellows who had brought me there in the first place pulled up.

"Take this man back to the Rainbow Hotel," the desk sergeant told them.

"Glad to," said the driver as I got in. "Good to see you again, kid. Where's your girlfriend?"

"I don't know," I told them. "They made me leave without her."

"We just saw her walking down the street toward the falls." As we drove out of town, we saw her walking along the side of the road. The police stopped the car and she and I got in the back seat and I gave her a hug. "What happened?"

"The detective made a pass at me and wanted me to go back with him," she said. "They tried to get me to tell them that you were a Rhodesian spy. Oh dear..."

It was cool and crisp in the back of the Landrover going out to the falls. After all that, I felt I had known Ann for years. The police let us off at my bungalow, I still had a lot of stories to tell her...

§§§

In 1967, a very scholarly and unusual article was published in *Nature* magazine (Number 216, pages 407-408).[33] Written by R.A. Dart and P. Beaumont, the article describes the "only ancient manganese mine yet recorded" and it is at Chowa, near Broken Hill, Zambia. Says the article, "Its rubble infilling contained crudely flaked mining tools chiefly of manganese, used as choppers, wedges and chisels, together with hammerstones, perforated stones and a single polished stone axe." Manganese is a hard, brittle, grayish-white, metallic element used chiefly as an alloy for steel to give it toughness. Such steel is generally known as Manganese Steel and it is also used in bronze. Manganese Bronze is an alloy of about 55% copper, 40% zinc, and up to 3.5% manganese. Was this ancient mine part of a steel manufacturing plant?

Before you say, "Why not?", I had better inform you about the rest of the article. According to Dart and Beaumont, workings at the entrance of the mine were dated by Dr. Minze Stuiver as 9640±80 B.P. or 7690±80 B.C. which is clearly in the "late Stone Age" say archaeologists. More incredibly, charcoal from the lower levels of the mine were carbon dated at Yale and Groningen laboratories as 22,280±400 B.P. or 20,330±400 B.C. from the Yale Laboratories and 28,130±260 B.P. or 26,180±260 B.C. from Groningen Laboratories. A manganese mine from 25,000 B.C.! The authors conclude that mining has been going on for 28,000 years on and off at Chowa, and mention that other mines in Southern Africa have been dated as ranging from 37,000 to 42,000 years Before Present (B.P.). Where is that civilization that was mining manganese, copper, zinc and iron 28,000 to 42,000 years ago? What were they doing with the metals? Where are their cities? Manganese is known as a strategic metal because of its strange properties, lightness and strength. For instance, today it is widely used in spacecraft alloys. Were Vimanas of ancient Rama and Atlantis using manganese alloys from the mines in Southern Africa for the hull? Such questions, naturally, are not brought up in the article. And, remarkably (or maybe, not remarkably), this earth-shattering evidence of advanced ancient culture in Southern Africa has not fazed the academic world one bit, and it is the rare book on African Civilizations that even makes a passing mention of it. After all, when the facts don't fit the prevailing theories, it is better to ignore them!

I hitched back to Lusaka a few days later. Then, I decided to hitchhike from Lusaka to Chipata in eastern Zambia and on to Malawi, a long thin country along Lake Malawi. The day before I was going to leave I met a young political student from West Berlin, a girl named Bubbles. I invited her to hitch with me to Malawi.

"You're the first other traveler I've met since I flew to Zambia from Berlin," she said. "I thought I'd meet lots of other travelers."

"Yeah, not too many travelers in Zambia these days; I suppose it's because of the war in Zimbabwe."

"All the whites should leave Africa," she blurted out hotly.

Suddenly, a black who was listening to us talk from the next table at the ice cream parlor said, "And all the blacks should leave America!"

I laughed so hard, I spilled my bowl of ice cream. Bubbles didn't find this very funny, however.

We decided to hitch out the next day along the road toward Chipata in eastern Zambia. This road was said to be dangerous as there are many Zimbabwe guerrilla training camps along it. We were warned, but for some reason I thought I'd gotten into enough trouble for one country. I was wrong.

We started out about midday, got a few short lifts out of town past the airport, and then an Englishman from India gave us a long lift, thirty miles or so, to a small truckstop. We got out and stood by the side of the road, waiting for a car or truck to come by.

Bubbles, with her short curly hair, cute button nose, and terribly pale skin, sat down on our packs. "We'll never get a ride!" she said.

"Come on, I replied, "we've only been waiting fifteen minutes." A large truck with a semi-trainer pulled up at the small bar next to a filling station. I explained how we were looking for a ride toward Chipata and followed him into the beerhall. We stepped up to a long wooden well-worn mahogany bar and each ordered a beer.

"Do you think you can give us a ride?" I asked. Suddenly someone tapped my shoulder, and, as I turned around, I became aware that I was the only white in a bar absolutely packed with blacks—that was no surprise, but now I realized that the crowded room had fallen into dead silence.

I found myself staring directly into the face of the biggest, meanest, most scarred and ugly black guy this side of the Congo gold mines. I could smell a strong scent of alcohol on his breath as he said gruffly in pretty good English, "Who are you and what are you doing here?"

Now there are times when a person's life is in great danger and his health and well-being may depend on his quick and rational thinking. There are other times when he should panic and run screaming in whatever direction his feet happen to be pointed at that time. I decided that this was not one of the latter times.

I glanced to my left at the driver, as if to say "What's going on?" He looked away and it dawned on me that I was face to face with a Zimbabwe guerrilla fighter. Not only was he a guerrilla, but I now realized that the whole room was full of guerrillas and this bar was right next to a ZANLA training camp. A powerful shiver shot straight up my spine to the cortex, electrifying my brain. Adrenalin surged into my stomach and I broke out into a cold sweat. There were probably half a dozen or more guns pointed at me right now and I was about to start leaking red stuff all over the cigarette-strewn floor if I didn't come up with some good answers mighty fast.

"I'm an American tourist and I'm coming from Tanzania and I'm going to Malawi and I've never been to Rhodesia and please don't shoot me, mister— here's my passport." In a jiffy I produced my trusty little American passport and handed it to him. Fortunately I always kept it in a pouch on my belt and it was handy to get at.

"See, I just came two weeks ago from Tanzania," I said nervously, pointing to the Zambian entrance stamp on my passport.

The bar was deadly silent; you could have heard a safety being flicked off an AK47 at fifty yards. The big, burly guerrilla looked over my passport for a minute, flipping through some of the pages. I knew that it didn't really mean a whole lot to him, but it might serve to squeeze me out of this alive.

Looking me right in the eye he handed the passport back to me. "You're OK," he said calmly, "but if you were from Africa..."—meaning from South Africa or Zimbabwe I supposed—he made a quick slice across his throat with his finger to indicate that I would be pushing up African violets in a bog in the back of the saloon.

I fairly melted back onto the mahogany bar. The truck driver reached over to support me. "Here, finish your beer," he said. With one big slobbery gulp I downed the last of my Mosi-O-Tunya Lager and made a beeline for the door.

Back across the road Bubbles said I looked like I'd just seen a ghost.

"I did," I told her, "my own!"

Fortunately it wasn't long after that that we were on our way to Chipata, cuddling up in the back of a little Japanese pick up driven by two rather strange African Jehovah's Witnesses. The sun was coloring the dry African forest in flames of orange, yellow and red. This was the end of one adventure, many more were sure to come. I didn't want to be negative, I thought, but Zambia had a long was to go before it was a tourist's paradise.

§§§

Those moments in the Zimbabwe guerrilla bar stayed with me for quite a while. I resolved that I'd have to straighten out my act and stop precipitating all these stressful situations in my life. Why was I attracting these things in my environment? I couldn't spend the rest of my trip in Africa in police stations and on the edge of danger.

"Perhaps you should stay out of war zones," said Bubbles, her brown curly hair flapping in the wind as we got a lift on the back of a truck into Malawi from Chipata.

I suppose this is just one of the dangers of traveling in Africa; after all, one third of the world is actually at war in some form or another, and Africa is especially ripe for a major war and the settling of old scores. It soon becomes apparent to the traveler that borders are quite haphazard and arbitrary. If the British had a fort here, the French a fort there, and the Portuguese or Germans a fort somewhere else, well, voila! you've got three African countries.

Entire African nations and tribes were divided by the colonial powers and remain divided today. But no matter what political philosophers say, Africans, as well as other people all over the world are tribally, not nationally, oriented. It's a mess, and unfortunately won't be sorted out without some bloodshed.

The month we spent in Malawi was like an idyllic vacation. Time was spent traveling down lazy country roads, sleeping on beaches and drinking Malawi-brewed Tuborg beer in friendly little Malawi pubs.

It had taken two days to get from Lusaka to the Malawi border a few miles east of Chipata. Bubbles and I then hitched to Lilongwe, the new capital of Malawi, that same day and went on the next day to Salima on the shores of Lake Malawi. Soaring mountains and the long inland sea of Lake Malawi have justly earned Malawi the nickname "The Switzerland of Africa." The resem- blance stops at the scenery, however, for Malawi is statistically one of the poorest and most underdeveloped countries in Africa, having very few natural resources to support a large population. At any one time, 300,000 Malawians are working as immigrant laborers in South Africa and Zimbabwe.

Malawi is one of the few African countries that feeds itself, however, and has some very rich farmland. Its exports, naturally, are agricultural: tobacco, tea, peanuts, corn, and cotton. Tourism has also become a major industry in recent years, largely due to Malawi's continued relations with South Africa and its picturesque 360-mile-long lake which has nice beaches and is one of the few freshwater lakes in Africa that is "officially" free of bilharzia, a blood fluke that infects the waters of Africa, making swimming hazardous to your health.

Malawians are an especially cheerful and friendly people. As Bubbles and I sat in the back of an army truck driven by several handsome soldiers, I couldn't help noticing the three pretty and winsome bank tellers who were also hitchhiking to Salima for the weekend. Ample and full of bright smiles, they immediately attracted me to Malawi and her people.

"We're going to have fun in Malawi!" I told Bubbles.

"I'll keep my eye on you," she returned, jabbing me in the ribs.

Farther up the lake is Nkotakota, a town already old when David Livingston arrived in 1859. Sitting beneath a giant elm tree in the spacious center of Nkotakota, having a shave, I decided that Nkotakota had not changed at all since 1859. Green tree-lined dirt roads stretched along the lake, lined with small wooden huts, shops and businesses. Dugout canoes were pulled up on the shore (they would be gone in the early mornings when the fishermen paddle out onto the lake with their nets). Women walked by, ever smiling in their bright African kanga skirts, carrying baskets of fish on their heads, shaking their hefty derrieres like the fish that flopped in their baskets.

And here, under the gigantic elm, which I privately named Livingston's tree, was a chair, a plastic washbasin, and a sign resting against the bottom of the chair, hand-painted by some budding young English student. It said "barber."

Standing by the chair was an elderly man. His black curly hair had long ago turned gray. Naturally he was clean-shaven; he was also barefoot and wore a piece of blue cotton cloth wrapped around his waist; his skinny chest was bare except for a few gray hairs. He stared wistfully off over the lake, all shiny and silver-bright in the midday sun. Even though I had already shaved that morning, I knew I had to, just had to, have a shave by this old African under Livingston's tree.

It was sheer pleasure looking out at the lake and eying the women passing by as he lathered my face with a real English boar bristle brush. "Did Livingston get shaved under this tree when he was here?" I wondered out loud.

A Scottish missionary and explorer, David Livingston was a fine person in my estimate. His travels around Malawi (which he named Nyasaland) brought the British attention to the horrors of the Arab slave trade in East Africa. His writings were instrumental in helping to halt the slave trade by the 1890's.

His revulsion of the Arab slave trade prompted Livingston to establish Scottish missions, and he promoted business groups to develop legitimate commerce as a substitute for the trade in human beings, a practice perpetuated by African chiefs and warring tribes as well as by the Arab slave traders. This allowed Britain to gain a foothold into the country and by 1891 Nyasaland was made a British protectorate.

Boarding the ship at night at Nkata Bay, I was once again traveling alone. Bubbles was traveling up to Tanzania to meet a friend and I was heading south. Sitting up on the top deck, talking with an English chap who had traveled down from Tanzania to visit his brother in southern Malawi, we were both slightly startled when a tall, slender Malawi soldier plopped down beside us and began to set up his wireless radio. He was slightly drunk and began telling us how much he loved his country and that Mozambique, the Marxist country across the lake, was wanting to invade Malawi.

"We have to keep our eye on them all the time," he said, patting me affectionately on the back. "They want our country."

"Why do they want your country?" asked the English fellow.

"Because of our women—very beautiful women in Malawi." He was friendly and eager to talk with us. I figured he was only joking about the women, but it was true that Malawians think that Mozambique would like to invade their country, though it seems unlikely to me, since Mozambique has plenty of problems of its own without invading Malawi. Still, nearly every African country indoctrinates its citizens with ideas that the country has some specific enemies in order to foster nationalism and a sense of national unity. Often the threat is imaginary, but it is a useful tool for controlling a populace. In actuality, said my English friend on the boat, every country does this, making use if its enemies, real or unreal, to mold the popular opinion of the citizens.

"Here, listen to this," the Malawi soldier said, handing me the headphones to his radio set. "Radio Mozambique!"

I listened for a while and heard someone speaking French; it sounded like Radio Madagascar to me. In Mozambique they would probably be speaking Portuguese. I handed the headset to the English fellow and the soldier handed me a can of corned beef and a can of mackerel in tomato sauce, which we all shared, and before you knew it we had a little party going!

"This is what traveling is all about," I thought, "meeting new friends and sharing their lives." It was really a joy to be listening to that radio and eating mackerel under the starry Malawian sky while the waves gently crashed against the hull of the mini-liner.

We were at Likoma Island by noon the next day. Here there is a rest house and a mission station established in 1885. It lies a few miles off shore from Mozambique and has a cathedral rivaling Westminster Abbey in size. At the end of the trip, two nights later, we arrived in Monkey Bay, the end of the line.

It was a further day's ride by bus to Zomba, the old capital, but I just passed through on my way to Blantyre, the largest city and business center of Malawi. That night, after I set up my tent at the Government Rest House, I suddenly felt chills all through my body and had a splitting headache. I put on a jacket, but was still shivering. I sat down at a table in the restaurant to wonder what was wrong. After a moment of thought, I realized that I must have contracted malaria, probably on the lake. I took a curative dose of chloraquine and two aspirin and got into my sleeping bag.

The next morning I felt better, but had a slight tingling sensation throughout the blood in my body. I was cured. I rested in Blantyre for a few days, regaining my strength. It would take more than malaria, prison and the Batmen of Africa to keep me down.

The Kongamato

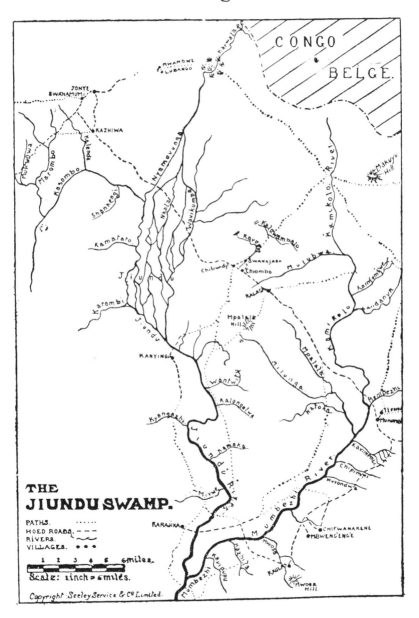

CONGO BELGE.

MWANOWI·LUBANGO
JONYE SWANAMUMU
KAZHIWA
Mukubwa
Kalemda
Kasombo
Masombo
Shankenji
Kamataro
Ntamovunga
Ntalai
Kavikumpu
Makuyu Hill
Kalowambalo
Kavo
Chibundji
SWANAJABA
Chiombo
Mukubwa
Kamikolo River
Jiundu
Karembi
KALAO
Karumunga
Kandanya
Mpalala Hill
Jiundu
Mpalala
KANYINGA
Wantwix
Kalangelwa
Kafola
Mambezhi
Kyangeshi
Jiundu River
Kamaka
Dilunga
Kavimunga
Chipimpi
Muvondwa
Mivuo
Mumbezhi River
KARAJIKA
Kazhibu
Miwoa
Muvondwa
CHIFWANAKENE
MBWENGENGE
Mumbezhi River
Kavupupu
Kaula
Miwoa Hill

THE JIUNDU SWAMP.

PATHS. . . . . .
HOED ROADS. - - -
RIVERS.
VILLAGES. • • •

1 2 3 4 5 6 miles.

Scale: 1inch = 6miles.

THE CRASH OF ITS FALL STARTLES THOSE WHO WALK IN THE MISTS OF THE LOST LANDS...

WITH A RUSH THAT SHAKES THE WORLD, THE GIGANTIC LIZARD HURTLES FORWARD. ITS GIANT JAWS GAPE WIDE, AND CRUNCH!

RAGE FLOODS THE JUNGLE MONSTER! HIS HEAD SWINGS AND HIS JAWS TIGHTEN! AS A DOG WOULD SHAKE A RAT, HE SHAKES THE BIG PLANE—AND A LIMP FIGURE DROPS EARTHWARD...

HISSS!

HISSSS

HOLY COW!

A Frank Frazetta comic book from the 50s about an American pilot who crashes in the African jungles. He faces such dangers as the Kongamato and the Mokele Mbembe.

Chapter 15

ZIMBABWE:
KING SOLOMON'S MINES AT OPHIR

If a man but had sense to see
Each time Death brushed his elbow
He would recognize himself to be but a Pilgrim
Traveling between clashing armies.
—*Beddoes, Death's Jest Book*

I headed back to Zambia on my way to the Zambian-Zimbabwe border. That meant a night in Chipata, sleeping outside on the lawn of the rest house because all the rooms were full. Not wanting to risk getting blown away by a nervous black nationalist guerrilla who might mistake me for a Rhodesian soldier, I decided to skip hitching and take the bus back to Lusaka. That was one day, and it was another back down to Livingston, where I spent the night.

It was a short morning from there to the Kazungula Ferry, the only point in the world where four international borders meet. The border consists of two small ferry stations, one on the Zambia side, and one on the Botswana side. This twenty feet or so, separated by the river, is the only place where the two countries have a common border. Also coming to a point at Kazungula are Namibia and Zimbabwe. The ferry isn't too impressive, really. The Zambesi River meanders along slowly through the cattails, papyrus and grass. I crossed on the ferry, and was stamped into Botswana. Later that afternoon I was crossing the Botswana-Zimbabwe border at Plumtree.

I gripped the straps of my backpack tighter as I started walking down the main road into Bulawayo, the second largest city of Zimbabwe-Rhodesia, as it was officially called at the time. I was interested in Zimbabwe-Rhodesia for a number of reasons. I've always had a healthy distrust of politicians; as Lord Acton said, "Power corrupts and absolute power corrupts absolutely." I don't see myself joining any political extreme, so being the Zen Cowboy that I am, I resolved to experience this country with an open mind and with as few preconceptions as possible. I also admit that a country involved in a civil war piqued my curiosity as I would probably never have a better chance of witnessing the birth throes of a nation. The conflict was reaching a crescendo, and both sides had said they were willing to negotiate.

I spent a night at the Youth Hostel in Bulawayo and caught a ride all the way to Harare (called Salisbury in those days) the next morning. In most areas of Zimbabwe at the time, people traveled in armed convoys, their passenger vehicles interspersed among "escort cars" which were usually Mazda pickups with a rotating machine gun manned by a helmeted soldier. Often several other soldiers with their automatic rifles sat in back with the machine gunner and his Browning 50-caliber machine gun. There were usually three such vehicles in a typical convoy, one placed at the rear, one center, and one in front.

It was only a six hour straight drive to Salisbury. Fortunately my lift was going straight through in convoy, and we made great time. The couple who had given me a ride, a dentist and his wife, gave me their address and told me to visit them at their comfortable suburban home, "and take a dip in our pool." They let me off at the Salisbury Youth Hostel, a large, rambling old wood frame building on the northwest side of the city.

I fished through my passport pouch for my Egyptian youth hostel card while a bald, elderly white man waited patiently behind the counter. This was no ordinary youth hostel full of friendly, fresh-faced Canadian girls with braces on their teeth. A sign at the reception desk said, "No machine guns, pistols, or explosives in the dormitories." No one paid much attention to that sign, however; the men's dormitory was a miniature arsenal. Almost everyone had a locker full of grenades, cartridges, pistols, and the occasional automatic rifle.

Once called Southern Rhodesia, Zimbabwe is an amazing country with a great deal to offer the traveler. A grassy, landlocked plateau with good mineral, manufacturing, and agricultural industries, Zimbabwe has one of the highest standards of living in Africa, and for the whites, it once had what has been called the highest material standard of living in the world. It was not uncommon for a salesman or garage mechanic to have an Olympic-sized swimming pool in his villa-type house. All made by the sweat of the blacks, of course, many of whom come from Malawi in order to find work, and are just happy to have a job. Still, for all that might be said about exploiting black labor, the standard of living and education was far higher than in any other African country; with independence the economy is expected to decline as the government socializes more and whites continue to leave the country, mostly going to South Africa. Civil war and tribalism among the different African groups in the country could take its toll in Zimbabwe, but if Zimbabwe can get it together, it has the most promise of any African State.

Zimbabwe was first opened to settlers in 1888 after David Livingston discovered Victoria Falls in 1855. Cecil Rhodes, a British financier, negotiated certain mineral rights for the British South African Company, and this eventually paved the way for white settlers in Northern and Southern "Rhodesia."

In 1952 Britain established an economic community involving Nyasaland (now Malawi), Northern Rhodesia (now Zambia), and Southern Rhodesia (Zimbabwe). This was dissolved in 1963 and Nyasaland and Northern Rhodesia were given independence. Britain wanted Southern Rhodesia to give the African population more representation, but the white government resisted. Eventually Southern Rhodesia made a "Unilateral Declaration of Independence" (UDI) and broke away from Britain in 1965. In an effort to force Rhodesia to its terms, Britain had sanctions placed on the country and the government declared illegal by the United Nations. A long and difficult guerrilla war followed, with several black nationalist guerrilla groups fighting a war of terrorism and harassment against the regime of Ian Smith, the white prime minister.

The economic sanctions placed on Rhodesia stimulated the economy to such a degree that the country attained a surprising amount of self-sufficiency and manufacturing capability, its imports coming from South Africa, the only nation that failed to comply with the United Nations' call for economic sanctions. Rhodesia, the colony in rebellion, was cast adrift by the world, its only friend South Africa, who shipped out the country's exports of tobacco and minerals.

The civil war reached its peak in 1979, one year after three moderate black leaders had agreed to be in an election which would give black majority rule, but guarantee the whites certain rights. Robert Mugabe and Joshua Nkomo, two opposing Nationalist leaders who often shot at each other as much as at the Government Security Forces, called the election a sham, and refused to participate, and instead intensifying their war against the new government. It was just at this time that I entered Zimbabwe, as the new government, headed by the African clergyman Bishop Muzorewa, was negotiating with the Patriotic Front, the alliance between Mugabe's Chinese-backed and Nkomo's Russian-backed guerrillas.

336

§§§

All sorts of travelers were hanging out at the hostels, mostly young males from English-speaking industrial countries, and a few other Europeans. They had come to find adventure and action, with a capital "A". As one fellow who showed up one day at the youth hostel said, "to kill somebody, white or black, I don't care." I was a bit aghast at the casualness of this statement. I certainly wasn't there to kill anyone. Like most of us, he was in his mid-twenties, youthful in appearance, and gave the impression that he could take care of himself. Still, there was a certain innocence (or was it naivete?) in his statements.

Typically there were about fifteen of us staying at the hostel, including five young women who came from Australia, New Zealand, and Britain. We'd sit around the hostel and drink beer, cook meals, and discuss the war or local and global politics, Africa in general, and swap traveler stories. There were plenty of things to do in Harare, though most of them centered around drinking beer.

One night a German fellow, named Otto, who was staying at the hostel came back about two in the morning in a state of shock. He kept saying over and over, "I don't believe it. He shot him right in the head! I don't believe it!" Finally he calmed down enough to tell us what had happened. It seems he had been sitting in a bar downtown with a black friend. The bar was essentially for blacks and there was only one other white man in the bar besides our friend Otto. Suddenly an African burst into the bar and shot the other white man, who happened to be sitting at the table next to Otto, point blank in the head. Otto figured he was next and dove in a panic under the table. But the African wasn't after Otto or any other whites. It seems the other white man had been fooling around with the African's girl friend and the murder was the result of jealousy and not racism.

Incidents of this sort happen all the time as weapons and explosives are very easy to obtain. There were other incidents where hand grenades were tossed into bars or fights were started in the toilets which ended in full-scale shoot-em-ups. Oddly, these fights were almost never racial in nature; they were more an expression of the frustration of a country at war, and the proliferation of weapons.

During the day it was nice to walk through the city of Harare, which is very spacious and attractive. During the months of September and October the many jacaranda trees are in full bloom—violet flowers cover the trees and color the streets as they drop their petals to the ground. The result is a rainbow city of purple trees and streets and parks glowing with pink, red, blue, and white flowers in bloom. There are many parks and gardens all over the city, as well as a number of beer gardens and sidewalk cafes to lounge in. In the center of town is a shopping mall with several streets closed off to allow leisurely shopping on foot.

The Rhodes National Gallery exhibits local and overseas art while the Queen Victoria Museum is mostly devoted to natural history. The nightlife in Harare-Salisbury is pretty wild, with some crazy clubs and discos to cruise at night. One night I returned to the hostel very late after drinking in town. I dozed for a bit and then had to get up about four o'clock and relieve the pressure on my bladder. As I lay back in my lower bunk, just about to fall asleep again, I heard the door open and caught the rays of a flashlight scanning the room. Hmmm, the night watchman possibly? The door started to close, and then he saw me looking at him and slowly the door opened again and the flashlight shot around the room. I was sleeping in the lower bunk closest to the door; I lay in my bunk and watched him warily.

"I kill you," he said softly; it was the voice of an African. "Give me your money." I could see a long slender black thing in his hand about the size of a rifle. "I kill you," he insisted again.

"Kill me? Jeez!" I thought, but kept my mouth shut. On the assumption that discretion is the better part of valor, I started carefully slipping out of bed and lay on the floor on the opposite side away from the door.

Suddenly, someone sat up in the back of the room. "Who's there, what do you want?" he said.

"Who said that?" stammered the thief, shining his flashlight toward the back of the room.

"He's a burglar," I warned from under my bed, "and he's got a gun!" My pack was sitting right by the door and my clothes were on top of it. The intruder grabbed my pants and some shirts from my pack, grabbed someone else's clothes from the top bunk and backed out swiftly, closing the door behind him.

I jumped up from under the bed, held the door closed in case he tried to return and flicked on the light. The other fellow who was awake, a Vietnam vet from Wisconsin, leaped out of his top bunk and grabbed an iron rod that was leaning on the far wall. A few other guys went for their lockers which covered the whole southern wall, and grabbed some pistols. When they heard what happened they asked me, "Why didn't you grab his gun?" "I would have filled him full of lead, could have had a great shot from my bunk."

Someone called the police, who made a quick and fruitless investigation. Since the thief had taken my best shirts and my only pair of pants, I was left without any clothes. In the morning I borrowed a pair of shorts and walked downtown to buy some pants. As an old African farmer had said to me a week earlier, "Life in the big city can be dangerous."

One afternoon I was sitting on the terrace of the Terescane Hotel having a beer when I noticed that the thin, balding man sitting next to me was reading *Soldier of Fortune,* an American magazine published in Boulder, Colorado, for "professional adventurers." He was wearing a T-shirt picturing a lot of people running in panic with the words "Look out, the Rhodesians are coming!" underneath it. This was a popular T-shirt in Zimbabwe at that time.

"Buy you a beer?" he asked when he saw me looking at him.

"Sure," I said, and moved to join him. He was looking at a blow gun ad in the magazine. He let me leaf through the magazine and I saw articles on "Southern Africa and guerrilla Wars," army training and lots of grim ads featuring tough-looking bearded brutes in camouflage gear lounging around with automatic rifles on their hips. In a grisly sort of way the magazine was funny, though it was obviously meant to appeal to people who dreamt of militaristic adventures.

Rhodesia was often accused of hiring "mercenaries," though the government denies this steadfastly, mostly on the grounds that they didn't pay enough to hire real mercenaries but had a regular army salary which didn't amount to a whole lot. Most mercenaries are paid quite a bit, anywhere from five hundred to three thousand dollars a month or more, depending on their expertise and the job. Looting and bank robbing is a preoccupation with "mercs." In the Congo during the sixties, "Mad Mike" Hoare, mercenary leader during the civil war, would put armed guards on the banks as soon as they entered a town, to keep his own men from blowing open the vault and robbing it.

Another famous "merc" was Rolf Steiner, a man who liked "cleanliness, beer, violence, and little else." He got his first taste of danger as a Hitler Youth skirmishing in Germany's last ditch defense against the advancing U.S. army. After the Germans surrendered, he enlisted in the French Foreign Legion and adopted their slogan, "Long live death, long live war!" Steiner was a colonel in the Biafran Army during the Nigerian civil war in 1968.

"Glory, hell, I fight for money," Steiner once said. Some mercenaries in Nigeria received four thousand dollars a month, plus looting. Mercenaries are said to "be looking for that strange 'something extra' in life" and find it for a few months or years living close to violent death, high on their own adrenalin.

As one merc said, "The smell of cordite (plastic explosives) is like perfume to me. I miss the sweat of jungle battles, the rat-a-tat of automatic weapons fire—and the prospect of knocking over a fat bank," Magazines like *Soldier of Fortune* would have us believe that this is the good life. What would they do if there weren't commies to kill?

My drinking companion was in a sorry state, but we talked intermittently between beers (we traded off rounds for the rest of the afternoon) and browsed through the magazine. He was of unusual parentage: he had a British-Rhodesian father and an American Indian mother. Apparently they had met in the U.S. during World War II. He was in the Rhodesian Special Forces, a group that lived in the bush with the "terrs" (slang for terrorists.) After he had described to me the horrible death of a buddy of his, he lamented the eventual end of the war.

"I don't know what I'll do. This war is my life. I love it. It's not the principle, it's the excitement."

"What will you do after the war?" I asked.

"I don't know; find another one somewhere, I guess."

After we mulled that over, we returned to the magazine. There were some T-shirts offered on the back page: "BE A MAN AMONG MEN: RHODESIAN ARMY FORCES." A large logo featuring a death's head wearing a beret was centered on the page. Underneath it was the slogan "LIVING BY CHANCE, LOVING BY CHOICE, KILLING BY PROFESSION." I wondered if any of my friends at the hostel had been lured to this country by *Soldier of Fortune* ads.

§§§

Drinking beer is the main recreation in Zimbabwe, and just about everyone goes at it with gusto. The gang at the Youth Hostel would either go out to a pub or just buy a couple of bottles and stay around the hostel. One day while quaffing a few beers around the kitchen table, a young Englishman, neat and clean-cut, came in. He looked like he'd just stepped out of a military academy. He was hailed in a friendly fashion by several of the guys at the hostel who knew him. His name was Mark, and he had come to Zimbabwe, like so many of us, to experience a real war. He was now a captain in the Guard Force, a minor division of the army used to control terrorist activity in farm areas. Primarily he went out on daily patrol with his small troop to police the farm areas in his district. Mark had a beer with us and then invited us up to his camp for the weekend. This was a fairly accepted way of recruiting people to join the army.

His base was an abandoned farmhouse surrounded by barbed wire with trenches inside the fence in case there should be an attack on the camp. There were about twenty men in his unit, all of whom were black except Mark and a West German sergeant. Three of us "hostlers" went on tour with him, and we were all a bit nervous. We sat in the back of an open armored troop carrier, with the sergeant and the black privates. As we drove out of the camp, everyone cocked his Belgian "Fabrique National" (FN) automatic rifle and buckled his seat belt. Seat belts were always used because one of the greatest dangers on patrol was the threat of hitting a mine. The floor of the troop carrier was mine-proof with solid steel plating, but there was still the possibility of a spinal compression from the shock of the explosion—the seat belt tended to lessen the shock. Another danger was that the engine could be blown right out of the hood and land in the back of the truck, a prospect that didn't thrill too many of us. The greatest danger of all was that a rocket propelled grenade (RPG) fired at us would hit the truck. These rockets or bazookas could devastate the entire truck, even with armoring and kill us all. For this reason, the troop carriers, as well as most armored vehicles, had metal shields and angled panels which were designed to deflect the rockets. With these thoughts in our minds, we rolled out on patrol.

339

Fortunately, that day, like most days, was uneventful. As we made our rounds to the farms down the country roads in our area, I couldn't help thinking about the black soldiers sitting with me in the back. They were draftees, conscripted into the army and forced to serve in the government troops. Probably they were neutral in the civil war, merely hoping the fighting would stop so they could pursue their normal work, get married, or go back to school. Perhaps all they really cared about was their next leave in Harare so they could whoop it up. At any rate, it was interesting that the great majority of Ian Smith's army was black. Of course, Bishop Muzorewa had been elected prime minister in elections run by Smith, but he was often thought of as a mere puppet of the white government by the foreign press. Still, it didn't seem to bother too many Africans. Many African nationalists actually stopped fighting at that time.

That evening I was just starting to feel safe as we guests and the two white officers sat around the farmhouse while all the black privates were in a separate bunkhouse about twenty yards away. A barbed wire fence separated us from the enemy. It was time to relax and have a beer. Mark told us how he though his own men had tried to kill him a few nights before as he sat at the dinner table. Suddenly the whole window next to where he was sitting had been blown out by a loud explosion; glass showered the room and narrowly missed him. Fortunately, the curtains were kept closed while he ate and it was impossible for the bomber to tell where he had been sitting. Assuming some of his own men had tried to kill him by firing at him through the window, he had all the troops assemble with their rifles and he personally sniffed the barrel of each one to see which had been fired, but none had been fired, apparently, as there was no smell of barrel powder.

The next morning a houseboy was cleaning up the room and found fragments of a "frag" grenade on the floor. Someone inside the camp had placed a grenade outside his window and detonated it, knowing that he would be having dinner at that time.

Hell, I thought, I not only had to worry about the guerrilla forces, but about the Government Security Forces as well. Feeling a bit uneasy, we all slept in the same room that night in our sleeping bags; I decided to sleep just beneath the window as it seemed the safest place in the room. If another grenade was detonated outside the window, I surmised that the fragments and glass would blow over me, hitting the opposite wall. As I dozed off into a fitful sleep, I remembered another T-shirt I'd seen in a bar in Harare, worn by a black soldier: "AIRBORNE: DEATH FROM ABOVE."

§§§

Eventually the dangers of war in Zimbabwe became a matter of everyday living. Talk of bombs, rockets, land mines, death and destruction became commonplace chitchat. When riding a convoy, you stopped scrunching down in your seat and tensing your rectum. When walking around a farm with a friend, you no longer kept your finger on the trigger of your ever-present rifle. I'd even grab my pack and check out of the hostel hit the road for a couple of days.

The people of Zimbabwe are essentially very open and friendly; it's fun to meet up with the different types that pick you up on the road. Once while hitchhiking from Harare to Bulawayo I got what was perhaps my most interesting ride in Zimbabwe. Standing on the road with my pack outside of some small town, I tried my skill at flagging passing cars. Eventually one stopped; it was a small Mazda pickup truck. I threw my pack in the back and got in the passenger seat.

The driver was African, and a very interesting fellow. Once a guerrilla fighter for one of the lesser-known liberation groups, he had lived in the bush for more than ten years. When the first elections were held the year before, he and his comrades had come out of the cold, so to speak, and been integrated with the government troops. He was now a colonel in the Zimbabwe-Rhodesian army. We talked for quite a while about what it was like being a guerrilla in the bush.

"We lived out in the bush most of the time. Hardly had any food. We'd just have a little nibble on some dried meat or a little fruit. It was pretty tough. And you know," he said, "we didn't hate the whites, we just wanted fair representation. We weren't like those other terrorists. We only attacked government troops, ambushing a convoy or shelling a police station. We never went in for this random violence stuff, killing women or children or even white farmers. Really, we need the whites; it would ruin the country if they all left. They have the knowledge, education, and drive that we need. They live here too, and we should all be brothers."

Just then he reached forward and said, "We also smoked a lot of *dagga* in the bush. When you're hungry and thirsty it gives you strength. If you smoke before some action, you're afraid of nothing—makes you very brave," he grinned at me.

"You mean you would smoke before you'd attack a convoy or something?" I asked him.

"Always. Everyone smokes. We may not have food, but we always have a good supply of dagga." I had to laugh at what he'd told me, and I realized it made perfect sense. Once while visiting a farm north of Harare we had heard on the inter-farm intercom that a farmer had been ambushed on his way home the night before and had been fired on for four or five minutes by seven or eight guerrillas with automatic weapons and not one bullet had hit his car! I laughed now as I realized why—they were stoned out of their everloving minds! The thought of a war being carried on by a bunch of stoned guerrillas really tickled me; what astonished me even more as I thought about it was that it was a moderately successful guerrilla campaign at that. I imagine that this is not the first time a war has been waged by a lot of stoned-out revolutionaries; it probably won't be the last.

"What would happen if we were suddenly stopped by guerrillas along this road?" I asked him after I'd stopped laughing. He stopped the car and got out. He was a tall man, powerfully built, with short hair, leaving just enough to stick a comb in the top of his hair and carry it there, as many Africans did. He had a kind face, but I got a bit worried at his abrupt manner of stopping. We both got out of the truck, and he flopped the seat forward. Reaching behind his seat he pulled out a G-3, the standard NATO issue automatic rifle, a common weapon among Zimbabwe government troops. I braced myself for a mad dash into the bush.

"Don't be afraid," he smiled. "You see, I always carry my G-3 with me wherever I go. I just wanted you to know that we're safe. Besides, all the guys out there know me. I'd love to run across a few of them right now," he said as we got back in the truck. "I'd tell them to give up their guns and join us. We'd give them all good positions in the army. They're my friends, really. Hell," he reminisced, "those were some wild days out there in the bush, but I'm glad to be back in civilization; I've got a wife and kids now!"

We came to the check point, a roadblock where cars were getting a cursory checking. He drove up to the checkpoint, and the white soldier at the roadblock, obviously recognizing my friend, waved him through. "You see," he said, "everyone knows me."

At this point, we picked up another hitchhiker, a young white soldier who was going up to the next town. He also recognized my friend and saluted him. Our new rider was blond and youthful, probably just out of high school. He was on his way to join his unit and wasn't too thrilled about it. We all talked for a while; there was a real feeling of camaraderie and hope. I sat between them, one a seasoned black guerrilla warrior and the other young, white and just beginning to see his future. They were working together for a new country they both could have a share in. As we rolled down the endless African highway, a glowing African sunset cast a rosy aura and I felt a sense of wonder that I should chance to be here and watch this nation's beginning.

It was good to be back in Bulawayo, the main city in western Zimbabwe and the second largest city after Harare. I had met an English traveler at the youth hostel in Bulawayo, a large and spacious home converted into a hostel that could hold some forty or more people. We were the only two staying at the Bulawayo Hostel. He was of medium height and handsome, with a thick, brown mustache and glasses. He was married and his wife was still in Britain. Like most people, he had come to Zimbabwe looking for adventure, and had come prepared with a Browning semi-automatic pistol which he brought with him from England. Walking the streets was a gas. It was like being in a western movie, walking down the street with a Texas Ranger bent on keeping law and order.

"Can't understand, mate, how you could come to this country without a gun. I'm surprised you're still alive!" he said, watching the shadows in an alley as we passed by.

"It's not that dangerous," I told him. "I've been here almost a month, and haven't been shot at yet."

"Don't worry," he said, "you'll get lucky." It was the earnest desire of many people here to get into a fire fight with "the enemy," the nebulous "them" out in the bush. It was just a thrill to carry guns and believe your life was in danger. Life could be so dull in Liverpool or Cleveland; this was where the Action was. But where was it? Why weren't they trying to kill us? We were white, weren't we?

"Let's go on into the dance hall here at the New Waverly," I suggested as we rounded the corner from the youth hostel. We had to check his Browning as we entered the hotel; a big black bouncer was at each door and frisked everyone as they went in, taking any potential weapons off them and handing them to the hat check girl behind him. This was very common in the black bars where fights and killing were most common. Those bars that catered mostly to whites, and there was no official segregation, didn't usually find it necessary to frisk the customers. Still, there were killings all the time.

After grabbing a cold Castle beer at the bar, we wandered over to the main dance hall. Walking through the two bars in the front, it seemed like it was mostly English men and younger African girls in their prime. The women found the older white men good providers and well mannered, and where else could an overweight clerk in his late fifties find a nubile twenty-year-old girl to marry?

The main dance hall was an absolute circus, totally packed to the brim with Africans jumping up and down to the rhythmic beat of a native "beat band."

"This is great!" I shouted over the music as we stood in the back and watched people dance. On stage was an attractive black woman belting out some familiar rock and roll song from the early seventies, shaking her breasts and bottom in time with the music. Nearly the entire crowd of about a hundred people were dancing with a fervor rarely seen in white bars anywhere.

"They sure do have rhythm, don't they?" said the Brit. Just then the music stopped, and the singer made her way through the crowd right up to the two of us.

"Hi!" she said with a big white smile.

"Howdy," I said.

"You are a super singer, dear!" said the Englishman.

"Thanks. My name's Anita. Want to buy me a drink?"

"Sure," we said, and then sat with her on the stage and chitchatted for a while until the band started to play again. Anita, the Englishman, a friend of Anita's and I then danced for a while. Between songs the Englishman suddenly took a look around the dance hall. It was packed with a hundred and fifty or more tall, sweating Africans. We were the only whites in the entire hall, and had been all night.

His hand suddenly went, involuntarily I suppose, to his hip, but his gun wasn't there. He looked at me with a slight expression of anxiety. "This would be a bloody bad time for a race riot!" he said, and we both had a good laugh.

§§§

After a visit to the Zimbabwe side of Victoria Falls, I hitched back along the highway to Harare, which takes a day, and spent another few days at a youth hostel there. After a few days, I decided to hitch over to the eastern part of the country on the Mozambique border, the main town being Umtali. It is only a hundred and fifty miles or so from Harare and can be easily traveled in an afternoon or morning. I caught the morning convoy out, which made things a lot easier.

I stayed for a few nights in Umtali, a town that would occasionally get shelled from ZANLA guerrillas in Mozambique, and then headed for Fort Victoria, the major town in the center of Zimbabwe and close to the country's mysterious ruins.

No one really knows when or by whom these tall stone structures and towers were built, but it is possible that they are three thousand years old or more. The first European in modern times to see the ruins was the German geologist, Karl Mauch, who was given to wandering, on his own and on foot, through the interior of the country. He came across the array of ruins in 1872, partly through the guidance of local natives, and partly through the help of a German-American ivory hunter known as Adam Renders who had heard of the ruins before but dismissed them until Mauch told him of their importance. Although the ruins were overgrown with vegetation, Mauch could distinguish various different constructions among them and was convinced that the building on the hillside was an imitation of King Solomon's temple at Mount Moriah in Jerusalem while the large rounded structure in the valley was a copy of the palace where the Queen of Sheba had stayed when she visited Solomon.

The discovery, once it was announced to the world, soon conjured up visions of gold and precious jewels, and fortune hunters of all kinds flocked to the district. By 1900 there were 114,000 gold claims registered in the surrounding countryside, most beside ancient workings, and the desecration of the many structures began in earnest.

In 1895 the famous novel *King Solomon's Mines* by H. Rider Haggard was published. The book drew on the discovery of Zimbabwe Ruins, and on the legend of King Solomon's mysterious gold source of Ophir. It was a popular concept at the time that Zimbabwe was the fabled land of Ophir, site of King Solomon's mines. Some cities have been found in other parts of Zimbabwe and Botswana and there are many legends of cities lost in the Kalahari someplace, seen by a few bushmen and a lonely prospector. Rider's hero, Allan Quartermain, is given a map to the lost mines, and after many adventures, discovers a lost city with connections to ancient Egypt and the lost mine full of gold and precious jewels. Quartermain narrowly escapes with his life when he is trapped in the mines by an African sorceress. The popular sequel, *She,* has Quartermain returning to the lost mines, where he battles the priestess again, and the book takes on an element of reincarnation and mysticism.

Against the background of the destruction of the ruins the first archaeologists tried to ascertain the date for the ruins and cleared much of area around the ruins away. One of the treasure hunting groups had the dubious name of "The Ancient Ruins Company" and while they may not have discovered a great deal of gold, they certainly left a lot of ruins behind them when they were done! The ruins, located about ten miles south of Fort Victoria, are quite impressive and worth a trip. The main gallery is especially impressive with its round, incredibly tall tower and high walls. A strong defensive structure, can be found on a 90-foot precipice and is known as the "Acropolis". All the structures are made of local granite; flat, brick-like stones chopped skillfully from wide 'leaves' of naturally split granite rock, and laid without any form of cement, mortar or other bonding.

Archaeologists were split into two main groups as to the origin of the structures. Basil Davidson says in *Old Africa Rediscovered,* [102] "Zimbabwe, thought the first school, had a 'minimum age of three millenniums': there were two main periods of building, the earlier being Sabaen of from 2000 to 1000 B.C., and the later being Phoenician 'somewhat anterior to 1100 B.C. down to some time before the Christian era'. This school of thought reflected the King Solomon's Mines' pioneers and was resolutely sure that no 'natives' had ever taken a hand in this building of a civilization. It evolved many variations; and there is scarcely a people of high antiquity whose influence was thought to have been absent here at one time or another.

"The second school of thought—the archaeological and scientific school—first made itself heard through David Randall-MacIver, an Egyptologist who examined the stone ruins of Southern Rhodesia in 1905 on behalf of the British Association. He concluded that those at Great Zimbabwe and other of their kind were African in origin and medieval or post-medieval in date, basing this on an investigation of seven from which no object, as he said, had been obtained by himself or others 'which can be shown to be more ancient than the fourteenth or fifteenth centuries'."[102]

The two viewpoints became the center of a heated controversy in the years which followed, and did not stop until 1929 when the British Association sent a second expedition, led by the skilled archaeologist Dr. Gertrude Caton-Thompson, to try and resolve the matter. In her fascinating and exhaustive survey *The Zimbabwe Report* in 1931, she came down firmly on the side of the second school and concluded, "Examination of all the existing evidence gathered from every quarter still con produce not one single item that is not in accordance with the claim of Bantu origin and medieval date."

§§§

Yet, Zimbabwe is not so easily explained, though the isolationist anthropologists who claim that all civilizations (especially remote ones such as Zimbabwe) were the result of isolated tribes who began building, and had little or no contact with the rest of the world. In *Archaeology Discoveries of 1960s* [103] Edward Bacon reports that as a result of the building of the Kariba Dam on the Zambesi revealed an ancient burial site called Ingombe Ilede. When the site was excavated in the early 1960s by the Rhodes-Livingston Museum, Carbon-14 datings placed the burials between A.D. 680 and A.D. 800, several hundred years earlier than the traditional dates given for Zimbabwe. The community of the site seems to have been a rich one, their prosperity apparently derived from agriculture, with a large number of imported goods found in some of the graves. In these rich burials, capper wire bangles encase at least one arm or leg or both. Quantities of beads were were found of glass (principally red, green, turquoise, and yellow), of seeds, of iron and of copper. Four of the bodies had gold beads plus there were also some bangles of fine gold wire; and in one burial, gold sheet had been hammered on to a *conus* shell. These conus shells were traditionally the perquisite of chiefs and had a tremendous trade value. In Livingston's time, the rate of exchange was two *conus* shells for a slave and five for a 100-pound elephant tusk.[103]

This helps disprove the medieval thesis on Zimbabwe culture, with Dr.

Caton-Thompson stating that archaeologists "can produce not one single item that is not in accordance with the claim of Bantu origin and medieval date." Such an item was produced, says Bacon, with the publication of *The Rock Paintings of Southern Rhodesia* in 1967 by the Singer-Polignac Foundation, Trianon Press, by the late Abbé Breuil. One the great authorities on cave art work, Breuil did transcripts of cave paintings at Chamavara and the Chehonondo shelter near Mapuwire in the Chibi reserve of Zimbabwe throughout the 1960s. These paintings are striking figures and do not appear to be Bantu. Indeed, they appear to be Phoenicians or other peoples from the Middle East. One particularly interesting rockpainting (reproduced here in the illustrations at the end of the chapter) is of "The Great man of Chamavara". He is wearing baggy pants, pointed shoes, possibly shoulder armor and a helmet very similar to those worn by the Greeks. Bacon states that the Abbé's support is for "the first school" of Zimbabwe thought and that he believes that southern Africa saw, several thousand years ago, an invasion' of 'foreigners', parallel with the Bushmen, but radically distinct from them and basically 'Nilotic' in origin. And in the term 'Nilotic', Breuil indicates peoples similar to those of the great civilizations of Crete and Egypt; and these peoples, it is supposed, were 'bred out' with the Bushmen.[103]

Still, if Africans are upset because some scholars claim that the ruins were built thousands of years ago by invaders, we need ask, who are the invaders and who are the present population? The mystery of the identity of Kalahari Bushmen comes into play here, though it will be taken up in a later chapter. Are the ancient invaders and the local tribesmen one and the same? Yes, says a Bantu witchdoctor named Wuzamazulu Mutwa who claimed in a book entitled *Indaba, My Children* published in South Africa in 1968. Mutwa claimed that the Bantu tribesmen of Zimbabwe and the surrounding area were the descendants of a race of people who had red hair, red skin (pink?) and knew all about space travel, robots and radioactivity. It was these early descendants, said the witchdoctor, who actually built the fortress of Zimbabwe, calling it Zima-mbje.

Orville L. Hope has taken Mutwa's tale and interpreted it his own way in his book *6000 Years of Seafaring.*[45] Hope says that the pink men with red hair arrived about 2000 B.C. and were Celts who had settled in Southern Arabia about 2500 B.C., forming the Sabean Civilization. Says Hope, "Native historians of Bantu tribes in southern Africa state that Sea People came to the Zambezi river. They are described as having blue, green or brown eyes with long flowing black, yellow or red hair. Women and children were with the strange men, making a total of about 200 on board the ship. They all had Pink Skin! Also, there were some smaller, darker men of a different race. They were Canaanite or Egyptian priests, guides and slaves. They arrived long before the hated Arabi came among the Bantu people.

"The Pink Men exchanged metal weapons for grain with a witch doctor chief, and encouraged him to conquer neighboring tribes. More Pink Men arrived from overseas. They then burned the witch doctor's village and established control over the entire surrounding country.

"Natives of the new Ma-Iti Empire, who did not escape, were taken as slaves. Some natives fled as far south as Swaziland. Their descendants retain words of the Pink Men's language and still practice the Celtic custom of bleaching and stiffening their hair with lime.

"Women and children were chained to carts and sleds, and driven like oxen. They hauled stones for fortifications, and ore-bearing material, containing iron, copper, and gold to the smelters. The iron mountain of Taba-Zimbi became a maze of tunnels where slaves died at their work. Hundreds of slaves were sold overseas. Elephants were slaughtered for their ivory and hippos for their bones and blubber.

"There can be little doubt Pink Men and their Sabaean allies built the great stone structures of Zimbabwe. The Temple was like a temple at Marib, the Sabaean capitol, and it is near the Sabi (Saba's) river....The Ma-Iti Empire acquired great riches. They imported silks and ceramics from China. They brought trees and plants from overseas. Several species of foreign plants are still growing near various ancient ruins in southern Africa. Foreign imports were bought with gold, diamonds and slaves. Most of the diamond trade was with India where there were skilled stone cutters and polishers. The chief port was Old Sofala on the coast of Mozambique near the mouth of the Sabi river. Gold was smelted in caves of the Fortress at Zimbabwe. It was poured into soapstone molds, shaped like an animal's hide.

"The original Pink Men were skilled in metal and woodworking. They and their children taught slaves many arts, such as making beautiful ornaments of gold and excellent bronze and iron weapons. The following generations learned the arts, but did no work. They degenerated into a life of idleness and debauchery. There was almost no overseas trade, the Ma-Iti had everything needed or desired. Great communities of slaves grew up around the walled cities. The gates were locked at night, when only servants were allowed inside.

"Then came the slave revolt, probably encouraged by a notion far away, anxious for the riches controlled by the Pink Men. The revolt was well planned, and executed with savage fury. In the predawn darkness, servants inside opened the gates of all principal cities. Slave communities outnumbered their masters more than two to one. No battles were fought, a few groups were able to barricade themselves in buildings. The furious mob dismantled the buildings piece by piece and stone by stone. No prisoners were taken, not a Pink Man remained alive in the empire by midmorning. Then, the total destruction of the cities began, even the stones from city wall were carried away.

"The fortress and Temple at Zimbabwe were taken in the same manner. It was plain, wanton destruction, as described by Rhodesian archaeologists. Golden jewelry was trampled into the mud, art objects were deliberately broken. Tables, holding Chinese ceramics, were smashed and left lying where they fell. It seems that the mob destroyed everything in sight, then moved on. No looting occurred; that came later, when tourists and fortune hunters discovered the ruins.

If Zimbabwe was first used by Sea People and the Sabaean allies, and if the slave revolt included Zimbabwe, the Fortress certainly was used later by other people for smelting gold. Archaeologists found caverns in the Fortress, containing smelting furnaces and soapstone molds for ingots. One mold still had clinging particles of gold from the last pouring. Chinese or Cambodians, Egyptians, Arabs, King Solomon's Hebrews with Phoenician sailors, and possibly another wave of Celts mined for gold, copper and diamonds, before Europeans 'discovered' the land during the Dark Ages. The capitol of Ma-Iti was on Lake Makarikari less than 100 miles from Plumtree, Rhodesia. The lake is gone, nothing remains but a large salt pan. The city is gone, only stone foundations remain, covered with soil and overgrowth. Nothing remains of the Pink Men, except scattered Ogam script, carved in the Canaanite language, and a few artifacts. Old swords, pieces of armor, battle axes and small trinkets or magic charms made of solid gold are in the possession of tribal historians and witchdoctors. The relics are brought from their hiding places on occasion and shown to the tribes. So far as is known, they have never been seen by outsiders."[45]

Hope pretty well sums up the diffusionist anthropological theory on Zimbabwe, and his last statements are particularly revealing. Rather than Zimbabwe being the isolated center of a Bantu empire, it was part of a world-wide trading network, where gold and ivory was shipped out and glass beads, chinaware and silk were brought from the Orient, and other goods from Arabia, the Horn of Africa and

Persia. Even iron gongs were brought to Zimbabwe from the iron mines of the Congo in the distant past. While critics may maintain that it is a racial slur to say that the Bantus did not build Zimbabwe, this is not the case. Africans of one sort or another must have built or at least occupied Zimbabwe at one time—it is hardly a matter of racial problems. Bantus came from West Africa and migrated down into Southern Africa, probably, after the building of Zimbabwe. The Bantus themselves were great travelers, and had colonized parts of Mexico several thousand years ago. There in Mexico, they carved gigantic heads, unmistakably Negroid, and are said to have been the originators of the Olmec civilization. East Africa was the realm of Azania, and the inhabitants were Ethiopians and Nilotic, not Bantus, as well as having a large Bushman and Hottentot population. These two later groups were eventually forced into the inhospitable deserts of South-West Africa by the invading Bantus. Was it they who built Zimbabwe?

Many atlases and archaeology books still print a date of 1300 or 1400 A.D. for Zimbabwe, when this has been conclusively proven to be wrong! The archaeologist J. Theodore Bent who excavated the ruins in 1891 and wrote *The Ruined Cities of Mashonaland* in 1892 said that a Roman coin of the reign of Antoninus Pius (138 A.D.) was found in a mine shaft at Umtali.[104] It was during the thirteenth and fourteenth centuries that the Bantus migrated down into the area from the Congo region. It was not until this time that the kingdom of Monomatapa was then set up, among the already deserted ruins. In 1950, S.D. Sandes, Warden of the Zimbabwe National Park, found a log used as a lintel over a drain in the Elliptical Building (or Temple, as it is called). Under carbon-14 tests, samples of this wood gave dates between 591 A.D. and 702 A.D. with a spread from 471 A.D. to 794 A.D. (this gives an indication of how inaccurate carbon dating really is; yet to date an artifact within a few hundred years is doing pretty well).[104]

Therefore, Zimbabwe ruins are at least as old as the eighth century, if not from the fifth century. They may well be older. Since stone itself cannot be dated, then artifacts, such as the wooden lintel, or a bone, or perhaps a Chinese saucer or cup, are dated. However, it cannot be proved that the structures were actually built at the time, but that they existed at the time, and could be much older! For instance, a Dutch Brandy bottle from the 1700s was discovered in one dig at the ruins[104]. One might infer that the structure was built in the seventeenth century. This is the very logic often used by professional archaeologists, who should know better. In the early 1950s, the American archaeologist Suggs was excavating the massive, megalithic platforms in the Marquesas Islands. He discovered a French brandy bottle and a Civil War musket inside one of the huge platforms. From these artifacts, he concluded that the platforms were built in the mid-1800's. Yet, Herman Melville had been on islands in the early 1800's, saw the same massive platforms, and was told by his guides that they were built in an era many thousands of years ago in a time before the islanders even arrived on the islands! Suggs ignored such evidence, and the idea that these clearly European items had been placed in the platforms at a later date than their construction did not even occur to Suggs (for more on this fascinating bit of archaeological foolishness, see *Lost Cities of Ancient Lemuria and the Pacific* [35]).

This may well be the case at Zimbabwe. Massive stone ruins have a tendency to remain standing for many thousands of years unless they are destroyed or covered in an earthquake, volcanic eruption or devastating war. The only thing that would keep some wandering tribe from moving into a perfectly habitable, but deserted, city would be fear of ghosts or protecting spirits. Therefore, we see that it is easily possible that the ruins at Zimbabwe, and indeed, all over the country, are many thousands of years old, even dating to 1000 B.C. or more. Yet, the only datable items (aside from the Roman coin) go back to only the fifth century. Gold cannot be dated, except by style or inscription, though it is possible to carbon-date iron.

It is historically recorded that in 500 B.C., a Carthaginian named Hanno explored a good portion of Africa with sixty vessels, sailing down the west coast and possibly back up the east coast. Arabian sailors were already aware of such far-flung islands as Madagascar and Ceylon in the fourth century B.C., and as has been previously discussed, the Indian Ocean, known as the Sea of Zinj, was a well traversed ocean-highway even in the second millennium B.C. The colonization of Madagascar from Indonesia must have taken place at least four or five thousand years ago, if not earlier.

That Southern Africa was a Phoenician colony circa 1000 B.C. is a fascinating and controversial theory that will be explored in the last chapter. Was Zimbabwe the Ophir of the Bible? The fact that the river that early voyagers to Zimbabwe must have used for transportation is called the Sabi River is a very interesting clue that cannot be overlooked. The name would indicate that the ancient kingdom of Saba or Sheba, extended from Yemen across the Red Sea (then known as the Erythrian Sea) and extended from present day Ethiopia all down the East African coast to Sofala in Mozambique and possibly to Zimbabwe. In this theory, we see Zimbabwe as an African empire built by Erythiotes and Nilotics (Africans of East Africa, Ethiopia and along the Nile) rather than by Central African Bantus.

Yet, even if Zimbabwe was part of the Sabean Empire, was it King Solomon's Ophir? Probably not. The reason why I say that it was not is because it is too close to Solomon's Red Sea Port of Ezion-Geber. It was a three-year journey to Ophir and back, a fact often overlooked by scholars because they cannot conceive of ancient peoples sailing any great distance in antiquity. They place King Solomon's mines in Saudi Arabia or the Horn of Africa because it is within reasonable distance of what cowardly, early seafarers could have managed to sail. Yet, even Zimbabwe is too close to the Red Sea for it to be necessary for a journey of three years. With the monsoons and East African coastline for easy navigation, a one year trip, there and back, would have been all that was necessary. Furthermore, with the sophistication of the local agricultural society, food would have been readily available and it would not have been necessary to grow a crop of food.

Indeed, Zimbabwe may well have been part of the Sabean Empire, yet it seems that Solomon and his Phoenician crew went elsewhere for their gold. As I have stated earlier, there is more evidence for Ophir having been in Sumatra or Australia, than in Africa. Still, H. Rider Haggard's exciting and provocative books would never have been written if it were not for the early theories, whatever their merits. As Ms. Caton-Thompson surmised in her book after excavation the ruins, "Zimbabwe is a mystery which lies in the still pulsating heart of native Africa."

No matter what the verdict on Zimbabwe ruins, sophisticated men have apparently been living in Southern Africa for many thousands of years. According to Rene Noorbergen in his book, *Secrets of the Lost Races,* [3] under the subtitle, *Who Shot Rhodesian Man?* Noorbergen states that at the Museum of Natural History in London there is an exhibit of a Neanderthal skull discovered near Broken Hill, Rhodesia, in 1921. "On the left side of the skull is a hole, perfectly round. There are none of the radial cracks that would have resulted had the hole been caused by a weapon such as an arrow or a spear," says Noorbergen. "Only a high-speed projectile such as a *bullet* could have made such a hole. The skull directly opposite the hole is shattered, having been blown out *from the inside.* This same feature is seen in victims of head wounds received from shots from a high-powered rifle. No slower projectile could have produced either the neat hole or the shattering effect. A German forensic authority from Berlin has positively stated that the cranial damage to Rhodesia man's skull could not have been caused by anything but a bullet. If a bullet was indeed fired at Rhodesian man, then we may have to evaluate this in the light of two possible conclusions: Either the

Rhodesian remains are not as old as claimed, at most two or three centuries, and he was shot by a European colonizer or explorer; or the bones are as old they are claimed to be, and he was shot by a hunter or a warrior belonging to a very ancient yet highly advanced culture.

"The second conclusion is the more plausible of the two, especially since the Rhodesian skull was found 60 feet below the surface. Only a period of several thousand years can account for a deposit of that depth. To assume that nature could have accumulated that much debris and soil over only two or three hundred years would be ridiculous."[3]

Noorbergen concludes by mentioning the skull of an auroch, a type of extinct ox, which was discovered west of the Lena River and has been judged as several thousand years old at the Paleontological Museum in Moscow. The curator of the museum, Professor Constantin Flerov, was curious about a small round hole piercing the forehead. The hole has a polished appearance, without radial cracks, indicating the projectile entered the skull at a very high velocity. The auroch survived the shot, as is evidenced by calcification around the hole. The auroch later died of other causes. Says Noorbergen, "Rhodesian man was shot by a high-velocity projectile, but the bullet that killed him must have been fired at an early period in human history."[3]

I walked around the ruins with wide eyes. They are truly remarkable, and it is not without reason that they are a major tourist destination in Southern Africa. From my travels in South America, I was amazed at the similarities between the Zimbabwe Bird, a soapstone carving of a bird, and identical carvings found in Peru. Similarly, Charles Berlitz points out in his book, *Atlantis, The Eighth Continent* [4] that the strange city of Kuelap in the jungles of Northern Peru is identical in design, construction and style as Zimbabwe! Indeed, even the chevron design on the top of the walls is duplicated! Kuelap was built at an unknown time by a mysterious group of people known as the Chachapoyos who were said by the Incas to have blue eyes, red hair and be very tall. These people have literally vanished, leaving only empty cities of the same design as Zimbabwe in the remote and high jungles of Northern Peru (see *Lost Cities & Ancient Mysteries of South America* [22] for more information on Kuelap). Berlitz also points out the similarity of the massive stone fort on the Aran Islands off Ireland to Zimbabwe and Kuelap. While Berlitz may feel that this is evidence for Atlantis, more probably it is evidence for the strange sea-going men sometimes known as the Atlantean League, a courageous and bold group of seafaring traders who sailed the entire world in the millenniums after the destruction of Atlantis, trading and building great cities around the world. These people would have have been the forerunners of the Phoenicians and Celts that later roamed the world in a similar manner.

The Atlantean League and early Phoenician Celts used a type of simplified writing known as Ogam. Ogam consists of a system of parallel lines, often with a slash through them, along with stars, squares, and other hieroglyphic symbols. It is written, like Hebrew or Arabic, without vowels. Ogam is typically found in Ireland, early Britain, Scandinavia, Iberia, North Africa and in fact, in New England and all over the world. An Ogam inscription carved on a rock at Driekops Eiland on the Riet River in the Republic of South Africa was translated by professor Barry Fell of the Epigraphic Society to read, "Under constant attack, we have quit this place to occupy a safe stronghold."[45]

In Wuzamazulu Mutwa's book, *Indaba, My Children* there is a photograph of the author holding a magic slate that has been handed down for centuries and supposedly tells the history of Zimbabwe and Mutwa's people. It is unmistakably Ogam writing on the tablet, though I am not aware of any current translations of this slate. The many mysteries of Zimbabwe have yet to be solved.

My time in Zimbabwe had pretty much come to an end. It was a wonderful country and I felt that it had a great deal to offer the traveler who searched for adventure, lost cities and ancient mysteries.

I slept out in a trailer park near the South African border that last night and the next morning found myself walking toward the border in the chilly early morning air. I reflected that I would have made a lousy mercenary. I had looted Idi Amin's house, which would score points with any soldier of fortune, but I just couldn't get into knocking off banks. Carrying a gold bar in my backpack, whether ancient or newly-heisted mercenary gold, just didn't seem like it was worth the weight.

H. Rider Haggard in Egypt (left) and (right) the 'route' to King Solomon's Mines from the 1902 edition of his famous book.

 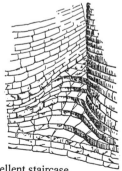

A section of Zimbabwe Ruins, and the excellent staircase.

A scene from Haggard's book, *King Solomon's Mines*.

1674 map of Africa. The empire of Abyssinia extends well into the interior of Africa. South of Abyssinia is the Empire of Mona, meaning the Empire of the

The great elliptical structure of the 'Temple' at Zimbabwe gave rise to many theories as to its origin: among them, that it was built by the same Sabaeans who built at Marib

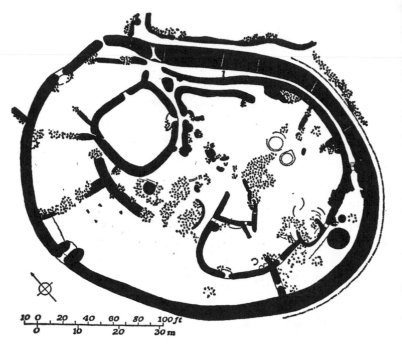

A plan of the Zimbabwe circular temple compared with the Temple of Marib in Yemen. Dotted line on the right side of the main wall indicates the extent of the chevron motif.

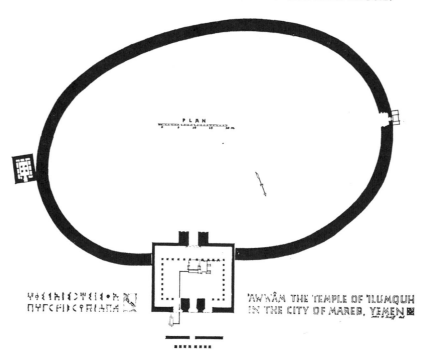

'AWWÂM THE TEMPLE OF 'ILUMQUH IN THE CITY OF MAREB, YEMEN

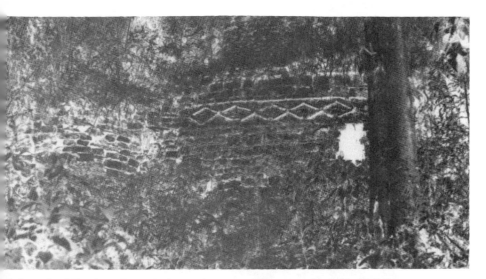

High wall of Kuelap, Peru. It is not known what race raised this building complex, but its similarity to Zimbabwe, even to the ornamental top of the wall, is notable. *Herbert Sawinski*

Ancient walls of Zimbabwe, the mysterious ruin that has given its name to a new nation, perhaps the most sincere compliment ever paid to an archaelogical monument. The high walls of this site and the apparent method of construction bear striking resemblance to other unidentified ruins on both sides of the Atlantic.

The Zimbabwe phoenix,
resembling the American
"thunderbird" totem.

The Zimbabwe monolith.

Credo Mutwa, Bantu witch doctor, tribal historian and author, holding his people's magic slate at Witwatersrand, Republic of South Africa. The slate contains Ogam script.

"Under constant attack, we have quit this place to occupy a safe stronghold."

The above inscription is carved on a rock at Driekops Eiland on the Riet River, Republic of South Africa. It is written in vowelless Ogam script in the Canaanite language, used in Libya and the Sabaean Kingdom of Arabia 3000 years ago, according to Barry Fell of Harvard, who deciphered it.

Look, two tribes come, great chief kill you

Men buried 1000 years on top of mountain

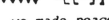

Yesterday we made peace, plenty in our land

Bantu inscription/hieroglyphs from the book, *Indaba, My Children* by Credo Mutwa.

"The Great Man of Chamavara", a rock painting in Zimbabwe depicting a man with a beard, pointed shoes, armor and a helmet, looking very much like a Phoenician explorer. From *The Rock Paintings of Southern Rhodesia* by Abbé Breuil, 1967.

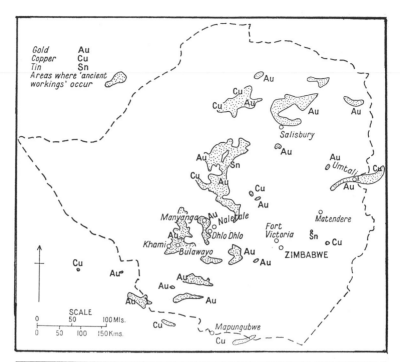

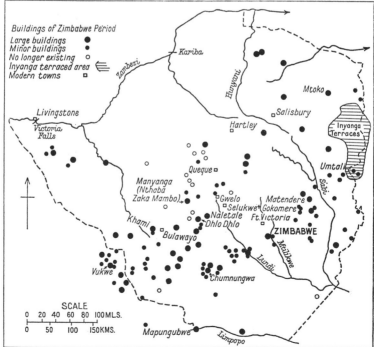

Two maps, the upper showing ancient mines (it is estimated that there were from 4,000 to 5,000 gold mines in ancient Zimbabwe) and the lower map showing ancient stone buildings (two to three hundred sites) in the same area.

Chapter 16

SOUTH AFRICA
& SWAZILAND:
GOLD RAIDERS OF AZANIA

He who knows others is clever,
but he who knows himself is enlightened.
He who overcomes others is strong,
but he who overcomes himself is mightier still.
—*Lao Tzu*

Once inside South Africa, I began walking down a small road heading south. I glanced to the west to catch a view of the afternoon sun; I was also looking for a ride to Johannesburg in the Transvaal. I hadn't a care in the world at the time; I felt I could just walk forever, absorbed in the freedom of the road and the eternal moment.

Roaring down the dirt road with a cloud of dust behind it came a small gray Peugeot pick-up. I waved and stood by the side of the road. My motions were friendly and definite, "Give me a ride, please." I pointed to the road at my feet, indicating that they should stop. I waved down the road to the south, indicating where I was going, and I waved at them just to let them know that I was a nice cheerful person. They kept their speed up, and didn't appear as if they were going to stop. I waved again; well, I might have to spend the night out here in the bush someplace, but that was all right. They drove by at 110 kilometers per hour and then suddenly slammed on the brakes and came grinding to a halt in a huge cloud of dust. I grabbed my pack and ran for the truck.

Two white men in their middle forties got out and took my pack, throwing it into the back of the truck. "Sorry, we almost didn't stop for ya, mon. Jump in!" they said in heavy South African accents, as we piled in the front, me sitting between them.

"Can't leave a white mon walking in this country! God, the heat will get to you if you're not careful. Where are you going anyway?" said the driver, a man of medium build in cotton work clothes and a typical South African bush hat made of well-washed cotton. They both had that odd Dutch-Afrikaans accent. I had heard it before and it was unmistakable.

"I'm on my way to Johannesburg. Where are you guys going?"

"Back home," said the taller and younger Afrikaaner sitting in the passenger seat. "We're foremen on a road crew and are heading back home to the other side of Potgietersrus for the weekend. Here, have some whisky and soda."

It was at this first meeting in South Africa that I discovered South Africans are, generally speaking, pretty heavy drinkers. I had a shot of whiskey just to be polite, and they finished off the bottle and started in on some blackberry brandy.

"Where're you from, anyway, lad?" the driver asked.

"America," I said, "I'm traveling through Africa; I've come all the way from Egypt to here so far."

"Ach, mon! You traveled through those bloody kafir countries?" said the younger guy on my right. "Kafir" is a slang expression for native Africans; it's derived from Arabic, meaning "heathen," or "unbeliever." The Arab traders and slavers originally called the blacks "kafirs."

"It wasn't so bad," I said. "In fact, I enjoyed it."

"Well, you're back in civilization again, lad. We know what it's like to be in those kafir countries; we work with 'em!"

This is a pretty typical conversation with a South African; many are racist, but fortunately, not all. In general I found South Africans friendly and hospitable—even the racists.

We stopped at a liquor store after an hour or so and they bought me a beer while they had another whiskey and soda and purchased a bottle of whiskey. We continued driving while they took nips from their bottle and I told them stories of Africa. It wasn't long before they were extremely drunk. As they carried on a one-way conversation about politics, mainly repeating over and over again that the President of the United States was an asshole, the car began weaving erratically from one side of the road to the other. Once I had to grab the wheel to keep us from going in a ditch. It's pretty hard to keep a *laissez faire* attitude when your transportation is swooping arabesques on a perfectly straight road. Their politics may have been bad, but their driving was worse. I was getting frankly scared. Fortunately we were on a back road in the bush and we hadn't seen another car the whole time. These guys were so drunk and happy to be going home to their wives after being out on the road crew for five days that they didn't really give a hoot about much else, much less traffic laws.

We finally stopped for a piss at my request. As we relieved ourselves in the bushes with a neon orange sunset to the west, I said heatedly, "You guys are crazy! You're stinking drunk and you're going to get us killed—I'm going to drive!"

"Thou shalt not worry," the driver told me with a pious wink. However, to my surprise, he did let me drive and I think they were grateful that I'd offered to do so.

I drove the last hour into town where we dined on fish and chips at a fast food drive-in. The older man said I should stay with him and his wife that night. That was perfect for me, so we dropped the younger man off and went straight home where I met his sweet wife—a lady in her late forties in a bathrobe and curlers who cooked us a dinner and showed me the family album while I had a quenching tall glass of South African beer.

"Now, here's Charlie," she'd say, pointing to a photo of their son in the album, "he's off to college now in Stellenbosh, studying engineering. And here's our oldest daughter Anne, she's married now and lives in Germiston. We have a grandchild on the way." Such a dear lady, filling us with desserts and telling her husband not to talk about blacks "that way."

"Thanks for driving him home; he gets in such a condition," she said, and showed me the bathroom. "Here's a towel for you, you'll stay in my son's room tonight." She could have been my own mother, I thought, as I was drifting away in the most comfortable bed I'd slept in for a long time.

§§§

The world's number one producer of gold, South Africa is a unique country both socially and politically. It is a country of startling contrasts, from steaming jungles and rich farmlands to rugged mountains, silvery endless beaches and desert wastelands. It is also a land of modern concrete skyscraper cities and shantytowns; of great wealth and bitter poverty; of freedom for some and artificial restrictions for others.

It is a major Zen challenge to come to South Africa with an objective viewpoint. One thing that startled me was the striking similarity to the U.S.A. If it weren't for the African villages between the western steel and glass cities, you could be anywhere in America. South Africa is like two countries in the same land—a traditional African country and a modern western state. Arriving in South Africa after hitchhiking through the continent gave me a different perspective on South Africa than someone coming straight from Europe or the United States, and I found I had a different perspective than did South Africans of all kinds, conservative or radical.

The history of South Africa is long, complicated and occasionally hazy. For instance, it is claimed that Cape Province was uninhabited when the Huguenots first settled there in 1652. Some historians question that. Whatever the case, the Huguenots came from Holland but were originally French, having fled France in the 1620's when Cardinal Richelieu began cracking down on the Calvinist Protestants. The Huguenots in South Africa are now called the Afrikaans Church and they have a great deal of power in the country. The church is very conservative: for example, no movies or sports can take place on Sundays and the whole country generally comes to a halt on this "day of rest."

Such events as the British acquisition of South Africa in the Napoleonic Wars, 1814, and the Boer War (Huguenots called themselves "Boers") at the turn of the century, are all well discussed in a great deal of literature.

Because of the huge diamond deposits found in Kimberly in 1867, and the vast gold deposits discovered in the Transvaal in 1886, the British wanted to annex the independent states of the Transvaal and the Orange Free State, formed by the Voortrekkers in the 1830's. The Afrikaaners had no outside help and were eventually subdued by the British who burned their farms and placed many of the people in concentration camps.

This struggle brought out the fierce Afrikaaner national consciousness that survives to this day. There is still a great gulf separating Afrikaaners- and English-speaking South Africans. They often refuse to learn each other's language and like to keep their heritage and traditions separate. The official languages are, therefore, English and Afrikaans (or Afrikaans and English, depending on who you are talking to).

Another important historical period encompassed the Zulu Wars, the most famous of which was led by Shaka, a brilliant military tactician who organized the Zulu nation and set about conquering southern Africa. Shaka was sort of the Ghengis Khan of South Africa, he reorganized his army and invented the *assegai*, a short-bladed sword that forced his men to fight at close range, instead of with spears. When he became chief of the Zulus, there were only some 1,500 in population. None of soldiers were allowed to marry until age 35, and by the time he was murdered by his two half-brothers twelve years after coming to power, the Zulus were the dominant tribe in the Natal area.

Shaka, by all accounts, ruled as a cruel, ruthless, and absolute dictator. The slightest opposition was met with execution. Any woman in his harem of 1,200 who became pregnant was killed. Shaka himself left no offspring, and was perhaps impotent. Upon the death of his mother, Shaka killed about 7,000 subjects out of grief, as well as thousands of cows. No crops were planted for a year, and any woman found pregnant was executed with her husband. Shaka's armies savagely annihilated all opposition, and repercussions of this war between Africans in the Natal area were felt all across the high veldt of the Transvaal. Some people fled as far as Malawi and Tanzania in their efforts to escape the Zulu. Some tribes held out in their mountain strongholds of Swazi and Sotho and were later protected by the British. These became the independent countries of Swaziland and Lesotho.

It is estimated that some two million people were killed in 1820's, leaving much of Natal and the veldt area to the north completely depopulated and unable to resist the the Boer "Great Trek" and mass white settlement of the 1830's. It was the forth Zulu king, Cetewayso, who revived the Zulu expansionism and battled the British with an army of 40,000. He was eventually defeated by the British in

1879 after many bloody battles, and Cetewayso was taken to England for a sensational visit with Queen Victoria. He died shortly after returning to Natal.[105]

Today, the historical event which probably concerns South Africans most is the national election of 1948, when the pro-British United Party led by Jan Smuts was defeated by the Afrikaans National Party—a political group with Nazi affiliations that had supported the Germans during the second World War. Apartheid, which, literally translated, means "separateness," now became official state policy. South Africa became the first and only state to take racial segregation (which exists unofficially all over the world) and legislate it. Suddenly, as if overnight, public facilities were designated either white or non-white universities were restricted to whites, and non-whites were no longer permitted to move freely around the country. "Job Reservation" was initiated; this means that certain jobs are reserved for certain races and other races are not allowed to hold them. This was done mainly to protect the colored (or "mixed race") population. They actually voted themselves out of the parliament in exchange for the job reservation policy so that they wouldn't lose their jobs to "blacks." Job reservation applies mostly to work that requires tools such as hammers and saws, which blacks are not allowed to use (at least officially) in the Republic of South Africa, so that "coloreds" may have these jobs.

Another regulation was the establishment of black Homelands. The Homeland plan was also begun in 1946; it created separate "Bantustans" for the Africans where they could have political autonomy and develop at their own rate. The first of these Homelands was Transkei, which gained independence in 1976 and broke diplomatic relations with South Africa in 1978, but South Africa was the only country to recognize its existence. A second Homeland, Bophuthatswana, gained independence in 1977 and a third, Venda, became independent more recently.

§§§

Johannesburg was the first place I wanted to see, and I hitched up there in one long day from Potgietersrus after the generous Afrikaans lady had stuffed me with boersvors sausage and eggs. Her husband drove me out to the main road that runs down to Pretoria and Johannesburg. Since we had been through Potgietersrus the night before, I didn't feel like hanging around.

After some hours and one long ride with a young white engineer in his snazzy Datsun which had a CB radio, I was left standing in downtown Hillbrow, the nightclub and "youth" district of Johannesburg.

South Africa's wealth is based mainly on the "golden reef" of the Whitwatersrand (Ridge of the White Waters). The golden reef is the South African term for the huge, long rand, or ridge, one long vein of gold ore. Johannesburg, with a population of one and a half million, is the center of this gold-rich area. It's interesting to note that "rand" is also the name of the South African dollar. The gold rand stretches to the east and west of Jo'burg, as the natives like to call it. It's traversed by enormous mine dumps which form a man-made range of hills dug up from some of the deepest mines in the world, going many miles underground, toward the center of the earth.

Strictly segregated, the western, eastern, and southern suburbs are mostly occupied by white artisans and middle classes. The northern suburbs have among them the finest and richest homes and gardens to be found anywhere in the world. Even the middle class whites, Asians, and coloreds typically have swimming pools and tennis courts at their homes. The majority of the Africans (three quarters of a million) live in the great southern city of Soweto. Soweto has no parallel in the world—crime in its most violent forms is commonplace, gangs roam the streets at night terrorizing the populace. It is illegal for whites to enter Soweto without a special permit, but it is possible to go on a tour of Soweto.

I spent a few nights in Hillbrow, going out to pubs and movies and to a multi-racial jazz club/disco called The Tower. The busy, bright, glittery world of Hillbrow was fun for awhile. I hadn't seen anything like it my whole trip (several years) in Africa. I might as well have been in Rio de Janeiro or Amsterdam. However, Hillbrow didn't hold my interest for long. Big cities have never appealed to me much; as the old saying goes, "They're nice to visit, but I wouldn't want to live there." So after a burly Afrikaans guy tried to pick a fight with me in the Beer Keller, a tiny beer stube in downtown Hillbrow, and later another guy actually tried to run me down in the street as I was jaywalking (I leapt over the fender of his car to avoid being mashed), I decided to split and travel down to Swaziland, 200 miles to the west.

Getting out of Johannesburg proved to be a major hassle. After a quick lift with an Austrian businessman a short way out of town, I got hopelessly lost in a maze of on and off ramps and freeways that branched off in a hundred different directions. I wandered about trying to figure out which highway to hitchhike on, cutting across busy freeways until I found what I distinctly thought was the right direction. Suddenly a car stopped on another off ramp just behind me and a greyhaired European in his fifties waved for me to come to his car. I ran over to him, cutting across the highway to get to his car.

"Where are you going?" he asked in a foreign accent.

"I'm trying to get to Swaziland, but I'm totally confused. Where are you going?"

"North. Here, hop in; I can give you a lift up to a highway that will take you right into Swaziland."

"Thanks a lot!" I said, "I was getting pretty exasperated out there. How confusing!" We talked for a while about ourselves. He was originally Dutch and had come to South Africa about twenty years ago to attend the London School of Economics which has a branch in South Africa; he had simply stayed on.

"South Africa seems to have a high standard of living," I said. "Everyone here is pretty well off materially, even the blacks. And your average middle class person has a large suburban home with a swimming pool. It's incredible!"

"Yes, the standard of living is high, but the quality of life is low," he replied. He was handsome and in very good physical shape for someone his age, I thought. In his three-piece suit, driving a large new Oldsmobile, he appeared quite prosperous.

"What South Africa is missing," he continued after a pause, "is a certain quality that adds meaning and direction to life. The people here have no philosophy or ideals to hang on to. They are merely satisfied with the pursuit of material things and sensory pleasure."

"I've heard that South Africa has a high rate of alcoholism and divorce. But in America it's the same—most people spend their lives chasing the dollar," I responded.

"Money isn't the cause of unhappiness, and it's important to have the basics of life without having to break your back all day to earn a decent living. It would be nice if we could find a balance between the physical and the spiritual."

"You don't sound like a typical economist," I said.

He laughed. "Economics is more philosophy than science—we're not all dry mathematicians."

We talked on about various economic theories, and then I told him of my research into lost cities, vanished civilizations and ancient mysteries. I also asked him what he though of the racial problem of South Africa. To my surprise he told me that, "The world as we know it has been destroyed many times over. Civilizations have risen from the dust and then declined. According to my teachers, the greatest of all civilizations was on a now-submerged continent in the

Pacific.

"While most people are unaware of it, Negroes are really not from Africa. Negroes, like all people, are originally from the vast continent in the Pacific, where all the races of man originated. In the later days of that great civilization, the different tribes of Mu, as we might call the continent, began to emigrate around the world. Negroes, as we know them, settled largely in Central America. Huge Negroid heads, many thousands of years old have been found in the jungles there. Other Negroes managed to survive in the New Guinea of South East Asia after Mu was destroyed in a cataclysmic pole shift.

"Africa was largely depopulated after the last poleshift about ten thousand years ago. Negroes from Central America migrated into the area around Nigeria, and began to populate that area of Africa. Meanwhile, Dravidians had settled in Abyssinia and Southern Arabia, while Asians began settling Madagascar and East Africa. The Bushmen and Hottentots are the remnants of these people. The world is one long history of migrations, and, unfortunately, of war and cruelty. Africa does not belong to blacks, or to whites, or to Asians. Like the entire world, it belongs to all mankind. Working together in harmony is the key.

"Negroes had extremely high civilizations in the past in Central America as well as in West Africa. Like all races, they have had to ride the rollercoaster of history. In Atlantis and other cultures, materialism and technology reached heights before they destroyed themselves. Our own civilization is coming to the same turning point in time." Suddenly he slowed the car at a crossroads. "Here's where I'm going to let you off." I grabbed my pack. "Take care," he said, "have a nice time in South Africa!"

"Thanks for the ride," I waved to him, "It's been interesting talking to you!" And it had been, too—he had given my body a lift and he'd also "lifted" my brain; I felt mentally stimulated by his ideas. I wondered where he had gotten his information. Though available in certain esoteric books, such ideas were hardly held commonly. James Churchward in his books on Mu[76] says largely the same thing. Perhaps more on the traditional academic level are the books edited by Ivan Van Sertima of Rutgers University in New Jersey. His books, *African Presence in Early America* [110] and *African Presence in Early Asia,* [111] are filled with articles and photos that show without a doubt that Negroes have lived, literally, all over the world, including the ancient Americas. While Van Sertima does not bring in such unorthodox theories as a lost continent in the Pacific, he is clearly of the belief that Negroes in ancient times developed many advanced civilizations and lived all over the world. However, he still describes Negroes as "Africans" which, at least by some accounts, is not really the proper term for such people (Bushmen are generally said to be the original inhabitants of Africa). However, in these times, Caucasians are "Europeans", Mongoloids are "Asians" and Negroes are "Africans". These are only words.

When you let the Tao flow through you, fate can toss you into someone's lap. Who knows what lesson or challenge you'll be facing? I thought about it while attracting my next ride. As car after car passed, I imagined myself riding a Zen wave through time and zeroed in on the next approaching vehicle. It stopped!

It was a brand new Mazda pick-up, silver gray and still gleaming with its showroom shine. Out jumped a small, rather weak, older white man. Nervously, he said, "Where are you going?"

"To Swaziland," I answered. "Thanks for stopping."

"Sure, hop in," he grimaced, "I don't know why I stopped for you anyway, I never pick up hitchhikers."

"That's interesting," I said, rolling down my window as the car started smoothly down the highway. "By the way, I'm from America; I've been traveling through Africa."

366

"God, who would want to do that?" He made a face expressing his disgust. "Those banana republics up north are crazy. I'd never go there." He was gripping the wheel like it was holding the car together. If he never picked up hitchhikers, he was probably nervous about his "first time."

"Don't worry, it won't hurt," I wanted to tell him. Instead I watched the scenery out the window. Bald, grimacing and basically a pessimist, my newest acquaintance was an English South African. I let my mind wander as we drove along, what mysteries and lost cities might be in South Africa?

South Africans generally believe that there was little in their country before the Boers came to the land, but this is decidedly untrue. In ancient times, the Limpopo River was a major highway into the High Veldt area of the Transvaal, and today, this river forms the border with Zimbabwe and South Africa. On the south side of the Limpopo, an important lost city was discovered was discovered only as recently as 1932. It was in this year that a farmer-prospector called van Graan convinced an African to show him the small table mountain of rough sandstone, precipitous on every side, which is known as Mapungubwe. To the local natives, Mapungubwe was "a place of fear", and even after whites had found it, Africans "would not so much as point to it, and when it was discussed with them they kept their backs carefully turned to it. To climb it meant certain death. It was sacred to the Great Ones among their ancestors, who had buried secret treasures there."[106]

Graan, with his son and three other men, were shown a concealed way to climb the hill by a courageous (or foolish) native. The hill had hundred foot high sheer cliffs and was a thousand feet long. It was a narrow "chimney" in the cliffside that was masked by trees. The explorers saw that the vanished dwellers on Mapungubwe had cut small holes inside it, facing one another, as though for the crossbars of a ladder. Without the rungs of the ladder, they made their way up the chimney to the top, to discover a low breastwork of stones as well as by large boulders balanced on smaller ones, as if poised to be rolled down on invaders.

They wandered over the summit and saw that its flat and rather narrow summit was littered with broken pottery. When they scratched in its loose, sandy soil they turned up beads and bits of iron and copper. Fortunately, a rainstorm a few weeks before had washed away some of the top soil and Graan saw some gleaming yellow metal in the soil, he picked it up and it was gold! The next day they returned and found more gold, including an exquisite gold rhinoceros. The gold turned out to be of exceptional purity and eventually the University of Pretoria began to excavate the site. The lost city of Mapungubwe then became an especially important site for two reasons: it was rich in skeletal remains as well as gold and other objects and since it had not been plundered by some "Ancient Ruins Company" like Zimbabwe Ruins, most things had not been disturbed and could be examined where they lay.[106]

On the strange rockcrop were skeletons with legs wrapped in over a hundred bangles of coiled wire, a number of pieces of beautifully worked gold plating, over twelve thousand gold beads, and over seventy ounces of gold buried with another skeleton. Yet, the most surprising discovery at Mapungubwe was that the skeletons discovered were not of Bantu Negroes, as was supposed, but were declared to be the skeletons of non-Negroid Hottentots. This ghrew something of a monkey wrench in the archaeological machine of academic South Africa, as it indicated that the Hottentots had once enjoyed a highly technical, metal-using society in the past, although they lived an essentially stone-age existence in South West Africa at the time of the arrival of Europeans. Basil Davidson says in *The Lost Cities of Africa* [106] that a date of 900 A.D. has been established for the city. He also surmises that at about this same time Bantu invaders from the north swept down into the area and intermarried with the Hottentots or forced them out and to the south (or both).

What is astonishing and important of these finds is that Mapungubwe must have had some affinity to Zimbabwe, Dhlo Dhlo and other cities only a hundred miles or so to the north in Zimbabwe. Were these cities, too, built and inhabited by Hottentots? What is also evident is that there has been a regression of culture in the Hottentots. It seems that they also went through a Dark Age, just as Europe did at approximately the same time. The final question is, then, who are the Hottentots?

§§§

The Mazda owner gradually relaxed his grip on the wheel and we fell into conversation about blacks and Afrikaaners. He didn't like either, but would take an Afrikaaner over a black any day. I read him part of an article that my sister had sent me from the University of Pennsylvania.

"This is a quote from some South African guy—he says, 'If things get rough, the South African government will bomb the blacks in Soweto and the other townships to pieces, and the whites will be a majority. The government isn't stupid by putting all the blacks together." I looked over at my driver. "What do you think about that?" I asked.

"Christ! I never heard that before!" he winced, frowning. "Actually, I'm fairly liberal, but that's not such a bad idea. Hmmm, maybe that IS their plan!"

"Oh, come on," I said, "The South African government wouldn't really do that, would they? Anyway, that wouldn't make a white majority."

"I wouldn't put anything past those people." A smile crept across his face for the first time. "Who knows what the government has planned for the blacks?" He came to the turnoff south to Swaziland and stopped the car.

"Thanks for the ride, mister. Drive carefully," I said as I got out of the truck.

"You too, kid," and he sped off east as I settled my pack to flag down another ride. South Africans have some pretty strange solutions to their complex problems, I thought; surely this one was a joke, but who knows...

I didn't want to let a negative conversation affect me, so I strove to keep my thoughts on a happy note. As I walked, it was a struggle not to get depressed, however. It was now overcast and getting darker. Rain clouds were looming to the north and I was walking down a rather lonely-looking road toward Swaziland, which was still 50 miles away. South Africa seemed a bitter place. So far I hadn't come across much brotherhood. The English and the Afrikaaners didn't like each other, and the other white minorities (mostly Jews and Portuguese, with some Greeks) just kept to themselves in their own neighborhoods. The Asians and coloreds too just kept to themselves. I had the impression they would even if they didn't have to. Meanwhile the blacks were killing themselves in the slums around the industrial complexes. I wanted to run back to Soweto and shout in the streets, "Leave this awful place; go back to your villages. Go back and live a healthy life in the country!"

§§§

South Africa has its share of ancient mysteries. In 1969, the book *The Carbon-14 Dating of Iron* [107] was published by the South African archaeologist Nikolaas J. van der Merwe. In this important, ground-breaking book, he says that the oldest piece of iron found in South Africa is from 770 A.D. and was discovered at twenty feet in a mine shaft at Loolekop, Palabora in the Transvaal. Yet, a year later, it was reported that a mine more then 40,000 years older was discovered in the Transvaal! According to the New York Times–Chicago Tribune

Service, date-lined Johannesburg, South Africa, February 8, 1970 and quoted by Ivan T. Sanderson in his book, *Investigating the Unexplained,* [61] the discovery was made of an iron mine that was radio-carbon dated by organic materials found in the mine at 41,000 B.C.! These incredibly ancient miners were apparently after not only after iron ore, but also a substance found in association with it named *specularitr,* which has been known from the very dawn of history as the basis for a most astonishingly vivid and durable cosmetic! That early stone-age man in South Africa was mining metals is bad enough (one has to ask how anyone mining metals can be called stone-age), but that they could separate two minerals like haematite (iron) and specularite should give us pause to consider and reassess a lot of our most cherished beliefs.[61]

One can only wonder if the "Rhodesian Man" with a bullet hole in his skull was one of the laborers at this ancient iron mine 41,000 years ago. The history of South Africa certainly takes a sharp turn into the world of the strange and bizarre past with that discovery!

Just as the first drops of rain started to fall on the tarred pavement, I saw a large, brown Mercedes coming toward me. "This is my ride," I said to myself, and started to wave the car down. I waved both arms over my head and did a little dance on the pavement. The car started to slow, then came to a stop just in front of me. To my surprise it was a black man. I was a little hesitant at first to take the ride. Should I? Was it safe? It did seem kind of strange that a black guy would pick up a white person out here.

"Hi! Where are you going?" I asked him.

"To Swaziland," he replied. "Want a ride?"

"Sure," I said, getting in and brushing my rain-wet hair out of my eyes. "Thanks for stopping."

He was a very large person, dressed in a brown suit and businessman's hat; his broad nose sat well on his broad face.

"I'm a businessman," he told me after a short introductory conversation. "I own about ten shops around Johannesburg and out this way near Swaziland. I start a new shop every year somewhere."

I looked at the glossy interior of his Mercedes. "Seems like you're doing very well as a businessman; do you find any restrictions on you as a black?" I asked.

"No, not really," he said, "just that we have to live in the black areas. Since I can afford a nice home, though, it isn't such a problem."

"What do you mean?"

"There are many wealthy Africans in South Africa. Our restrictions are mostly political; economically we can do almost anything we want, just so we don't get into politics."

"What about job reservations?" I asked.

"Oh, that's no big deal, that's just to keep the coloreds happy and secure in their jobs. We still learn whatever trade we want, nobody stops us."

This African was obviously well off; I tried to question him about what blacks wanted in South Africa and what was going to happen to them. He clammed up though and I realized I had been a bit too inquisitive. We rode on in silence for a while until we came to a small gas station and store in the hilly, green country on the outskirts of the Swaziland border. Pulling in, he said, "This is my place. How do you like it?"

It looked pretty rundown and scuzzy. Some black men in greasy overalls were hanging around outside. No one was really doing anything. "Looks great," I said.

We waited there for a while as he checked things out at his store; then we headed over the mountains in to Swaziland. We went through customs and immigration and then wound our way through the green misty hills of Southern Africa. There was a soft, wispy spring fog curling around the valleys, lending a romantic, fairytale atmosphere to the land.

It was only another half hour into Mbabane, the capital of Swaziland. Because of a friend's recommendation, I asked my lift to let me off at the Happy Valley Motel a few miles south of the city. At nine rand a night, it was pretty cheap for Swaziland, but my image of this country as a fairytale came to a halt when I stepped into the bar. In the corners were all kinds of gogo dancers in various stages of undress, a porno movie was being shown on the wall, and the place was full of women of "easy virtue," as the Victorians used to say. Swaziland is South Africa's Sin City. All the South Africans come to Swaziland to gamble at the casinos, watch pornographic movies and have sex with blacks and coloreds —all of which is illegal in South Africa and legal in Swaziland!

Still, Mbabane is a pleasant and friendly place. I spent a day looking around the capital and then hitched around the Usuthu Forest in the high veldt, a heavily forested park area with some spectacular views of the countryside and mountain range of Swaziland.

The next day I started hitching south to Durban on the Natal Coast. I never expected to make it all the way in one day since it's a good two hundred and fifty miles and part of the route is unpaved and poorly trafficked. I had been thinking of going to Mozambique while I was in Mbabane, but was told it was out of the question at that time.

Anyway, I was lucky to get a lift with a colored South African who took me a hundred and fifty miles down the coast. He was a nice guy and interesting to talk to. A mechanic in his early thirties, he wore brown denim overalls and had a knit cap to cover a bald spot on the top of his head. We talked of the freedom of traveling and the thrill of meeting new people.

"Life is good," he said with a smile, "we have the power to make or break our lives. We can shape our own destinies."

"Right on!" I grinned back. This was largely my own philosophy, that we are the makers of our own destiny, not the victims of it.

He also told me something interesting about South African politics. "South Africans are generally a lot more liberal than it appears to the outside world. They use a system called gerrymandering—I believe you have that word in America. The political districts are all gerrymandered so that the rural Afrikaaners will have more voting power than they would have if the vote were actually a straightforward popular vote. So the most conservative element in our country has the most political power. I think if we had a popular vote in South Africa, apartheid would be taken out."

"Yeah, the politicians use that trick in America quite a bit, but I never thought about it happening in South Africa as well," I said.

My lift suddenly yelled, "Hey! There's a car with Durban license plates." He pulled up next to the car and shouted out the window, "Hey, mister, are you going to Durban?"

The fiftyish white man at the wheel looked slightly startled, but hollered back, "Yes!"

"I have an American here who needs a lift, will you take him?"

"Sure!" The man in the Impala yelled back, and they both pulled over.

"Thanks a lot!" I said to my lift.

"Anytime!" He waved back and sped away. I got in with the greyhaired energetic white. He was friendly like so many South Africans, and we had a nice talk as we drove the last hundred miles into Durban. I had made it in one day, thanks to the colored mechanic.

The Natal coast is the most popular holiday area in South Africa. The Indian Ocean waters are warm all year round for swimming and surfing. South Africa has some of the best surfing in the world and is full of "surfies" and bikini-clad "surfettes" that make hanging out on the beach an eye-filling experience. To the north of Durban is the Zulu Homeland, known as Kwazulu. Lots of nice scenery around as well as Umfolozi, one of South Africa's best game parks, famous for the square-lipped rhinoceros. This is one of the few game reserves in Africa that can be traversed on foot.

I have spoken about the fierce Zulu wars of the 1800's, but the origin of the Zulu is mystery with some tantalizing clues. An anonymous article in *American Antiquarian* in 1885 (7:52)[33] says, "A paper received from Natal Africa contains an article by Rev. Josiah Tyler on the similarity of Jewish and Zulu customs. Among them we mention several: The feast of first fruits, rejection of swine's flesh, right of circumcision, the slayer of the king not allowed to live, Zulu girls go upon the mountains and mourn days and nights, saying, 'Hoi! Hoi!' like Jepthah's daughter, traditions of the universal deluge, and of the passage of the Red Sea; great men have servants to pour water on their hands; and throwing stones into a pile; blood sprinkled on houses. The authors' belief is that the Zulus were cradled in the land of the Bible. Certain customs are mentioned which may be ascribed to the primitive tribal organism. These are as follows: Marriages commonly among their own tribe; uncle called father, nephew a son, niece a daughter; inheritance descends from father to eldest son. If there are no sons it goes to the paternal uncle. A surmise has been advanced by some that the relics of the Queen of Sheba's palace may be found in certain ancient ruins described Peterman, Baines and others, and the Ophir of scripture has been located at Sofala, an African port."[33]

The Reverend Tyler is referring to the recently discovered Zimbabwe ruins as the palace of the Queen of Sheba, and the port of Sofala in Mozambique as the Biblical Ophir. Indeed, Tyler's letter may have been an important influence on H. Rider Haggard who was to write *King Solomon's Mines* ten years later. Tyler essentially would call the Zulu one of the lost tribes of Israel, an interesting concept. Sofala was indeed the main Indian Ocean port for the export of the metals coming out of the plateau that was reached by the Sabi River, Limpopo River, or Zambesi. Everyone who was anyone in the ancient world around the Indian Ocean must have visited Sofala at one time or another, even Chinese traders. This would of course include King Solomon's fleet from Ezion-Geber as well. While the Phoenicians working for Solomon may well have visited and traded here, and even established some of the cities and mines in league with Sabeans, it is still doubtful that Sofala or Zimbabwe was Ophir. Yet, it is possible (though no doubt hotly contested) that the Zulu are descended from ancient Jewish colonists, though the evidence is circumstantial. Another custom that is not mentioned, but very important, is the circumcism rites of manhood, a custom carefully practiced by the ancient Egyptians as well as the Hebrews. It is also interesting to consider Wuzamazulu Mutwa's book *Indaba, My Children* in light of this article.

The eastern influence is strong in Durban—lots of Hindu temples and Muslim mosques. There is a colorful Indian market full of Oriental curios, tropical spices, and exotic fruits. Most people come to Durban because of its beaches, and since I hadn't seen the ocean since I left Dar Es Salaam in Tanzania more than three months earlier, I was ready to just kick back and hang out on the beach for a while. I thought I might even rent a surf board. Incidentally, most beaches around Durban are guarded by shark nets; its wise to avoid those that are not!

In about a week I was ready to head down the coast toward my ultimate destination in South Africa, Cape Town. In one day I hitched southeast along the coast to Transkei, one of the independent Homelands in South Africa. This area is considered the wildest part of South Africa; in fact it's called the "Wild Coast." The Bomvana tribesmen of this area had the curious habit of fighting to the finish with battleaxes. Every Sunday in each district the young men with a score to settle would meet and hack each other to death. I learned this from an English South African who was giving me a ride.

"That's horrible," I said. "Do they still have these fights today?" I was wondering if I was going to get into trouble traveling through this tribe's territory.

"Not so much any more. I used to see a lot as a kid. In those days rival groups of ten to twenty would clash together, attacking bystanders only if they gave the impression they were favoring one side or the other."

"Being a spectator at those kind of team sports sounds kind of hazardous," I joked.

"I never felt in any danger," he replied, "there's more danger in the rivers. I remember one time when I was crossing a river in the upper Natal with some Zulu guides. I looked up the river and saw this big black thing floating straight down river at us. 'Is that a crocodile?' I asked my guides. 'No,' they said, 'It is not a crocodile.' I looked at it again. We were standing chest deep in the middle of the river, wading across. This thing was coming straight for us; if it was a crocodile we were in big, I mean BIG trouble!"

"Was it a crocodile?" I asked.

"Well, I'll tell ya—I said to those boys, 'Are you sure that isn't a crocodile?' They looked up the river again and told me, 'No, Bwana, that's no crocodile.' And you know, those bastards were right; it wasn't a crocodile— it was the half-eaten body of some tribesman upriver that was floating down the river without any arms or legs. Came floating by just like that. Yessir, those crocs, they'll get you..."

South Africans love to tell foreigners ghastly stories of life in Africa. I didn't mind, in the Rocky Mountains we have our grizzly bear stories. Guess it's the same the world over.

Port St. Johns with its high, wooded hills and rugged, enclosed bays is a spot of delicious beauty and great angling. I couldn't resist spending a couple of nights here in a shelter down by the river. I felt fine just sleeping out under the thatched shelter by the river. I met a young South African and an English traveler who were camping out in tents a few miles away down by the beach. They were eking out a living by collecting mussels along the beach and steaming them.

I enjoyed the beach for two days, but couldn't swim because of sharks, so I hitched with a vacationing couple in a Winnebago-type camper through Umtata, the capital of Transkei. It looked like a rather dull and uninteresting place, though a couple of Englishmen at the bar in St. Johns had told me there was some pretty wild night life there. I didn't stop to find out.

The retired couple with the Winnebago were a conservative pair—stern, grey-haired and righteous. "You know," said the husband, "the blacks will take over the world. They'll do it through sex. Look—whites and blacks have sex and what do they get? A black! That's how they'll take over the world. But you know who is going to stop them? The Chinese, that's who! Yessir, they'll stop the blacks!"

"That's crazy," I wanted to say, but didn't. A good traveler keeps his mouth shut and asks the right questions; he is nice and friendly and tries to learn something from everyone. Occasionally he might be able to help his "ride" see something in a new light—but not when he starts by telling the "ride" he's crazy.

I made it to East London that day, spending the night in one of the nicest youth hostels I've ever stayed in. That was probably due to the labors of the dear, elderly lady who ran it. Wandering around the town, I found a number of restaurants, some night clubs, and to my surprise, even a strip joint or two. Not much to keep me in East London, I decided as I walked back to the youth hostel in a fierce wind that whipped up along the beach. The next morning as I was packing up to go, the old lady cooked me the best breakfast of bacon, eggs and toast I'd had in months because I had gone grocery shopping for her the day before. "Take care, darling," she said as I was leaving, and asked me why travelers always stole her toilet paper.

"An old traveler's habit from the Middle East and North Africa," I told her, though I hadn't taken any of her toilet paper. But like all travelers, I had occasionally stolen a roll from a hotel when I'd been desperate enough and couldn't find any to buy.

The East London Museum has one of the strangest exhibits to be found anywhere in the world. On December 22, 1938, a South African trawler pulled out of the depths of the South African waters a prehistoric fish that was supposed to be extinct for more than 70 million years, the bizarre coelacanth, a five-foot fish with leg-like paired fins. Four more coelacanths turned up over the next thirty years, and while the zoological world should have been considerably rocked by such a discovery, it still maintains that while coelacanths may have survived the 70 million years they were supposedly extinct, other dinosaurs and prehistoric animals could not. Coelacanths are that big slap in the face to those stuffy scientists who have all the answers and scoff at tales of sea monsters or living dinosaurs. The mystery of fossilization comes into play here, and raises the question of why there are no fossils of a coelacanth in any strata since 70 million years ago. Can it be that our geological dating is way off? What other extinct animals will show up in a fishing net one day?

I decided to hitch to Jeffries Bay via Grahamstown the next day; it's a famous surfing spot just outside Port Elizabeth. Jeffries Bay is a great place to stay, especially during the summer when it's full of surfers. It's popular with surfers because the bay has one of the longest left-handed waves in the world, but it's not the place to learn how to surf. The water is shallow with plenty of jagged coral to slice up swimming trunks and their wearers along with them.

As I hitched that one long day to Cape Town with a sign that said "Montana to Cape Town," I couldn't help thinking with some degree of sadness that my overland through Africa trip was almost over. It had taken me two and a half years to do it (counting working time in Sudan), but I'd had a mighty exciting and fascinating time. For some reason I thought of Cape Town and Table Mountain as a sort of Mecca for the African traveler. When I got to Cape Town, I would probably collapse in the center of downtown, my thumb out, frozen to the pavement like some glacier on top of Mt. Kilimanjaro.

Now I was bombing down the coastal highway toward this Mecca, my destination. An Indian had picked me up outside of Riversdale. He was a nice guy, a projectionist at a cinema in Cape Town. He was forty-five, brown and handsome. He told me all about his kids and the affair he had had with a Swedish lady in India when he had gone back there for a visit.

"You must meet my daughter," he said with a soft smile. "I'll introduce you to her when we get to Cape Town. She was dating an American for a while. You'll like her."

"Gee, I'd love to meet your daughter," I said, "but isn't it illegal to date like that, I mean Indians and Europeans?"

"Oh, nobody really cares about that in Cape Town. Besides, that's just sex. I always watch my daughter very carefully; she's in before midnight every night. No hanky panky."

Oh well, it sounded like he wanted her married off—he didn't know us maverick archaeologists, I guess. We talked about life in South Africa, and about apartheid. We pulled into a gas station with a small cafe along the road. "Let's go into this cafe and grab a bite to eat," I suggested.

"Oh, no, we can't go in there; they'll never let me in."

"Really, you mean you couldn't get a sandwich in there?" I eyed the slightly rundown building—it was hardly the Ritz.

"Certainly not," he replied, "go on in and ask them yourself if you don't believe me."

While he was getting gas, I walked into the cafe and addressed the rotund white cook behind the counter. It was a typical greasy spoon—hamburgers, french fries and coffee—nothing special.

"Excuse me," I asked, "Could my father-in-law and I come in and get a sandwich?"

The cook took a half-hearted wipe of the counter with a graying dishcloth and wiped his hands on his apron. "Sure," he said, perking up a bit, seeing that I was a foreigner.

"Great," I said, "but he's an Indian."

A look of sadness came over his tired face. "Sorry, kid, but he can't come in. I could make you some sandwiches and you could take them with you."

"No, thanks," I said, turning away. "We'll do without."

"Sorry, kid," said the cook again as I walked out. At that point the horrible degrading essence of apartheid hit home for me. I was crushed. This decent respectable man couldn't eat with me, a lousy five dollar a day traveler, in a greasy spoon in the middle of nowhere. God damn!

"I told you," said the Indian, handing me a beer from a six-pack he'd bought while I was in the cafe. "That's just the way it is."

In Cape Town he took me home to introduce me to his children. His daughter turned out to be extremely beautiful and delightful; it was a pleasure to make her acquaintance. While at their house I called some students of the University of Cape Town whom some friends had told me about, and it was arranged that I would stay with them. After dinner with his family and an arranged date with his twenty-year-old daughter, he drove me to a student quarter of Cape Town, where I met Andreis and Gaby. They were to become my best friends in South Africa.

Cape Town is a beautiful city, often rated as one of the most beautiful cities in the world. Founded in 1652 by the Dutch as a way station to the East Indies, Cape Town is a fine port with a sheltered bay, sandy beaches on two oceans, old trees and beautiful gardens nestled beneath craggy Table Mountain and Lion's Head Peak. The Cape Town Castle dates back to 1666, and it houses plenty of relics from the early colonial period. The slopes of Table Mountain itself can be ascended by cable-car from the station at Kloop Nek.

There are beaches galore around Cape Town. One can choose between warm Indian Ocean beaches at the resort towns of Muizenberg or Simonstown, or the colder beaches on the Atlantic side of Sea Point, Clifton or Llandudno.

I found a job selling camping gear and back packs at a local sporting goods store and settled into a cooperative student house with Andreis and Gaby and three other young people, students and teachers. Pretty, fun-loving Gaby was studying to be an elementary school teacher. Tall, handsome, bearded Andreis and another fellow were working illegally in the black township of Googaletto, teaching carpentry to the immigrant blacks there at a church school. They told me that our phone was tapped and that the police knew that they were working in the township, where whites couldn't even go without a permit, much less work. They certainly didn't have any kind of permit; however, they were never arrested.

One grand night everyone in the house and a few select friends were invited to the township for the ritual circumcision of one of the carpentry school's students. Freely we moved into Googoletto on a few motor bikes and cars, breaking South African law as we did so.

At twenty-one it would be rather painful to be circumcised, but this is Bantu tradition, at least in South Africa. The person fasts for a week, is then circumcised, and doesn't see anyone for a week afterward, just living by himself in a hut in the bush somewhere, where he is visited only by the witch doctor.

We sat around a campfire in the back of a pretty rundown house, occupied by a family of six or seven. Relatively speaking it was considerably better than housing in other African countries; in Cape Town it was what you could call crowded but by no means horrible. Many a time in my travels I'd been happy with a place of less comfort to stay in.

Huge bowls of African beer were passed around. We visited the guy about to be circumcised, who was in a small hut made of sticks especially for the occasion. We were allowed to talk with him, but we were not to touch him. After briefly socializing with just about all the Africans living in the township, one of the fellows from our house and I decided to go out and explore Googaletto a bit. We walked around the block. The roads were dirt, rather than paved; there were some street lights at least, but no sidewalks. All the houses looked the same; old rundown, four- or five-room-thirties-style houses.

One black guy from the party took us to a shabine where they served beer illegally. We sat on a couch with a dozen or so blacks and talked very reasonably about life. They gave their grievances about the system but they didn't seem hostile in the slightest bit, and we both felt quite safe there.

Suddenly Andreis came running in. "Oh, I've found you! Thank God! I was so worried. You two shouldn't go off like that. It's dangerous! You could get killed! People are killed out here all the time!"

We thanked the gentlemen we had been chatting with and walked back to the party with Andreis. We told him we hadn't felt in any danger. It's so hard to tell, really. I suppose everything can be hazardous. As one of the fellows back in the shabine had told me over a beer, "Live your life as if you knew you would die tomorrow. If you would change anything, then you are living your life wrong." I could buy that.

Two rare photos of ancient ruins near Vaco in the the state of Cela in Angola. **Above:** Used as a tomb, this structure had two chambers and contained a Bantu mummy. The man was mummified by having his body sitting with the head tilted back. Hot palm oil was then poured down the throat for ten days in a row. Note the Zimbabwe motif of chevrons, also used at Kuelap in Peru. Arrow points to pyramidical stones on roof of the chamber. **Below:** A structure, use as a tomb at Quibala, near Cela. Note how this structure uses megalithic menhirs as part of the wall. Photos from *Arquelogia Angolana* (1980, Lisbon) by Carlos Ervedosa courtesy of Dr. Aurelio deAbreu of Sao Paulo, Brazil.

The ancient city of Sofala in Mozambique, from an old print. In the foreground can be seen the newly built Portuguese fort. Portuguese explorers to the Indian Ocean believed Sofala to be King Solomon's Ophir.

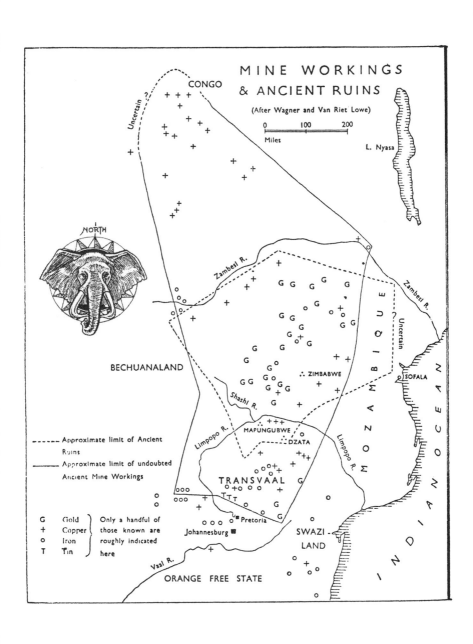

MINE WORKINGS & ANCIENT RUINS

(After Wagner and Van Riet Lowe)

0   100   200
Miles

L. Nyasa

NORTH

CONGO

Uncertain ?

Zambesi R.

Zambesi R.

MOZAMBIQUE

Uncertain

BECHUANALAND

G G G G
G
G G G
G G
G G G
G
G G G
∴ ZIMBABWE
G G G G
G G G
G

SOFALA

Shashi R.

Limpopo R.

+ + +
MAPUNGUBWE
DZATA

Limpopo R.

TRANSVAAL

T T T

Pretoria

Johannesburg

SWAZI-
LAND

INDIAN OCEAN

Vaal R.

ORANGE FREE STATE

- - - - Approximate limit of Ancient
         Ruins
———— Approximate limit of undoubted
         Ancient Mine Workings

G   Gold   } Only a handful of
+   Copper   } those known are
o   Iron   } roughly indicated
T   Tin   } here

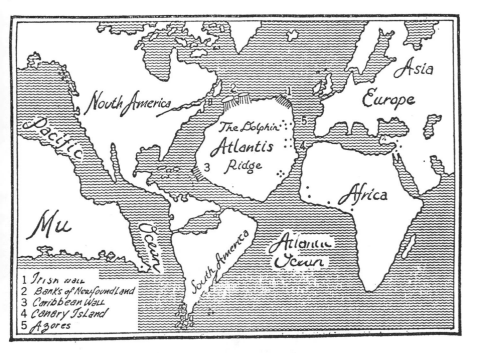

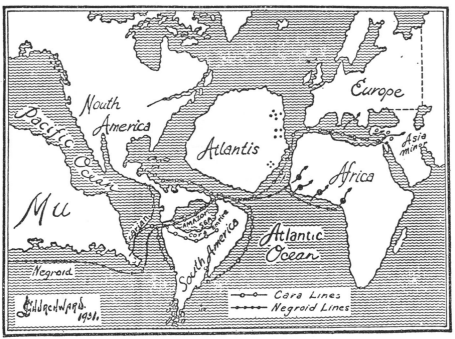

Maps of Mu, Atlantis, and the rest of the world
by James Churchward. He believed that the
world was populated by different races coming
from a lost continent in the Pacific.

One of many "Olmec Heads" from La Venta in Mexico. With obvious Negroid features, and many thousands of years old, the question is, were Negroes going to West Africa, or coming from it?

Olmec head excavated in San Lorenzo, Mexico, when small part of right forehead was noticed under burned over jungle. These enormous heads, in appearance more African than Indian, are one of the New World's greatest and oldest mysteries. *(Photo: Dr. Gordon Ekholm. Courtesy of American Museum of Natural History)*

AFRICA'S STRANGE AARDVARK IS LIKE A
LIVING MOSAIC MADE OF PARTS FROM MANY
ANIMALS. IT HAS A POWERFUL KANGAROO-LIKE
TAIL, DIGGING CLAWS OF AN ANT BEAR, BRISTLES
LIKE A PIG, AN INCREDIBLY LONG SNOUT AND
ENORMOUSLY OVERSIZED EARS. THE THREE KNOWN
SPECIES LIVE IN BURROWS AND FEED ON TERMITES.
AARDVARK FOSSILS, 10 TO 15 MILLION YEARS OLD, HAVE BEEN FOUND
IN FRANCE AND GREECE SHOWING THE ANIMAL HAS REMAINED
UNCHANGED THROUGH THE AGES. MOST ZOOLOGISTS AGREE ITS
ANCESTORS WILL NEVER BE IDENTIFIED.

THE STRANGE, BLUISH COELACANTH, OR TASSEL
FISH, APPEARED ON EARTH 350 MILLION YEARS
AGO. ENORMOUS SCHOOLS SPREAD THROUGH
EVERY OCEAN. THEN THEY VANISHED LIKE
THE DINOSAURS. SCIENTISTS CONCLUDED THE
SPECIES BECAME EXTINCT 70 MILLION YEARS
AGO. BUT IN 1938, OFF SOUTH AFRICA, A LIVE
COELACANTH WAS NETTED IN 250 FEET OF
WATER. SINCE THEN THE REGION HAS
YIELDED SIX MORE. HOW THEY SURVIVED IS
ONE OF NATURE'S GREATEST MYSTERIES.

Chapter 17

NAMIBIA & BOTSWANA: ATLANTIS IN THE KALAHARI

...the stars there have heart in plenty and are great hunters.
She is asking them to take from her little child
his little heart and to give him the heart of a hunter...
Surely you must know that the stars are great hunters?
Can't you hear them?
Do listen to what they are crying!...
You are not so deaf that you cannot hear them.
—*Kalahari Bushman to Laurens Van der Post;*
The Heart of the Hunter

I had heard the strange tales of lost cities in the Kalahari, and naturally, I wanted to investigate them. Both Namibia and Botswana are still unexplored to large extent, and all kinds of ancient mysteries were said to be present. One weekend some friends were heading north up along the coast, and it was a good chance to a good start on the long road up to the Orange River where the border of Namibia was. Piling into an VW Microbus, I began the last leg of my two-and-a-half-year African odyssey. It was sad to leave my friends after living for six months in Capetown, but I was ready to hit the road again.

After spending the night at a family holiday cabin with my Capetown friends at Langebaan Bay, I began walking down the main road north. After a few good rides up the coast, one with a Swiss chemical salesman who had immigrated to South Africa, I got a ride with a young man from Capetown on his way to Windhoek in a large Ford truck full of gun safes that he was going to sell to a local gun dealer. It was good sitting up front with him, chatting about the adventure of travel in this vast desert country.

Namibia is vast, rugged, beautiful, varied and virgin. Some of the best vagabonding in Africa is offered by this incredible country. The Namib desert, said to be the oldest in the world, stretches for one thousand miles along the coast. There are only three towns along this coast; it's one of the most desolate and lonely coastlines in the world (though parts of Western Australia could compete). Inland is the central high plateau, where rainfall increases slightly. This plateau slopes down to the Kalahari in the east. Rainfall increases to the north, where are found the swamps of the Caprivi Strip, the farmland of Ovamboland, and the largely unexplored Kaokoveld Mountains to the northwest.

Rich in minerals and the world's largest producer of diamonds, Namibia was annexed by Germany as Southwest Africa in 1883. Fierce opposition to the Germans by the Nama and Herero natives of Namibia led the Germans to virtually exterminate some 65,000 Hereros, and the Hereros have still not recovered from this. South African military forces, as part of the British Empire, drove out the Germans during World War I and a League of Nations mandate was given to South Africa in 1920 to supervise the former German colony. The United Nations attempted to take control of Namibia in 1945, but the South Africans refused to relinquish it, claiming that the UN was not the legal successor to the League of

Nations. Meanwhile Namibia was virtually treated as a sixth province of South Africa and continued to have its independence frustrated by South Africa. At least, in 1975, the all-white assembly of Namibia passed legislation ending all racial discrimination and "petty apartheid."

It was a long ride after we crossed the border of the Orange River and began the drive to the first major town in Namibia, Keetmanshoop. To the west we passed the Fish River Canyon, which Namibians claim is an even more impressive sight that Arizona's Grand Canyon. Keetmanshoop was the site of an extraordinary series of events a number of years ago. Roy Mackal reports in his book *Searching For Hidden Animals* [97] that Dr. Courtenay-Latimer, one of the South African zoologists responsible for examining the coelacanth, investigated a strange sighting of a large "flying lizard" in the area of Keetmanshoop. Farms in the are usually quite extensive, as much as 100 square miles (160 square kilometers) large. The main livelihood is derived from a flowering plant named kana and colossal flocks of sheep. The natives of one particular farm around Keetmanshoop eventually left because the owner would not take seriously the allegation that there was a very large snake in the mountains where the sheep grazed and were shepherded. Having lost all of his hired help, the farmer sent his son of sixteen to tend the flocks of animals. When the boy did not return that evening, a party went into the mountains to search for him. They found him unconscious and brought him home. According to the attending doctor the boy was unable to speak for three days because of shock after he regained consciousness.

When questioned the boy related that he had been sitting under a shade tree quietly carving animals from pieces of soft wood when he heard a great roaring noise—like a strong current of wind. His immediate thought was that a "wind devil" (whirlwind) was responsible, since such air disturbances are common in this mountainous area. Looking up, he observed what appeared to be a huge snake hurling itself down from a mountain ridge. As the creature approached the sound was terrific, combined now with that of sheep scattering in all directions out of the creature's path. The creature landed, raising a cloud of dust and a smell he described like burned brass. At that point he apparently lost consciousness from shock and fright. The incident was investigated by police and farmers some of whom actually saw the creature disappear into a crevice in the mountain. Sticks of dynamite were fused and hurled into the crevice. After the detonations a low moaning sound was heard for a short while and then silence.[97]

Dr. Courtenay-Latimer arrived shortly afterward and studied photographs of the marks and tracks of the creature and theorized that it was possibly an injured python. Yet, Mackal does not buy this hypothesis, especially since the creature appeared to have wings and genuinely flew, in some form or another, creating an air disturbance with its wings. Was the dragon of Keetmanshoop a living pterodactyl that had found an easy lunch with the nearby shepherd? Maybe the Batmen of the Jiundu Swamps roamed the remote deserts of Namibia and Botswana as well. According to Dr. Courtenay-Latimer, literally anything is possible in the remote parts of South-West Africa.[97]

§§§

It was two in the morning when we arrived in Keetmanshoop, the center of the sheep ranching district and a fair-sized town for Namibia. Since there were no flying lizards in sight, I spent the night on the porch of the Union Hotel while the driver from Capetown slept in the cab of his truck.

The next day, after breakfast, he then drove into Windhoek, situated in the rugged Khomas-Hochland Mountains at the center of the country. Windhoek, with

61,000 people, is Namibia's capital and largest city. Modern, international, and expensive, Windhoek has plenty of German cafes, fashion shops, delicatessens, restaurants and hotels with swinging nightclubs. Its major tourist attractions are the old Fort, which houses the Cultural History Museum, and the three German castles at Klein Windhoek.

I found Windhoek a very pleasant place, full of German-style pubs where sausage and sauerkraut are served, and enough diversions to keep a traveler busy for a few days. I visited the local biltong (beef jerky) factory one day and met with a friend I had met in Capetown on another. As the center of Namibia, I knew I would be returning to Windhoek at least once, and so after a few days headed out to the west.

I left early in the morning and decided to visit a part of the Namib desert called Sossuvlei. Hitching a hundred miles through some pretty desolate country to a place in the middle of nowhere, appropriately called Solitaire, I was once again well stocked for camping several days in the desert. Solitaire, while on most maps of Namibia, is just a gas station and general store. From there I got a ride with some South African tourists on their way to Sossuvlei, an oasis of sorts amongst the highest sand dunes in the world, curled like sleeping dinosaurs. It is fun to climb up them and slide down, and they make for excellent photography.

At certain times of the year, April through September of so, there is a small water pan where animals, springbok, gazelles and even lions may come in out of the desert to drink. We camped for two nights at the dunes, telling stories around the campfire and watching the moonlit dunes curl away into the distance.

The South African family left me at a dusty desert crossroads where, after several hours, I got a ride to Walvis Bay, one of the three towns on the desolate Skeleton Coast. Walvis Bay is the major port in Namibia, but strangely, it is politically part of the Republic of South Africa, rather than part of South West Africa, and even in an independent Namibia, Walvis Bay will remain a part of South Africa. After checking the town out, I spent the night camping out in a trailer park down by the docks.

The next day I took a bus the few miles north to the quaint German port of Swakopmund. The German legacy of Namibia can best be seen at Swakopmund, the country's main seaside resort. It's a pleasant holiday town with a relaxing, touristy feel. As I walked about town, I met a young student from Germany who was in Namibia to visit a relative and was now hitching around the country a little. Later, I reflected that she was the only other traveler I ever met in Namibia. She was surprised to meet me as well, with my backpack and foolish grin of delight, so after lunch in a cafe, we decided to travel together into the Erongo Mountains and other remote areas in the north of the country.

We spent the night on the floor of a friend of hers, a German/Namibian woman who had studied for awhile in Munich. Then the next morning we headed for Usokos and then Otjiwarongo farther north. We began walking down the dry road out of Usokos, with low brown and green bushes covering the rolling plains after our first good ride. The German girl's name was Sabrina. She was short and stocky, with curly brown hair and glasses; not what you would expect to find out in the middle of Namibia hitchhiking, but then, come to think of it, what would you expect to find?

Our next ride was with a German-Namibian farmer who spoke to Sabrina most of the time in German. Having been born here, he had gone back to Germany in 1927, "because the time was right." He then fought for Germany during the war and left to come back to Southwest "because the time was right." I figured he was probably some die-hard Nazi, as Namibia is full of right wing-extremists, or so I was told. On the other hand, they were supposedly fighting a bunch of left-wing extremists up on Ovamboland, so I guess it all balances out. I dozed most of the time, not being able to understand German, waking up just as we got to Outjo. We didn't stay there long but headed back out on the road to Khorixas in Damaraland. We got a ride with a colored gentleman, a Southern African of mixed race, who was taking a load of supplies to his store in Khorixas in his big Chevy pickup. It was a long ride through wild, beautiful country. We saw kudu antelope and springbok along the way and then were treated to a super sunset, slow and easy with just the right number of clouds to catch the glow and tint of the reds, oranges, and yellows. Being a genuine sunset nut, I was overwhelmed by the colors and movement while I stood in the back of the vehicle, as we roared over hill after hill. With the wind in my hair and the glow of the west calling me forward, I stood up in back, holding on to a railing on the side of the truck and just whooped and yelled for ten minutes. Sabrina thought I'd been in Africa too long and tried to get me to sit down, and the shrieks of joy from the back made the driver look back more than once.

Just as the last glowing embers of sunset faded away, we pulled into Khorixas. A small town with a few shops, some run-down wooden buildings, and a government rest house, it was the administrative center of Damaraland. After some bacon and eggs for dinner, we camped out on the grass behind the rest house.

The lady who ran the place, a charming colored lady from the Cape Province, gave us a lift outside of town after breakfast. No sooner had she let us off than we got a ride with a white doctor making his rounds to the remote villages to the west. We were on our way to the rock paintings of Tyfelfontaine and the petrified forest, as well as the adventure of camping out in the desert for a few nights. We drove around with him for the day, visiting a school and a village started by a bunch of local Africans with the aid of the South African Government. We even chased a lost zebra along a big stretch of flat desert for a while. Such was the excitement of the Namib.

The doctor then took us down the infrequently-used road toward Tyfelfontaine, a strange area of petrified forests and bizarre "bushman" paintings that are quite controversial. He let us off near the cave paintings and then set off back the way he had come to his clinic. It was hot and barren, the only shade was a road sign indicating that the paintings were nearby. The paintings at Tyfelfontaine are interesting, but it is the famous "White Lady of the Brandenberg" that is so controversial. In 1917, during a topographical mission, an engineer named Maak discovered the side of Mount Brandenberg a painted shelter. Today it is known as Maak Shelter.

The paintings depict a procession of people, led by women. They are of Mediterranean type and their elegance and posture are similar to those of the figures depicted in Egyptian frescoes. "The figures have nothing common with the aborigines of southern Africa," says Reader's Digest in their book *The World's Last Mysteries.* [23] The procession is dominated by a woman, known as the White Lady, dressed in skin-tight shorts like the Cretan bull-fighting girls, and carrying a lotus flower. She and her companions had bows, and their wrists are protected by gauntlets. They all wear boots and some have red hair and fair complexions.

Says Reader's Digest, "In Rhodesia (Zimbabwe) there are similar paintings. It would seem that travelers or invaders may have come from the north-east of Africa to the southern areas. In one of the decorated shelters archaeological material has been uncovered enabling the painting to be dated to about 5000 B.C. which was still in the age of prehistory in Egypt. However, the Egyptians are considered by some authorities to be a branch of the ancient Libyan peoples. Perhaps another branch emigrated southwards?"[23]

Indeed, it has long been surmised that Phoenicians may have attempted to colonize parts of Southern Africa, and that such rock paintings of men in armor or white women as that at Maak Shelter were representations of Phoenician explorers or colonists. However, a date of 5000 B.C. throws such theories totally out the door! In mundane academic histories, mankind was hardly getting out of his caves in 5000 B.C., much less having holidays in remote Southern African deserts. Today, Maak Shelter is an extremely remote and difficult place to reach, and this is in the era of motorcars and roads!

The surrounding territory is unusual indeed. There is the extinct volcano of Erongo, around which there are a large number of "bushman" paintings, there is the Burnt Mountain, completely bare of plant life and composed of rainbow-colored lava rock, the petrified forest on the road to Torra Bay, and further east, near Kalkfeld, are dinosaur footprints in rock.

All of the remote "tourist attractions" (probably the number of tourists to actually visit these remote sights can be counted on two hands) are indicative that Namibia enjoyed a different climate then it has now, one where forests once covered the land and where large herds of wild animals roamed, providing sufficient game for hunters. Today, the land is barren, waterless and uninhabited except for a few hardy, lone animals. If the dating of some of the cave paintings is correct, a mere seven thousand years ago Namibia was quite different. This would have been a similar time period as when the Sahara was part sea and part savanna. Further clues to the mysterious past of the area is to be found at the gigantic and mysterious "lost cities".

§§§

After viewing the rock paintings we waited by the road for a lift back to Khorixas. No cars came that day, and so we decided to make camp. The land was like a moonscape, full of strange rocky spires and craters, absolutely devoid of plants, except for some dead, dried bushes scattered about. The only plant that grows all year round is *Welsitchis mirababilis,* an unusual prehistoric plant which grows low to the ground and has two big leaves which get lacerated by the wind. When the Namib was turned into a desert, at some time in remote prehistory, only this plant could survive the climatic change. The reason it could survive is that it extracts water not from the ground but from the fog that rolls in from the coast. Most of these plants are form 1500 to 2000 years old! When it does rain, the desert suddenly blooms with bushes and flowers. The infrequent rains had left enough remains of bushes for us to make a small fire.

We spent a restless night on the sand by the side of the road. I had set up my tent because of the possibility of lions, but Sabrina had refused to sleep in it. Therefore we both slept outside, but I knew that there were lions out there in that desert roaming around, and in that desert, they were bound to be hungry! After I set up my tent we watched the sunset from a hill overlooking the desert to the west. The view was breathtaking; a desolate area of hills, eclipsed by more hills with a haze flowing in the valleys between them. In the distance were tall mountains, rising up in the brown and red tint of the late evening desert. It was like the dawn of time, a primordial prehistoric beginning. Not a sound to be heard, not a thing stirring.

"Truly, it seems like some ancient, time-forgotten desert, and buried deep inside it a secret of some ancient civilization," said Sabrina suddenly.

"Like Lemuria?" I asked. But she had never heard of Lemuria, and it didn't matter.

Fortunately, we survived the night, and then waited all morning for a car to come by, but none came. We decided to walk back toward the main road, as we only had water enough for one more day, and it might take that to get back to Khorixas.

Just as it seemed that we would not get a lift back to Khorixas, a car came along the road toward us and we stood in front of it waving our arms desperately. Rarely had it been so important to get a ride. This was surely the first and only car along this road today, and there might night be another one for several days. We would never make it back to Khorixas if we didn't get a ride with him. Of course, he stopped and took us back to Khorixas. He was a colored guy (of mixed race), going to some research station on the Skeleton Coast a hundred miles or so away. We spent another night on the lawn at the rest house and late the next day at Otjiwarongo, Sabrina and I split up, she going back to Windhoek.

In true adventurer spirit, I decided to hitch up to some of the remotest spots in Namibia. No sooner had I stepped onto the road on the way out of town the next morning than I hailed an oncoming VW bus in much the same fashion that one would hail a taxi, and to my surprise it stopped. It was an Afrikaans family on their way up to Tsambo in the north to buy a house and "shoot terrs" (meaning, of course, to shoot the terrorists) of SWAPO, the outlawed Black Nationalist organization of the Ovambo tribe along the Angolan border. We drove all day up to Tsumeb, a small little town in the farming district. I had dinner in a local pub and then slept out in a park in the center of town.

The next day I headed down the road to Rundu, which I began to think of as the "road to nowhere"—300 kilometers straight through the empty bush to the Angolan border. I didn't know what I'd find there. I got a ride with some black fellows in an old rundown Ford Fairlane station wagon all the way to Rundu where they let me off at the Rundu Club, the only accommodation in town. Naturally I ended up at the bar having a beer and met a couple of South African whites from Capetown who described themselves as the "biggest piss cats in the South West." I took that to mean that they were alcoholics living in Namibia. They called themselves "piss cats" and soon had a nickname for me—"The Biltong Kid". Biltong was dried meat jerky and I always had several strips of beef, ostrich or kudu biltong to chew on. As I had my first beer in the club at Rundu, I'd pulled out a strip of ostrich biltong to chew—and the name was given to me then and there.

Before I knew it, these two piss cats, Neil and Rheinhard, whisked me off to their place to drink more beer and watch an old John Wayne movie called "Hatari" about some cowboys rustling rhinos and zebras in East Africa. Neil and Rheinhard often had movie parties at their house, renting a projector and a 16mm film and inviting the odd nurse and doctor or two from the local hospital. There were also some contractors and surveyors who worked on the same road crew as Neil and Rheinhard, and a Portuguese guy with a shifty manner who would periodically disappear for two months at a time into Angola for the South African army.

These guys were all lonely, desperate souls who for some reason had volunteered to work at the end of the world. Neil, bitter and jolly at the same time, was forty and somewhat overweight. Twice divorced, he had settled into a life of alcoholism in the far reaches of the empire, content with screwing nurses from the mission hospital and going back to Cape Town once a year.

Rheinhard was far more promising. In his mid-twenties, he was doing his engineering internship for the South African Government in Rundu. Tall and handsome, he still had his whole life in front of him and hadn't reached the point of having to forget the past. Still he drank a lot; there wasn't anything else to do in Rundu anyway.

They let me sleep on a cot in their living room for two days, happy to have some fresh blood around to drink with. "Let me tell you," said Neil, opening another Castle beer with a *pphhtt,* "about South-West Africa, er, Namibia to you. One interesting thing is that the second largest ethnic group in Southwest is whites.

Yep, that's right. The biggest ethnic group is the Ovambos just west of here in Ovamboland. However, they live on just a small percentage of the land along the Angolan border. SWAPO, South West Africa's People's Organization, is the political wing of that tribe. And did you know," he asked, opening another beer, "that the UN has recognized SWAPO as the legitimate voice of the Namibian people? If we had a democratic election here, who do you think would win? The Ovambos, of course, and the other tribes in Southwest would not like it at all. Africans are tribal!"

"I guess that is true," I said. "Democracy is often called the tyranny of the majority. Just because 51% of a population wants something to be so doesn't mean that it is right. There are many cases of democratic governments where what really happened was the most populous group gained control of the country and began wiping out the other groups. I've read of this happening to several African countries since the colonial powers began granting independence."

"That's why we're reluctant to just turn the country over to the Ovambos," said Neil. "You know, there are white political groups in Namibia that are so conservative they think that the South African government, the Afrikaaners, are a lot of Communists! Really! They blew up a Masonic lodge in Windhoek this year and have threatened a terrorist campaign of their own."

"Democracy is a farce," said Reinhard. "People have to be mature enough to vote and take responsibility for themselves and other people, in an enlightened fashion. That leaves out most of the whites and the blacks in South Africa!"

So much for politics in the hinterlands of white-ruled Africa. I did agree with Rheinhard that a democracy of immature people was worse then a benevolent dictatorship. It's amusing to think of the South African government as a bunch of radicals. In fact many Afrikaaners in the Republic have expressed their opinion that the government is too lenient and progressive in their reforms. It seemed impossible not to get into arguments over politics when talking and drinking with these fellows in Southern Africa. And my being an American made it all the worse somehow.

"You Yanks should just keep your mouths shut, and stop butting into our affairs!" someone would say. "You think you own the bloody world!"

It's a great lesson in tolerance, but tongues loosen up after a few beers and I was likely to be kicked out into the rain if I said what I really felt. I escaped Rundu after two days of non-stop beer drinking. In a morning-after haze, Rheinhard drove me out to the dirt road that would take me back to Tsumeb and the area populated by white farmers. I walked down the road for a mile or two, feeling relieved that I was getting out, and feeling sorry for all those people who couldn't throw a pack on their back and hitch out of town like the "Biltong kid."

I caught a couple of lifts on Army trucks up to Tsumeb again, driven by pimply-faced South African kids who were doing their compulsory military service up here in Namibia. This was South Africa's hot spot and it seemed as if all high school boys ended up here sooner or later. After spending another night in the park in Tsumeb, I thought I would hitch to Oshakati, the capital of Ovamboland and the center of all the terrorist activity in Namibia. I stood out on the main road to Oshakati and had no sooner started hitching when a truckload of Ovambo natives pulled up and I piled in. They were friendly and curious, wanting to know where I was from and where I was going, and most of all, how did I like their country, Ovamboland? It was great, I told them, although we hadn't even got there yet. We cruised along at a steady 90 kilometers an hour with the radio blasting away all our "favorite hits on Radio Ovambo," and reached Oshakati in about three hours.

Oshakati was quite a disappointment. I searched all over for a town center or a market, but all I could find was a deserted petrol station. "This is it," said the old man in brown clothes and rubber sandals. Otherwise it was just a mass of fences, barriers, army camp, and the like.

Standing on the road, hitching back south, I was glad to have satisfied my curiosity about the northern area of Namibia. At least I could say I had been there. I got a ride with four black dudes in a run-down white Chevy of some kind and they asked me to drive the 250 km back to Tsumeb, perhaps because they thought it would be less suspicious and safer for them if a white guy was driving. After all, the area was crawling with South African Army personnel. The guy who owned the car, a tall skinny fellow with a big gold watch and a hearty smile, told me he was a "consultant" for the Orangemund Diamond Mines down in the south, but it later turned out that he was a driver of one of their trucks, which seemed more likely. At the time it did not occur to me that they might be smuggling diamonds, and that was why they wanted me drive, but it did occur to me later. Along the way we passed what I would call a "pack" of fifteen South African army commandos on scrambler motorcycles. Both they and their dirt bikes were camouflaged, and they had backpacks and rifles on their shoulders. They reminded me of some sort of legalized motorcycle gang, heading down the highway, looking for trouble and adventure in a deadly kind of way. To the Namibian natives in the car it was no big deal, just part of everyday life in Ovamboland.

The black Ovambo natives and I tried stopping at the Etosha Pan on our way to Tsumeb, but it was closed for the season. Etosha Pan is one of the largest game parks in the world with some good game viewing to be had around the waterholes when the animals gather during the dry season, from May to December. We barreled on through the night, as the guys were on their way to Windhoek, and they let me off in Otjiwarongo, where I spent a rainy night in my tent at the local caravan site.

It was mid-afternoon the next day when I arrived back in Windhoek. I spent a night at a friend's house and left the next morning heading east toward the border of Botswana and the Kalahari Desert. In the late afternoon I arrived in Gobabis, and started once again my nightly search for a spot to sleep. "You'll stay at my place, you can sleep on the couch," said a young German-Namibian I met on the street. And so it was that we went back to his place, listened to the newest Pink Floyd album that he had just bought, and then went bar-hopping for the rest of the night.

As I walked down the road the next morning out of Gobabis with a slight hang-over, I felt a certain sadness about leaving Namibia. What a great country: wild, hospitable, and yet inhospitable at the same time. It was a frontier by any definition, and I felt that it was country with a great deal of promise, yet it would have to have its political problems sorted out, and that would not be easy.

Before heading out of town, I bought supplies for the trip: some tunafish and cans of pork and beans, some dried ostrich jerky, a flagon of wine, a six-pack of beer, a loaf of bread and a gallon of water. This, I figured, would get me to Ghanzi, about 250 miles to the east. With my pack full of supplies, I felt confident that crossing the Kalahari would not be so tough. A lost city in the desert awaited me, as well other adventures.

I was on my way back to Zambia, where I would get a flight to the Seychelle Islands in the Indian Ocean. Crossing Botswana was my last main sojourn in Africa and that meant one of the great thrills of my trip, the opportunity to hitch across the Kalahari desert. Like climbing Kilimanjaro, crossing the Kalahari while in Africa seemed like something one just had to do, not for any reason I could name, but simply, like Mount Everest, because it was there. It is this kind of illogical bullheadedness that makes the best explorers, I suppose.

Known as Bechuanaland until its independence from Britain in 1965, Botswana is famous for its Kalahari Desert, which isn't a real desert, according to many Botswanans. The majority of the country is a tableland, with the Kalahari covering most of the center and southwest and the Okavango Swamps and Chobe Game Park taking up the northwest. Most of the people, about seven hundred thousand of them, live along the eastern border with South Africa and Zimbabwe where cattle ranching is the main industry. Totally landlocked and semi-arid except for the Okavango Swamps, Botswana is about the size of Texas or France. Botswana is also known as the only multi-party democracy in Africa, all other African countries being dominated by the party that happens to be in power at the time—and that party generally outlaws all other political parties. Still, like most African countries, Botswana seems to have its president for life. Sir Seretse Khama, an Oxford-educated lawyer married to a rather reclusive English lady, has been reelected to the presidency over and over again since 1965. With his recent death, the Botswana legislature ruled that only a pure-blooded black could be President, which pretty much stopped his son, part English, part Tswana, from becoming President. Instead, he is the head of the military. Still, Botswana remains one of the most stable and progressive of all black African nations.

It is said that the original inhabitants of Botswana were the Bushmen. They are also said to be the earliest-known inhabitants of Africa; but frankly I doubt that this is true. At any rate, there are only about 4,000 left. The mystery of their origin is still unknown, but it is a little-known fact about the Bushmen is that they are not racially classified as "Negroes" in the five major racial classifications, but are known anthropologically as "Caucasians". Generally, anthropologists like to classify mankind into five main groups; the white man or Caucasian; the black man or Negro; the yellow man or Mongoloid; the red man or Amerindian; and the Australoids of Australia whom are in a class of their own. The bushman or Hottentot is not a Negro, all anthropologists admit that. But what is he? Generally they are classed as whites, though another popular theory is that they are Asians or Mongoloids. Indeed, Bushmen look a great deal like certain hill tribes in Laos, Vietnam, Thailand, Burma and China.

I personally tend to this view, that the Bushmen are part of an Asian invasion of Eastern and Southern Africa. "What!" ask the critics. We must remember that Madagascar, an island just off Mozambique was settled by Asians from Malaysia and Indonesia many thousands of years ago. In a migration that preceded the mass immigration of Europeans to North and South America by six or seven thousand years, so-called stone-age people navigated an empty expanse of ocean that is twice as far as from Africa to Brazil. At Mapungabwe in the Transvaal, excavations of skeletons at the site proved that the inhabitants were bushmen or Hottentots.[106] Now these people roam in small numbers their last refuge of the unpopulated waste of the Kalahari, where they track animals and follow the seasonal pans of water. What seems evident is that the Asian invasion of Southern Africa included not just Madagascar, but also the high plateau of Zimbabwe and other areas of Southern Africa that could be reached by the river systems. This would make the Bushmen relatives of the Malagasy.

Even more tantalizing is the mystery of lost cities in the Kalahari; megalithic stone cities that must predate even the arrival of the Bushmen.

§§§

As I stood with my pack on the outskirts of Gobabis, a large cattle truck suddenly drove toward me and came to a stop. I piled in back with a dozen natives, and we were off through the desert. To my surprise, I met a Swiss guy on the truck, who was heading to Ghanzi to take a part for his Volkswagen bus that had broken down on the road. He had been to Gobabis to get the part. As we drove that first day, through the low bushes of the Kalahari sandveldt, I saw the entire desert blooming with a small yellow flower called the Devil's Thorn, but except for an occasional ostrich, there was almost no wildlife visible. The Kalahari is not uninhabited, for there are small cattle ranches scattered throughout, except in the very central part of the Kalahari which was now east of us. Just at dusk, the driver let the Swiss and me off at an absolutely desolate crossroads in the middle of the desert somewhere, as he took off south down an overgrown ranch road in his rusty old cattle truck. The Swiss, myself, and a few natives then settled down at the remote desert crossroads where the two tracks met, and made a camp beneath the tree.

The Swiss guy and I sat beneath the thorn tree with about ten other traveling Botswanans, Hereros, and Tswana who were on their way to see relatives or something. Like us, they were waiting for a ride to Ghanzi. Our supplies of food and water came in handy now. We cooked up some soup with our precious water and then I brought out my sixpack of beer. As we enjoyed the now lukewarm brew, I told him of the lost cities of the Kalahari.

The bushmen and Hottentots have many legends and tales of lost cities in the Kalahari. These cities are not built by them, but by the *ancients.* [108] In the *Sunday Times* of Johannesburg for March 15, 1931 and included by A.J. Clement, in his book *The Kalahari and Its Lost City,* [108] it is said that a Mr. T. H. Howard met with a Mr. A. A. Anderson near the Nossob River in the Kalahari during the summer of 1873-4. Anderson told Howard that he had seen ruins of a lost city "up north", but that he had not approached the spot, for his Bushmen had warned him that the Ovambos would kill anybody who actually entered the city. The building of the city had never been completed, and it was still possible to unearth tools from the debris. Other newspaper articles at the time told such popular reports of the rumors of lost cities and the expeditions that searched for them. A Mr. R. Craill said that "stories more strongly authenticated than those told in the past pointed to the possible existence of mysterious ruins in the desert. Bushmen told tales of the existence of strongly fortified walls, now crumbling into the sand, which had been seen in the Kalahari, and several pieces of stone carving stated to have come the 'lost city' were actually produced."[108]

These mysterious ruins were apparently to found in different parts of the Kalahari, even north near the Caprivi Strip and Ovamboland (as in the first report). Dr. Gustav Prelude, a well-known historian, reported that a party of bushmen in the Kalahari who had camped near his expedition had told him of a terrible drought and of large ruins somewhere to the north, and expressed their willingness to lead him there once the rains had fallen. The Bushmen also claimed that precious stones had been found in the desert farther north.[108]

The *Kimberley Diamond Fields Advertiser* newspaper carried a story on August 24, 1949 about an expedition into the Kalahari led by a Dr. van Zyl in search of a lost city. Dr. van Zyl was prompted to search for the city because of a Bushman legend about a ruined city—one of a chain of forts said to have been built between Zimbabwe and the coast by past civilizations in order to safeguard the transport of gold and silver. A Dr. G. C. Coetzee, who had been a member of the expedition, said, "Nomadics swear to having seen the city, but searches have all proved vain. The heavy sand moves with the wind and landmarks disappear after every sandstorm."[108]

In 1956 an expedition left Natal which included the well-known South African author Alan Patton (author of *Cry the Beloved Country*) and a Major D.C. Flower. They had heard encouraging news from a man in Gobabis, who claimed that Bushmen had told him of ruins between the Toanakha Hills and Mount Aha. Other Bushmen rumors reported in the press were of ruins between Kang and

Kaotwe, as well as between a place called Na Na (Nxau Nxau) and the Caprivi Strip. In the Nxau Nxau district an explorer named Dr. Haldeman was told by a Bushman that his father knew where there were some ruins. Dr. Haldeman went with the Bushman to a place called Donkey, but when they arrived the Bushman's father refused to guide them to the site. The father claimed that he was the sole survivor of nine Bushmen who had seen the ruins.[108]

Hottentots, relatives of the Bushmen, but living a more settled lifestyle, told a Dr. W. M. Borcherds about a lost city along the course of a dry tributary which joined the Nossob River somewhere between Kwang Pan and Rooikop. The ruins were supposed to be close to a fairly high sand dune and a "large kaal (bare) pan". A French explorer named Francois Balsan who crossed the Kalahari in 1948 told of a Hottentot living at Lehututu in the Central Kalahari who was supposed to have knowledge of the whereabouts of a lost city. Balsan had employed this Hottentot as a guide three years earlier on a previous expedition to Bechuanaland (Botswana, as it was called before independence); on this occasion Balsan thought he had trekked to within 35 miles of the lost city. After Balsan's expedition failed to find the fabled lost city, the Johannesburg *Star* said, "South Africans resent this skepticism about the fairies at the bottom of their garden... Rider Haggard himself, a court official in his day and an honest South African has shown that the drier parts of the sub-continent abound in hidden cities and lost treasure. But you will never find them if you cross the desert by motor-car....The least the country can do for the lost city of Kalahari is to proclaim it a national monument, 'whereabouts at present unknown'." (10 October, 1951)[108]

"Well, it's a fascinating story," said the Swiss, as I handed him another beer, and tossed a stick onto the fire. "But has anyone ever seen this city? It seems that it is just legend."

"Ah, but that is part of the mystery," I replied, "In 1885, an American explorer did see the ruins, and he took photos of them and wrote a book!" And with that I told the Swiss the story of Farini the Great. Born William Leonard Hunt in New York City, he became a showman, best known for tightrope walking across Niagara Falls, he changed his name to Gilarmi A. Farini and his show name was Farini the Great. Farini exhibited shows at Coney Island, and when a show did well, it was taken to London. One of the shows in London that was popular was entitled *Farini's African Pygmies or Dwarf Earthmen from the Interior of South Africa*. Essentially a tableaux of Bushman life in the Kalahari, it included six live Bushmen. Fariniwas very interested in the Bushmen and became even more so when they showed him diamonds they said had come from the Kalahari. Farini, his son Lulu, and a colored South African showman named Gert Louw who had brought the Bushmen to London, sailed from London to Capetown on January 7, 1895 and arrived on January 29. After meeting certain dignitaries in Capetown, in early February, the party departed for the Northern Cape and the Kalahari. Six months later they returned to Capetown, claimed to have discovered a lost city of colossal proportions, and departed for London on July 22, 1895.

Back in London, Farini addressed the Royal Geographical Society and later the Berlin Geographical Society. His book, *Through the Kalahari* [109] was published at about the same time, and it contained photographs of the lost city taken by Farini's son Lulu. Farini even staged a *Lost City Exhibition* in London, which included photos of the city. These photos showed the city to be of huge, massive stones stacked on top of each other, and of extremely ancient construction.[108]

Farini described the megalithic city as a long line of stone laid out in the shape of an arc and resembling the Chinese Wall after an earthquake. The ruins were quite extensive, partly buried beneath the sand at some points, and fully exposed to view in other. They could be traced for nearly a mile, and consisted mainly of huge flat-sided stone. In some places the cement was in perfect condition and plainly visible between the various layers of the heaps. The top row of stones was weathered and abraded by the drifting sand. Some of the uppermost stones were grotesquely worn away on the underside so that they resembled a small center table supported by a short leg.[108,109] In his Royal Geographical Society report he described the stones as "cyclopean".[108]

Heaps of masonry, each about eighteen inches high, were spaced at intervals of about forty feet inside the wall. The heaps were shaped in the form of ovals or obtuse ellipses; they had flat bases and were hollowed out at the sides for about twelve inches from the edge. Some of them consisted of solid rock, while others were formed from one or more pieces of stone accurately fitted together. Where they had been protected from the sand the joints were perfect. Most of the heaps were more or less covered with sand, and it took his local colored guides almost a day to uncover the largest of them.

The following day, with no assistance from his guides, who apparently felt it was all a waste of time, Farini and his companions dug the sand away from the middle arc and exposed a pavement structure built of large stones. The pavement was about twenty feet wide, and so designed that the longer, outer stones were laid at right-angles to the inner ones. A similar pavement intersected it at right angles, and the whole structure resembled a Maltese Cross.

Farini visualized an altar, column or some other kind of monument at the intersection of the two pavements. The remains of the base, which were clearly visible at the junction of the pavements, consisted of loose pieces of fluted masonry. There were no inscriptions or markings of any kind. He concluded that the ruins were probably thousands of years old, and they must be of a city, a place of worship or the burial ground of a great nation. His son Lulu sketched the ruins and took several photographs. [108,109] Farini composed this poem for his lost city:

> A half-buried ruin—a huge wreck of stones
> On a lone and desolate spot;
> A temple—or tomb for human bones
> Left by man to decay and rot.
>
> Rude sculpted blocks from the red sand project,
> And shapeless uncouth stones appear,
> Some great man's ashes designed to protect,
> Buried many a thousand year.
>
> A relic, maybe, of a glorious past,
> A city once grand and sublime,
> Destroyed by earthquake, defaced by the blast,
> Swept away by the hand of time.[108]

In his book, *The Kalahari and Its Lost City*, Clement does an exhaustive study of Farini's book, his route and the inconsistencies to be found in the publication (and there are many). Clement is to be commended for his research, though his final conclusions are to be questioned, as we shall shortly see. Farini's book caused a brief sensation at the time, and was published in both German and French. But then the whole business of a lost city in the Kalahari was generally forgotten until 1923 when the story was revived by Professor E. H. L. Schwartz of Rhodes University. Farini himself died at his ranch in Ontario in 1929. From the 1920's up through the 1950's, many expeditions set out in search of the incredible lost city, many using aircraft. No one was able to find it, largely due to Farini's wildly inaccurate maps to the spot. Many began to feel that the whole thing was really just a natural limestone formation, yet Farini had photos of the city in his book, and no one had yet come up with a suitable natural formation that fit Farini's description. Clement also shows in his book that Farini almost certainly did not travel up to Lake Ngami afterward, as he claimed in his book. Clement believes that he turned back after discovering the city, and used details supplied by his secretary W. A. Healey who had visited Lake Ngami the year before collecting items, as well as Bushmen, for the London exhibit.

After pouring through Farini's book, Clement finally concluded that Farini's lost city must actually lie near the small town of Mier, now called Rietfontein. With partial sponsorship from the *Sunday Chronicle* newspaper, Clement set out with his 77-year old father, a reporter from the newspaper and a professional photographer, on Easter Monday of 1964. At Rietfontein they were shown an extremely unusual "rock formation" known to the locals as Eierdop Koppies (Eggshell Hills). Says Clement, "The unmistakable outline of a large, oval-shaped amphitheatre, perhaps a third of a mile in length, was the predominant feature. In numerous places there was striking resemblance to a double wall built from large, glistening black rocks, and it was obvious that many of the individual boulders could easily be confused with square building blocks. There were several examples of flat slabs of rock perched precariously like table-tops on underlying boulders, and one of them—more impressive than the rest—closely matched the one appearing in Farini's illustration. One or two of the rocks showed a kind of fluting, several were encrusted with a mortar-like substance, and a few were shaped like a basin. To use the phraseology employed by Farini in his lecture before the Royal Geographical Society: 'The masonry was of a cyclopean character...'."108

Clement showed one of Farini's photos of his lost city to the oldest man in town who agreed that it seemed to show the same place. Clement, it seems, had genuinely rediscovered Farini's lost city—known all the time to local residents—but concluded that it was no city at all, merely a highly unusual natural rock formation of dolerite, a hard igneous rock. After showing his photos to a geologist, the geologist suggested that the "ruins" were the product of the weathering of dolerite, in this case, magma intrusions forced their way in the form of sills or sheet along the bedding planes of sediments (some 180-190 million years ago guesses the geologist) forming the level planes or flat sheets of rock found at the site. As the magma cooled, it formed cracks and splits, making it seem as if the rock had been carefully cut and dressed, then stacked up on top of each other. Part of composition of dolerite is pyroxene and over time a chemical reaction takes place that precipitates a brownish "desert varnish" and kind of cement.

Clement concludes his book by saying, "Like all legends, that of the Lost City will be a long time a-dying, and doubtless there will still be some who are disinclined to let the matter rest in spite of all the contrary evidence. And possibly this is just as well, for there is something rather sad about the destruction of a legend."108

I wonder if Clement wrote those words for me? Certainly, I am disinclined to accept his conclusions, and proof against them is given in his own book. Even before setting out, Clement was convinced that Farini's city was a natural formation. He could not conceive of a "cyclopean" structure in the Kalahari that was not natural. Says he, "The climatological history of the Kalahari does not appear to have undergone any marked change for several thousand years, and it is obvious that no settlement of the size indicated by Farini could exist without perennial rivers or lakes in the vicinity."(p.145) And, "...suitable conditions for the establishment of a 'city' cannot have existed along any of the river courses for tens of thousands of years. Furthermore, if the age of the Lost City is assessed in relation to Zimbabwe and the ancient ruins of Persia, it is impossible conceive of

any 'city' in the Kalahari having been built more than 15,000 years ago."
(p.147-148)

It is painfully obvious that the "city" has nothing to do with Zimbabwe ruins, and his mention of ancient ruins in Persia seems totally beside the point, indicating that Clements, for all his excellent research on Farini, knows little about ancient history, and had probably never seen a cyclopean wall before. Cyclopean architecture, which can be found in Egypt, Turkey, Greece, Malta, Peru, Bolivia and other areas, is indeed an astonishing sight! To this very day the method for building such walls still confounds architects and engineers. Farini had traveled a great deal in Europe and the Mediterranean and had probably seen cyclopean walls in the Peloponesse in Greece or at Abydos or the Valley Temple of Chephren in Egypt. Farini was also a Mason, and, depending on his initiatory status within the Masons, had probably been exposed to the Masonic beliefs in Atlantis, cataclysms, Mystery Schools and such. His poem about the city indicates as much.

Other clues to the non-natural origin of the rocks can be found in photos taken both by Clement and Farini. The rocks are all neatly squared and the lines of "masonry" are parallel and at right angles. Some igneous formations such as basalt do indeed crystallize in regular patterns, but not like the dolerite rocks at Rietfontein. The final proof is Clement's own photo (reproduced at the end of this chapter) of one of the massive blocks with a series of four parallel, horizontal grooves on it. Clearly these are not natural! Even Clement admits that they could not be natural. In the photo caption he asks, "Are they natural, or were they made by Farini?"

I would venture to guess that they are neither. Farini's lost city is probably just as he believed it to be, a cyclopean structure from another era, destroyed in a cataclysmic shift of the earth's crust, possibly 15,000 years ago, but probably more recently, such as about 10,000 years ago. It has been suggested that a shift of the earth's crust about this time sent Africa moving to the south, causing a huge tidal wave to wash over all of Southern Africa. Any cities, such as Farini's, would have been destroyed and depopulated during such an event.

If the Kalahari was some sort of Atlantean colony 15,000 years ago or more, then I surmised that there must be other cyclopean ruins around that were also being mistaken for natural formations. And indeed there are. Clement mentions "massive stone blocks on Mr. Guy Braithwaite's farm, 'Gesond', in the Tuli Block of north-east Bechuanaland (Botswana). Similar stone blocks are also present in the sandy bed of the Amacloutsie River. The so-called 'Solomon's Wall' is 'between fifty and one hundred feet high (imagine from three to five double-decker buses piled on top of each other), it is between ten and fifteen feet thick and dominates the countryside for a couple of miles... on the exposed faces of Solomon's Wall, chemical erosion has produced a network of lines which give the appearance of building blocks. Some of these blocks are several feet square while others are only a matter of inches....The local Ngwato tribe claim that it was built by the 'old people', and it is suggested that something similar to 'Solomon's Wall' may have been seen by Farini in the Kalahari."[108]

As I drifted off to sleep that night, the Swiss and I having finished my sixpack of beer, I wished I could somehow make it down to Rietfontein and see this city first hand, but it was too far to the south, and traffic, like roads, were non-existent. I would have to be content to see this Atlantis in the Kalahari on some other trip.

§§§

It was only one and a half days before another truck came along. When we saw it, we all ran out onto the rutty tracks, and waved and jumped up and down like mad. You can bet he stopped. There was no way he was not going to give us a ride! Happily, all twelve of us climbed in the back of this empty cattle truck and headed for civilization again. So far the trip hadn't been so bad, I decided. We spend a nice night in the driver's brother's house in Kalkfontein, a very small town between Gobabis and Ghanzi, the site of an especially deep well.

Once we chased three ostriches down the road for five minutes, their tail feathers up, their mighty haunches bulging as they sprinted down the dirt tracks east toward our destination: the central Kalahari town and commercial center of Ghanzi. We arrived just at sunset, my favorite time of the day. I don't think I will ever tire of sunsets, especially African ones—a flock of birds on the horizon, and the deep reds and oranges of the equatorial sun.

The main attraction in Ghanzi is the people and atmosphere of the place. I guess that's the whole idea of traveling around Arabia and Africa; meeting people and growing from the experiences. I sat on the porch of Ghanzi's main spot of interest, the Oasis Store. It was a general store in the traditional sense of selling everything the Bushman could ever want. At the same time, it seemed to be the social center of the entire Kalahari Desert. All kinds of people would come and go; thin haggard Bushmen, in to buy some supplies, what little they needed and could afford.

I sat with them on the porch of the Oasis Hotel and drank cans of Lion Lager beer. Sometimes a Tswana, in cowboy hat, wool jacket, and boots, would join us. They were the cattle ranchers of the Kalahari. Usually they spoke better English than the Bushmen, and after a few beers would tell about the lost cities and old mines in the desert, about which there are a number of legends. One grizzled, wrinkled old rancher even talked about a lost continent, Lemuria, and how the people had fled to all parts of the world, even South Africa. But that "was a long time ago."

After two days of hanging out in Ghanzi, I arranged a ride out of town toward Maun in the Okavango Swamps on a truck, driven by an old Tswana guy who spoke hardly a word of English, but could go on in Afrikaans all day. His truck was an old British Bedford flatbed truck, empty except for an old bed frame in back and me. His tires were as bald as he; a curly gray beard kept the flies out of nose and mouth.

An hour after leaving Ghanzi we were stuck in swamps around Lake Ngami, "the great lake of the unknown region." In the past, only the Bushmen knew of this lake, telling of it in awe at their evening campfires in the desert, "a lake with waves that throw hippos ashore, roaring like thunder. A lake of many, many fish..." Eventually the baYei hunters stumbled onto the lake and sent word back to their tribe—that was only in the 1750s. By the 1890s, the more aggressive baTawana moved in with their cattle. There was only one way to find the mythical lake of the unexplored southern desert of southern Africa, and that was across the Kalahari itself. It was on August 1, 1849 that the intrepid explorer, David Livingston, first set eyes on the lake. At that time, it was 240 kilometers wide. At times it completely dries up!

It certainly wasn't dried up the year I was there, but seemed to be covering a lot more than its usual boundaries. We were slipping and sliding all over the place, usually in several feet of water that was supposed to be the road. We had neither four wheel drive nor even good tires. Eventually we got stuck and there seemed no way out of the mud and water this time. "It is times like these," I reflected, "that I'm glad I'm hitchhiking in Africa. I can walk to Maun if I have to."

Fortunately, just then, a very large tractor came by and pulled up out, and even pulled us to a road camp where the tractor driver and a bunch of others were spending the night. We decided to spend the night here also. I heated up a can of pork and beans over the fire and finished off the last of my beer from the Oasis Store in Ghanzi, before spending a fitful night in the front seat being eaten alive by mosquitoes while it rained outside. The driver slept in the tents of the road camp.

Early the next morning we were off again, around the back side of Lake Ngami, each tree quivering and trilling with thousands of Quelea finches. It was all day through arid grassland, passing one or two small farms near the lake. I drank half a flagon of South African wine, fell asleep in the hot afternoon sun and got terribly sunburned.

Though I could have been riding inside with the driver, I preferred to stand up in the back, the wind blowing through my tangled hair. I had a great feeling of openness and communion with the countryside as we passed through it. Standing in the back has its dangers too; you have to keep a wary eye out for low branches and watch out for bushes whipping your face as the truck plows on its own road through the underbrush. As we stopped in a small African village just after the lake, all the kids in the village came running when they saw a European traveler standing in the back of the truck.

The kids gathered around the truck, laughing and staring at this crazy guy. "Where are you going?" asked one kid.

"To Maun," I replied, pulling a small green chameleon off my pants and putting him on my shirt where he clung for some time. These charming, slow-moving lizards are utterly harmless, having two feet on the end of each leg and eyes that are like cones, the eyeball at the tip, the cones moving independently of each other. The children, to my surprise, where quite frightened by this little critter, and were shocked to see me let it cling to my shirt. I surmised that he must have gotten knocked off a tree as we drove through the branches and landed on me. The driver was also amazed that I wasn't afraid of the lizard and asked me to throw it out, so I placed him high up on a tree, where he could go his slow and merry way.

It was evening when we arrived in Maun, a sort of safari town with lots of Great White Hunters, all well over the hill and leading safaris of telephotoing tourists by canoe into the crocodile- and bird-infested swamps of the Okavango Delta.

I met John on the streets of Maun. He was a Peace Corps volunteer who was teaching English in town. I also met Patrick, a traveler in his early twenties from California, who was on his way from Lesotho in South Africa after visiting his mother who was also a Peace Corps volunteer. We all hit it off well, John invited to sleep at his place and then we all went out to a small restaurant called "The Bistro," run by a French world traveler, who like a lot of people, came to Maun and couldn't get out. Perhaps like one type, he was running away from the army and couldn't return to France, or else, like another type of traveler, he had to get away for some other reason.

As John, Patrick, and I sat at a small table drinking beer, waiting for our hamburgers and french fries, I knew that we fit into the latter category. However, I did not know what we were running from.

Patrick and I were fortunate enough to get a ride from Maun to Kasane on the Zambia border with John, the crazy surfer from California, who had a Toyota Landcruiser he had to drive up to Kasane, nearly two hundred miles away. The drive is mostly though Chobe National Park on parallel tire tracks through a dirt road that changes from season to season as mud holes and other obstacles get more and more difficult to drive through and have to be circumvented. I was still headed for Lusaka, so I could get my flight to the Seychelles.

By midmorning, we were whipping along the African bush past hordes of zebra, springbok, and giraffe. There would also be the occasional herd of elephant and a lioness or small spotted cerval cat crossing the road. Chobe National Park, it occurred to me, must be one of the most underrated game parks in Africa, absolutely teeming with game and with virtually no tourists. Nevertheless, we managed to have a head-on collision with what was probably the only other vehicle for a hundred square miles.

We were barreling along down the dirt road that was no more than two dirt strips, each big enough for a tire, with elephant grass as tall as nine feet growing on either side of the road, making turns a bit on the blind side. As we came around a sharp, grass-obstructed corner, Patrick and I, who were both sitting in the back, were thrown forward to the front of the truck as Surfer John slammed on his brakes. Doubtless the other driver was doing the same, and we both hit head on, our truck bouncing off theirs and jumping straight back four feet with a neck-snapping lurch. Fortunately no one in either truck was hurt.

We continued on until just before dusk and spent the night in a small village about sixty miles from Maun. John commandeered the local school room for us to sleep in and set off looking for a "shabine" or local African bar that sold Kadi, the home beer of the locals. An hour later, when Patrick and I had our mosquito nets all set up and a fire going in the yard just outside the school, John turned up, drunk as a hyena in heat. Singing something about California girls and the surf in the moonlight, he insisted that we go with him back to the shabine as he had left his sandals there. The three of us tripped on off into the night, weaving our way between the grass huts that were all scattered along the bank of the Linyanti River. Then we were sitting around the fire in a rather large grass hut drinking thick, sour Kadi like some good draft beer. To our surprise, (though not to John's) the place was full of nubile young women, five or six of them and all quite smashed. They couldn't keep their hands off John, but Patrick and I were too shy to get involved with these rather forward young ladies. Besides, Patrick had mentioned to me that Botswana had something of a VD epidemic, one of the highest rates in the world. We therefore tore ourselves away before it was too late and we were too drunk, leaving John to close the bar, so to speak.

§§§

We arrived just after noon the next day in Kasane, a small town on the Botswana side of the Kazungula Ferry. John found another school room for us to sleep in and then told us about a dance at the local hotel in Kasane, we thought it would be a blast to go.

Later that afternoon, after freshing up, John, Patrick and I were walking along the river when there was the deafening drone of gasoline engines and the beat of whirling metal blades came from the skies. Having already been in two wars on this trip, I dove instinctively for a bush. John, too, hit the dirt, and Patrick dove down the bank of the river toward the shallow reeds along the edge. A helicopter, probably a West German Leopard attack chopper, came flying along the river at the level of the tree tops, flying east toward the South African military bases in the Caprivi Strip, which sticks out between Zambia and Botswana. The noise and speed of the chopper's arrival was startling and rather frightening. I could imagine what the Africans must think of these loud toys of destruction. From my position flat on the ground, I noticed machine guns mounted on the underbelly of the chopper, each capable of devastating a village; this was death and destruction at its most sophisticated. As fast as it had come, it was gone, speeding down the river on its border patrol.

"Holy shit!" cried John. I was helping Patrick out of the reeds. He was coughing, wet and confused.

"That was some dragonfly," said John. We all had a good laugh and headed for the Safari Lodge. Soon we were guzzling beer, as people are known to do in the afternoon sun.

As we sat on the terrace watching the river, I mused over the mystery of the Bushmen and Hottentots and the solution to the ancient history of Africa. From the way I was able to piece it together from various sources, a massive continent in the Pacific went down in a catastrophic pole shift circa 22,000 B.C. The twelve different races of mankind were said to come from that continent, though they had begun migrating around the world many thousands of years before that.

South America's continental plate was raised at that time, and an island archipelago in the Atlantic became a small island continent named Poseid, Atlan, or Atlantis. Ancient India, China, the Gobi Desert, the Andes and Mexico were highly developed civilizations. The North Pole was in the Hudson Bay. Atlantis was destroyed in an isolated geological upheaval, possibly of its own making circa 8,000 B.C.

Shortly after the sinking of Atlantis, the earth experienced another pole shift, causing the North Pole to be shifted from the Hudson Bay to where it is now. Much of Antarctica was ice -ree but now began to accumulate the awesome amount of ice it currently has. Today the lopsided ice cap in Antarctica is more than a mile thick and contains 90% of the fresh water in the world!

Southern Africa was completely depopulated at this time, as the African Plate slid in an essentially southward direction. A gigantic tidal wave washed over the entire southern portion of Africa, destroying all civilization. With a sudden tilt of the plate, parts of the ocean bottom, for instance around the Namib, were tilted up, while the Triton Sea began to drain, eventually leaving only a few oasis' and Lake Chad. Only the vast swamps of Central Africa were left, and even those began to slowly drain. Giant saurians of the recent, but prehistoric, past slowly retreated to the remotest of swamps and river tributaries.

The Mediterranean, if it had not been flooded earlier at the time of the sinking of Atlantis, began flooding. The climate of Egypt changed, and in the end it was the only civilization left from ancient Osiris.

Ancient Rama was gone, but small kingdoms in India and the Middle East survived the pole shift somewhat intact. In the wake of the devastation, a bold class of ancient seafarer, men who were scientists, architects and navigators, and who worshiped the sun; the amazing Atlantean League; sailed the earth, exploring new worlds, colonizing far-flung areas, and remapping the earth. It was these people who later became the Phoenicians and Carthaginians.

Since much of Africa was depopulated in the last pole shift, it began to be slowly repopulated. Egyptians lived in the northeast corner, Ethiopian/Sabeans in East Africa, originally from Southern India. Negroes lived in Mexico, Cameroon and New Guinea, and Tuaregs in the Tibesti Mountains.

Asians probably began migrating across the Indian Ocean to Southern Africa about 5,000 B.C. They populated Madagascar and the coastal area of Mozambique and inland to the high, mineral rich plateau of Zimbabwe. It was these people, Asians, along with Sabeans, who then began remining southern Africa, an area that had not been populated since before the last poleshift several thousand years before. Later, they built Zimbabwe and some of the other cities. The area was probably an international trading and manufacturing center, visited frequently by Phoenicians, Chinese, Persians, Indians, Arabs, and nearby Sabean traders who were just up the coast. Sabeans from Azania in East Africa then began encroaching in their territory and finally some disaster, either from within, such as a slave rebellion, or from without, such as an invasion, ended forever the

Ma-Iti Empire.

The Asians, those that survived the slaughter, were forced to scatter and were pushed out of the Zimbabwe Highlands and High Veldt, into the desolate deserts of the Kalahari and Namib. Today, these people are the Bushmen and Hottentots. Madagascar remained the Asian stronghold, though there were incursions in the northwest by Arabs and Negroes. Bantus had only settled as far as Natal or Transkei, though Hottentots and Bushmen must have roamed through the Orange Free State and Cape Province at that time. Though the High Veldt of the Transvaal was heavily populated at the time, the Zulu Wars in the early 1800s depopulated much of the country. The Portuguese and Dutch began to settle Southern Africa, and so history comes up to the present—and some geologists tell us that we are on the verge of another pole shift, and it is about to happen all over again!

As the sun went down and palm trees were reflected in the silvery water, music started to play in the upstairs disco of the lodge. A disc jockey was spinning his favorite tunes, a mixture of American disco, Jamaican reggae, and African "juju" music. By the time we staggered upstairs, the whole place was packed to the balcony with twisting, shaking Africans, each with a can of Lion Lager beer in his hand, thus giving the name to their dance, "The Kasane Beer Can Boogie."

I danced for awhile, obligatory beer can in my hand, and then walked out into the night, the party behind me. Relieving myself of part of the evening's beverages, I looked up into the sky. I thought of the Oriental bushmen, lost cities, neo-dinosaurs, and continents in cataclysm that surrounded me. A planet that breathed in and breathed out spun about me in a dizzying motion. Suddenly I lost my balance and fell backward! I landed on my back in the sand, fortunately not hitting a rock. Many stars twinkled in the sky—in another few hours a new day would begin. Many adventures lay behind me. Many more lay ahead me, but on that starry night, it was only then and there that mattered.

Left: Rock drawing from Zimbabwe showing a reclining figure with a beard and hat that looks very Middle Eastern. Right: One of the strange rock paintings in the Namib desert of bearded white explorers. Phoenicians?

This prehistoric rock painting near Brandberg,
South-West Africa, depicts young Amazons
whose European faces are painted with a light tint and
the hair shown in red or yellow. Who were these
European girls in a remote part of Africa?

Maak Shelter. Wall-painting showing the White Lady on the right.

Farini the Great (G.A. Farini) in the frontpiece of his book, *Through the Kalahari Desert*, 1896.

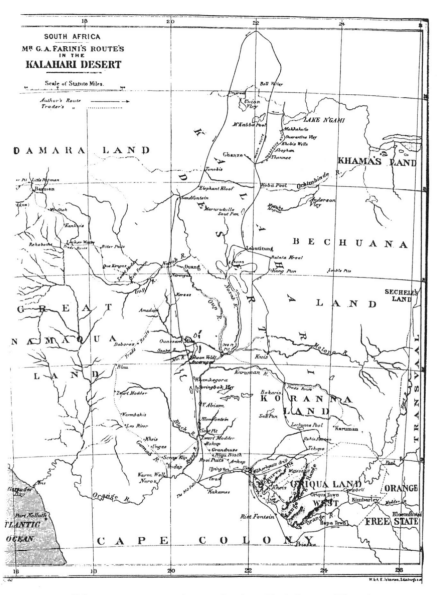

Farini's map of his journeys through the Kalahari. The lost city is marked in the center of the map, just near the letter "E" in DESERT.

Top: Farini's "Lost City" seen from a distance. The huge dolerit[e] stones dwarf the person on the right. Bottom: A closer view of th[e] huge stone blocks, notice the corners at right angles and straigh[t] parallel lines between the stones. Photos from Clement's book, *Th[e] Kalahari and Its Lost City.*

Proof that the dolerite blocks are not natural can be seen in this photo of parallel, horizontal grooves obviously cut into the stones. Since this cannot be the product of nature, Clements suggests that perhaps Farini cut the lines into the rock!

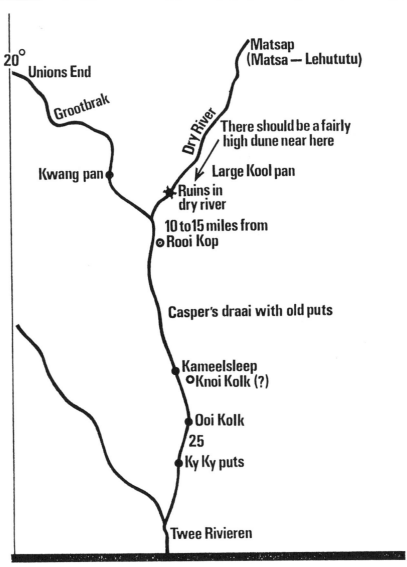

The Simon Kooper Hottentott Map. Supplied by the late Dr W. M. Borcherds.
(Unions End, Grootbrak and the 20° line of longitude have been added for clarity.)

A map to the massive "ruins" in the southern Kalahari. Fro[
Clement's book, *The Kalahari and Its Lost City*.

A. J. Clements examines some of the blocks. The white streaks are the result of chemical weathering. The right-angle corners and squaring of the blocks can clearly be seen. Incredibly, this appears to the remains of a megalithic structure built before the last cataclysmic pole shift that shifted the African continent "south."

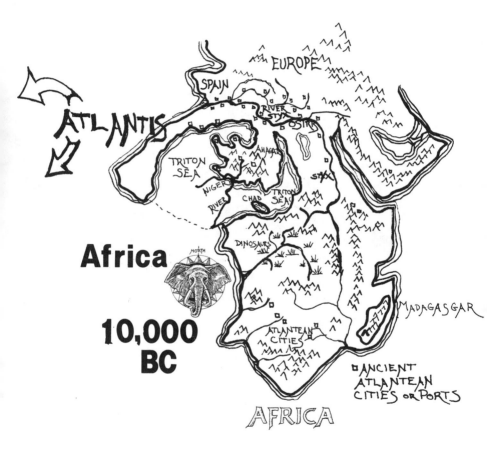

FOOTNOTES & BIBLIOGRAPHY

1. *Africa*, edited by Phyllis Martin & Patrick O'Meara, 1977, University of Indiana Press, Indianapolis.
2. *Kingdoms of the Asia*, the Middle East & Africa, Gene Gurney, 1986, Crown Publishers, Inc. NYC.
3. *Secrets of the Lost Races*, Rene Noorbergen, 1977, Harper & Row, NYC.
4. *Atlantis the Eighth Continent*, Charles Berlitz, 1984, G.P. Putnam's Sons, NYC
5. *The Atlas of Mysterious Places*, edited by Jennifer Westwood, 1987, Marshall Editions Limited, London.
6. *Petra*, E. Raymond Capt, 1987, Artisan Sales, Thousand Oaks, CA.
7. *Vanished Cities*, Hermann & Georg Schreiber, 1957, Alfred A. Knopf, NYC.
8. *Explorers Extraordinary*, John Keay, 1986, Jeremy Tarcher Inc, LosAngeles.
9. *Lost Cities*, Leonard Cottrell, 1957, Robert Hale & Co. London.
10. *The Search For Lost Cities*, James Wellard, 1980, Constable & Co. London.
11. *Archaeology Discoveries in the 1960's*, Edward Bacon, 1971, Praeger Publishers, NYC.
12. *Atlas of Ancient History*, Michael Grant, 1971, Dorset Press, Dorset, England.
13. *Atlas of Ancient Archaeology*, edited by Jacquetta Hawkes, 1974, McGraw-Hill Book Company, NYC.
14. *Atlas of the Ancient World*, edited by Frances Clapham, 1979, Crescent Books, NYC.
15. *Lost Cities of China, Central Asia & India*, David Hatcher Childress, 1985, AUP, Stelle, IL.
16. *The Phoenicians*, Gerhard Herm, 1975, William Morrow & Co. NYC.
17. *Peoples of the Sea*, Immanuel Velikovsky, 1977, Doubleday & Co. Garden City, NY.
18. *The Atlas of Archaeology*, consultant editor Professor K. Branigan, 1982, St. Martin's Press, NYC.
19. *Early Man and the Ocean*, Thor Heyerdahl, 1979, Doubleday & Co. Garden City, NY.
20. *America B.C.*, Barry Fell, 1976, Simon & Schuster, NYC.
21. *Saga America*, Barry Fell, 1983, Times Books, NYC.
22. *Lost Cities & Ancient Mysteries of South America*, 1986, David Hatcher Childress, AUP, Stelle, IL
23. *The World's Last Mysteries*, 1979, Reader's Digest, Pleasantville, NY.
24. *Dictionary of Philosophy and Religion*, W.L. Reese, 1980, Humanities Press, Atlantic Highlands, NJ.
25. *Discovery of Lost Worlds*, edited by Joseph Thorndike, Jr., 1979, American Heritage Publications, NYC.
26. *INFO Journal*, No. 55, "Unbelievable Baalbek", Jim Theisen, 1988.
27. *The Bible As History*, Werner Keller, 1956, William Morrow & Co. NYC.
28. *The Anti-Gravity Handbook*, David Hatcher Childress, 1985, AUP, Stelle, IL
29. *Anti-Gravity & the World Grid*, David Hatcher Childress, 1987, AUP, Stelle, Illinois.
30. *The Bridge To Infinity*, Bruce Cathie, 1983, Quark Enterprises, Auckland.
31. *Lost Discoveries*, Colin Ronan, 1976, Bonanza Books, NYC

32. *Mysteries of the Lost Lands*, Eleanor Van Zandt & Roy Stemman, 1978, Aldus Books, London.
33. *Ancient Man: A Handbook of Puzzling Artifacts*, William Corliss, 1978, Sourcebook Project, Glen Arm, MD.
34. *Jawa, Lost City of the Black Desert*, Svend Helms, 1981, Cornell University Press, Ithica, NY.
35. *Lost Cities of Ancient Lemuria & the Pacific*, David Hatcher Childress, 1988, AUP, Stelle, IL.
36. *Lemurian Fellowship Lessons*, 1936, Lemurian Fellowship, Ramona, CA.
37. *The Ultimate Frontier*, Eklal Kueshana, 1962, The Stelle Group, Quinlan, TX.
38. *Treasures of the Lost Races*, Rene Noorbergen, 1982, Bobbs-Merrill Co. NYC
39. *Biblical Archaeological Review*, May-June 1983
40. *Elijah, Rothschilds and the Ark of the Covenant*, Jeremiah Patrick, 1983, Restoration Press, Frankston, TX.
41. *The Unseen Hand*, Ralph Epperson, 1986, Publius Press, Tuscon, AZ
42. *Biblical Archaeological Review*, July-August, 1983.
43. *What the Bible Really Says*, Manfred Barthel, 1982, Souvinir Press, London.
44. *The Treasure of the Copper Scroll*, John Allegro, 1960, Doubleday, NYC.
45. *6000 Years of Seafaring*, Orville Hope, 1983, Hope Associates, Gastonia, NC.
46. *The Alexandria Project*, Stephan Schwartz, 1983, Dell Books, NYC.
47. *The Bible As History*, Werner Keller, 1956, William Morrow, NYC.
48. *The Treasures of Time*, Leo Deuel, 1961, World Publishing Co., Cleveland, OH.
49. *The Maldive Mystery*, Thor Heyerdahl, 1986, Adler-Adler, Bethesda, MD.
50. Egypt & the Sudan, Scott Wayne, 1987, Lonely Planet, Oakland, CA.
51. *Alexandria: A History & a Guide*, E.M. Forster, 1922, Whitehead Morris Ltd. Alexandria, Egypt.
52. *The Alexandria Project*, Stephan Schwartz, 1983, Dell Publishing, NYC.
53. *The Traveler's Key to Ancient Egypt*, John Anthony West, 1985, Alfred Knopf, NYC
54. *The Magic of Obelisks*, Peter Tompkins, 1981, Harper & Row, NYC.
55. *The Serpent in the Sky*, John Anthony West, , Harper & Row, NYC.
56. *Migdar–The Secret of the Sphinx*, F.L. Oscott, Neville Spearman, Suffolk, UK
57. *The Great Pyramid*, Piazzi Smyth, 1880, Bell Publishing Co. NYC.
58. *The Riddle of the Pyramids*, Kurt Mendelssohn, Thames & Hudson, London.
59. *Nefertiti*, Philipp Vandenberg, 1978, Hodder and Stoughton, London.
60. *Behind the Mask of Tutankhamen*, Barry Wynne, 1972, Souvineer Press, London.
61. *Investigating the Unexplained*, Ivan T. Sanderson, 1972, Prentice Hall, Englewood Cliffs, NJ.
62. *The Curse of the Pharaohs*, Philipp Vandenberg, 1975, J.B. Lippincott Co. Philadelphia, PA.
63. *Secrets of the Great Pyramid*, Peter Tompkins, 1971, Harper & Row, NYC.
64. *Egypt Before the Pharaohs*, Michael A. Hoffman, 1979, Alfred Knopf, NYC.
65. *The Shaping of Western Civilization, Vol. 1*, 1970, Holt, Rinehart & Winston, NYC.
66. *South From Khartoum*, Alan Caillou, 1974, Hawthorn Books, Inc. NYC.
67. *The Lost Pharaohs*, Leonard Cottrell, 1950, Pan Books, London.
68. *The Sphinx and the Megaliths*, John Ivimy, 1974, Sphere Books, London.
69. *The Great Pyramid: Man's Monument to Man*, Tom Valentine, 1975, Pinnacle, NYC.
70. *Solomon's New Men*, E.W. Heaton, 1974, Pica Press, NYC.

71. *Pyramidology, Books I & II,* Adam Rutherford, 1957,1962, Institute of Pyramidology, Hertfordshire, Great Britain.
72. *Egyptian Myth and Legend,* Donald Mackenzie, 1907, Bell Publishers, NYC.
73. *The Sirius Mystery,* Robert Temple, 1976, Harper & Row, NYC.
74. *The World Atlas of Archaeology,* 1985, Portland House, NYC.
75. *The Ancient Atlantic,* L. Taylor Hansen, 1969, Palmer Publications, Amherst, Wisconsin.
76. *The Children of Mu,* James Churchward, 1931, Ives Washburn, NYC.
77. *The Search for the Tassili Frescoes,* Henri Lhote, 1959,Hutchinson & Co., London.
78. *In Quest of Lost Worlds,* Count Byron de Prorok, 1935, E.P. Dutton & Co, NYC.
79. *Vanished Civilizations,* 1963, Thames & Hudson, London.
80. *The World's Last Mysteries,* Nigel Blundell, 1980, Octopus Books, London.
81. *Citta Perdute nel Deserto,* Cino Boccazzi, 1977, SugarCo Edizioni, Milan, Italy.
82. *Yemen A Travel Survival Kit,* Pertti Hamalainen, 1988, Lonely Planet, Oakland,CA.
83. *Looking For Dilmun,* Geoffrey Bibby, 1969, Alfred Knopf, NYC.
84. *Arabia,* Jonathan Raban, 1979, Picador, London.
85. *Arabian Sands,* Wilfred Thesiger, 1959, Penguin Books, NYC.
86. *Oman,* John Whelan, 1981, MEED Books, London.
87. *Qataban and Sheba,* Wendell Phillips, 1955, Harcourt, Brace & Co., NYC.
88. *The Queen of Sheba,* H. St. John Philby, 1981, Quartet Books, London.
89. *The Bible Came from Arabia,* Kamal Salibi, 1985, Pan Books, London.
90. *Seas of Sand,* Encyclopdia of Discovery & Exploration, Vol.11, 1971, Aldus Books, London.
91. *World's Beyond the Horizon,* Joachim G. Leithauser, 1955, Alfred A. Knopf, NYC.
92. *Picture-Writing of the American Indians,* Garrick Mallery, 1889, Smithsonian Institute, republished by Dover Books, NYC.
93. *Kebra Negast,* translated by Sir E. A. Wallis Budge, 1932, Dover, London.
94. *On the Track of Unknown Animals,* Bernard Heuvelmans, 1959, Hill & Wang, NYC.
95. *Living Wonders,* John Michell & Robert Rickard, 1982, Thames & Hudson, NYC.
96. *A Living Dinosaur,* Roy Mackal, 1987, E. J. Brill Publisher, Holland/NYC.
97. *Searching For Hidden Animals,* Roy Mackal, 1980, Doubleday & Co., Garden City, NY.
98. *More Things,* Ivan T. Sanderson, 1969, Pyramid Books, NYC.
99. *Project Kenya,* David Round-Turner, 1986, Kenway Publications, Kenya, distributed in the U.S.A. by Adventures Unlimited Press, Stelle, Illinois.
100. *Timeless Earth,* Peter Kolosimo, 1973, Bantam, NYC.
101. *In Witch-Bound Africa,* Frank Melland, 1923, J.B. Lippincott Co., Philadelphia.
102. *Old Africa Rediscovered,* Basil Davidson, 1961, London
103. *Archaeology Discoveries of the 1960s,* Edward Bacon, 1971,Praeger Publishers, NYC.
104. *Citadels of Mystery,* L. Sprague & Cathrine De Camp, 1964, Ballantine Books, NYC.
105. *The Second Book of the Strange,* 1981, World Almanac Publications, NYC.

106. *The Lost Cities of Africa*, Basil Davidson, 1959, Atlantic-Little, Brown Co., Boston.
107. *The Carbon Dating 14 Dating of Iron*, Nikolaas van der Merwe, 1969, University of Chicago Press, Chicago.
108. *The Kalahari and Its Lost City*, A. J. Clement, 1967, Longmans, Capetown.
109. *Through the Kalahari*, G.A. Farini, 1896, Sampson, Low, Marston & Co., London.
110. *African Presence In Early America*, edited by Ivan Van Sertima, 1987, Transaction Books, Rutgers University, New Brunswick, NJ.
111. *African Presence in Early Asia*, edited by Sertima & Rashidi, 1988, Transaction Books, Rutgers University, New Brunswick, NJ.
112. *Ancient African Kingdoms*, Margaret Shinnie, 1965, St. Martin's Press, NYC
113. *Shanidar, The First Flower People*, Ralph Solecki, 1971, Knopf, NYC.
114. *Akhenaten The Heretic King*, Donald B. Redford, 1984, Princeton University Press, Princton, NJ.
115. *Ships and Seamanship in the Ancient World*, Lionel Casson, 1971, Princeton University Press, Princton, NJ.
116. *African Mythology*, Geoffrey Parrinder, 1967, Hamlyn Publishing, NYC.

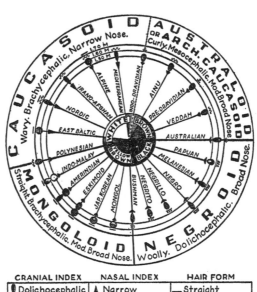

CRANIAL INDEX	NASAL INDEX	HAIR FORM
Dolichocephalic	Narrow	— Straight
Mesocephalic	Moderate	Wavy Curly
Brachycephalic	Flat or Broad	Woolly Spiral

Lost Cities

LOST CITIES & ANCIENT MYSTERIES OF AFRICA & ARABIA
By David Hatcher Childress
Across ancient deserts, dusty plains and steaming jungles, maverick archaeologist David Hatcher Childress travels through Africa and Arabia, to lost cities in the Empty Quarter, Atlantean ruins in Egypt and the Kalahari. , a port city in the Sahara, and more. This is an extraordinary life on the road: across war-torn countries, Childress searches for King Solomon's Mines, the Ark of the Covenant and the solutions to the the fantastic mysteries of the past.
ISBN 0-932813-06-2
420 Pages 110 Rare photographs Maps and Drawings
6 x 9 Paperback, $12.95.

LOST CITIES OF ANCIENT LEMURIA & THE PACIFIC
By David Hatcher Childress
Was there once a lost continent in the Pacific? Did ancient Egyptian, Peruvian, Chinese and other sailors continually cross the Pacific in ancient times? Who built the mysterious megaliths throughout the Pacific? In the forth book chronologically, in the Lost Cities Series, Maverick Archaeologist Childress searches the Indian Ocean, Australia and the Pacific in search of lost cities and ancient mysteries. Includes nearly every map drawn of lost civilizations & continents of the Pacific.
ISBN 0-932813-04-6
120 Maps Prints and Photographs
382 pp, 6 x 9 Paperback. $12.95
Portuguese edition, $12.95.

LOST CITIES & ANCIENT MYSTERIES OF SOUTH AMERICA
by David Hatcher Childress
Archaeologist/Adventurer Childress takes the reader on unforgettable journeys deep into windswept mountains in quest for cities of gold, deadly jungles searching for living dinosaurs, and scorching deserts on the track of Egyptian gold mines. Translated into both Spanish and Portuguese, and a top ten best seller in Brazil, (as well as Cuzco) this book is quickly becoming a classic on the mysteries of South America.
ISBN 0-932813-02-X
100 Maps Prints and Photos
375 pp , 6 x 9 Paperback. $12.95
Portuguese edition, $12.95.

LOST CITIES OF CHINA, CENTRAL ASIA & INDIA
By David Hatcher Childress
Search for mysterious tunnels leading to supernatural realms, forgotten cities in the Gobi Desert, hidden monasteries high in the Himalayas while Childress spins amazing stories of travel, history and the mysteries of the past while searching for clues to the fabulous Rama Empire of India and other advanced, lost civilizations. Chronologically first in the Lost Cities series.
ISBN 0-932813-00-3
80 Photographs Maps Drawings
447 pp, 6 x 9 Paperback. $12.95
Portuguese edition, $12.95.

Lost Science and Anti-Gravity

ANTI-GRAVITY AND THE UNIFIED FIELD
edited by D H. Childress.

The latest book in our best selling Anti-Gravity series. We delve deeply into the secrets of Einsteins theory to help explain the propulsion systems of UFO's and Gravity Control. Is Crystal Harmonic Propulsion the key to unlocking the mystery of Space - Time? Are the Earth's energy Vortex's actually classrooms to explain Time ? This book continues on where *Anti-Gravity and the World Grid* leaves off and takes you farther into the depths of the borderlands than ever before.
ISBN 0-932813-10-0
307 pp 100's of Photo's & Drawings 7 X 10 Tradepaper. $14.95

TAPPING THE ZERO POINT ENERGY
by Moray B. King.

The author, a well-known researcher, explains how "free energy" and "anti-gravity" might be possible with todays physics. The theories of the zero point energy show there are tremendous fluctuations of electrical field energy imbedded within the fabric of space and how in the 1930s the inventor T. Henry Moray could produce a fifty kilowatt "free energy" machine; how the Pons / Fleischmann "Cold Fusion" experiment could produce tremendous heat without fusion; how certain experiments might produce a gravitational anomaly. 6x9 tradepaper, 170pp, illustrations, diagrams, bibliography, ISBN 0-9623356-0-6, $9.95.

ANTI-GRAVITY AND THE WORLD GRID
edited by D.H. Childress.

Is the earth surrounded by an intricate electromagnetic grid network offering free energy? This complex pattern of the earth's energies, researchers believe, if properly understood, can shed light on the nature of gravity, UFO's, vortex areas, power spots, ley lines, and even the placement of ancient megalithic structures. One of our best selling books—fascinating and visual.
ISBN 0-932813-03-8
267 pp 100's of Photos & Drawings 7x10 tradepaper. $12.95

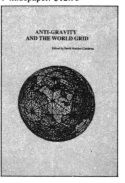

THE ANTI-GRAVITY HANDBOOK
edited by D. H. Childress

Now into several printings, this fascinating compilation of material, some of it humorous, explores the theme of gravity control and the theoretical propulsion technique used by UFOs. Chapters include "How To Build a Flying Saucer", Quartz Crystals and Anti-Gravity, Arthur C. Clarke on Gravity Control, NASA, the Moon and Anti-Gravity, & flying saucer patents. There is also a rare article by Nikola Tesla entitled "A Machine to End All War".
ISBN 0-932813-01-1
195 pp 100's of Photos Drawings & Patents 7x10 tradepaper, $12.95

THE MANUAL OF FREE ENERGY DEVICES AND SYSTEMS
compiled by D.A. Kelly.

In this manual which combines volumes I and II, D.A. Kelly describes the viability and progress of each device from Nikola Tesla to present. Also mentioned are various spin-off inventions as a result of "free energy" research. Included are chapters on Joseph Newman, "N" Field Machines, Viktor Schauberger, Rudolf Steiner, Wilhelm Reich, John Searle, and more. ISBN 0-932298-59-5, 123 Pages, 100's of Photographs, Diagrams & Patents, 9 x 11 Paperback, $12.95.

THE BRIDGE TO INFINITY: Harmonic 371244
by Captain Bruce Cathie.

Cathie's fourth and latest book on his popular theory that the earth is crisscrossed by an electromagnetic grid system. The book includes a new analysis of the harmonic nature of our physical reality, acoustic levitation, harmonic reciever towers, UFO propulsion, and demonstrates that today's scientists may have at their command a fantastic store of knowledge with which to advance the welfare of the human race. 200 pp, dozens of photos, maps & illustrations, 6x9 tradepaper, $12.95.

BRUCE L. CATHIE
THE BRIDGE
TO INFINITY

HARMONIC 371244

FURY ON EARTH: The Biography of Wilhelm Reich
by Myron Sharaf.

Free energy, cloud busters, deadly orgone, and UFOs are all part of the strange and ultimately tragic world of Wilhelm Reich. Said the TIME obituary on Nov. 3, 1957 when Reich died in prison: "Died. Wilhelm Reich, 60, once famed psychoanalyst, associate, and follower of Sigmund Freud, founder of the Wilhelm Reich Foundation, lately better known for unorthodox sex and energy theories, of a heart attack in Lewisburg Federal Penetentiary, Pa; where he was serving a two-year term for distributing his invention, the "orgone energy accumulator", a telephone-booth-sized device which supposedly gathered energy from the atmosphere, and could cure, while the patient sat inside, common colds, cancer and impotence." 6x9 trade paper, 546 pp, ISBN 0-09-158731-X, $10.95.

THE AWESOME LIFE FORCE

JOSEPH H. CATER

THE AWESOME LIFE FORCE
by Joseph Cater

Here is a book that purports to solve all of the mysteries of life, including, but not limited to, UFO phenomena, gravity, Wilhelm Reich and orgone energy, teleportation, time travel, materializations, and just about every other strange occurrence one can think of. A must for anyone seriously exploring the strange nature of reality. ISBN 0-86540-374-0, 472 Pages, 30 Diagrams and Line drawings, 5 x 8 Paperback, $15.95.

ETHER TECHNOLOGY: A RATIONAL APPROACH TO GRAVITY CONTROL
by Rho Sigma

Written by a well-known American scientist under the pseudonym of Rho Sigma, this brief book discusses in detail international efforts at Gravity Control. Included are chapters on Searle Discs, T. Townsend Brown, Ether-Vortex- Turbines, and more. Forward by Astronaut Edgar Mitchell.
108 Pages 20 Photographs Diagrams 6 x 9 Paperback. $9.95

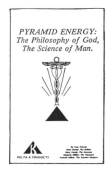

PYRAMID ENERGY:
The Philosophy of God,
The Science of Man.

DELTA-K PRODUCTS

PYRAMID ENERGY: THE PHILOSOPHY OF GOD, THE SCIENCE OF MAN
by D & M Hardy and M & K Killick

This book is far more than the title proclaims. It is an exhaustive study of the many energy fields around us, including the purpose of ley lines, megaliths and pyramids, vortex energy, Nikola Tesla's coil energy, tachion energy, levitation, the meaning of the Ark of the Covenant, & more. ISBN 0-9322298-58-7, 266 Pages 100's of Photographs & Diagrams, 6 x 9 Paperback, $15.95.

EXTRA-TERRESTRIAL ARCHAEOLOGY

THE MONUMENTS OF MARS
by Richard C. Hoagland

In 1976, NASA sent four Viking spacecraft to Mars, to photograph the planet and set landers on its surface to test for the presence of life. On July 25, a lander photographed a mile long peculiar-looking mesa resembling a human face. Through careful analysis of the Cydonia region, the author has uncovered other monuments and structures, including what is possibly an underground city. Does the "face" summon us to Mars? Why is it our face? Why has NASA denied the existance of the artifact? What are the implications of a Soviet-US manned mission to Cydonia? Who were the "Martians?" How does the existance of the "face" change our entire history? The Monument of Mars covers these topics and many more and should be one of the most controversial and talked about books of the century. ISBN 0-938190-78-4, 432 Pages, Photos & Illustrations, 7 x 10 tradepaper, $14.95.

ALIEN BASES ON THE MOON
by Fred Steckling.

This is a pretty strange book! As the title says, this is a book about weird stuff on the moon. It is jam packed with official NASA Apollo photographs and area maps of the Moon that show various craft, robot vehicles and their tracks, pyramids. Readers can examine the full page photographs themselves and decide whether or not there are structures on the Moon or not. Steckling analysed over 10,000 photos, and came up with some startling conclusions. Who is on the Moon? This book can help you decide for yourself. ISBN 0-942176-00-6, 191 Pages, 125 Photographs in color & b/w, 5 x 8 tradepaper. $12.95.

PLANETARY MYSTERIES
by Richard Grossinger

This anthology opens with an illustrated guide to a bizzare and controversial discovery - a mile long face on the surface of Mars, a massive statue flanked by pyramids and a wall. Dan Noel's discussion of megaliths and mythic signifigance of the space age follows. Richard Grossinger reports on a visit of an Australian Aborigininal holy man to a New Age retreat and asks how we can tell the difference between authentic and fake spritual practice. Rob Brezsny presents the atomic warhead as an archtype and spritual shadow. José Argüelles explores the numerical and fractal mysteries of the Mayan calendar in a long illustrated piece. Also included are *The Recollection of Osiris* by Normandi Ellis, *Artic Explorations* by Elisha Kent Kane and more. 272 pp 6 x9 Tradepaper, illustrated with photos and diagrams, $12.95.

MOONGATE:
Suppressed Findings of the U.S Space Program
by William Brian II

Documented evidence, NASA photo and official government papers ar used in this incredible book to expos the real nature, method and goal of th Apollo missions. Includes informatio on suppressed gravity research, secre discoveries on the moon by astronaut "The Great Energy Cover-up", an more. A must for anyone interested i advanced technology, gravity contro NASA and the "Space Race" Profusely illustrated with color & b/ photos. ISBN 0-941292-00-2, 23 Pages, Color Photos & Illustration 6 x 9 Paperback, $15.95.

THE LOST MILLENNIUM
by Walt & Leigh Richmond

How did the Atlanteans tap the fr energy of the Universe? What mac the continents break apart? Why d some people feel the last earth up heaval was only 5200 years ago? Di Atlanteans have free energy? Did th science cause their destruction? Wa the Great Pyramid of Egypt built as "solar tap"? All this questions an more! 172 pp Glossary, 6 x 9 Tradepaper. $9.95.

MYSTERIES OF THE CRYSTAL SKULLS REVEALED

by Bowen, Nocerino, Shapiro. An interesting collection of material on the many crystal skulls that have been found in lost cites or secret libraries around the world. Especially interesting are the pychic impressions on some of the Mayan and Tibetan skulls, and ofcourse that of the Mitchell-Hedges skull found at the lost city of Lubaantam in Belize. 8x10 tradepaper, illustrated, 298 pp, index, ISBN 0-929781-26-0, $17.95. **NEW PUBLICATION**

THE DESTRUCTION OF ATLANTIS

by Francis Joseph.
The first book from Atlantis researcher and co-discoverer of the underwater pyramids in Wisconsin, Frank Joseph. "You have obviously read a lot and accumulated much interesting material on a topic that is so controversial that most scholars turn their backs on it for fear of being ridiculed." –Thor Heyerdahl, world-famous explorer. A fascinating, well researched and unusual book about the legendary continent of Atlantis. ISBN 0-1421-5656-5, 109 pp, Photographs Illustrations, 6 x 9 Tradepaper, $9.95.

OUR COSMIC ANCESTORS

by Maurice Chatelain. Chatelain, a former NASA space expert, compiles compelling evidence to show that a highly evolved civilization existed on Earth approximately 65,000 years ago. Included are chapters on Tiahuanaco, the Egyptian Pyramids, the Mayan Calendar, the Maltese Cross, and the Sumerian Zodiac. 6x9 tradepaper, 216 pp, illustrated, ISBN 0-929686-00-4, $9.95.

THE 6000 YEAR OLD SPACE SUIT

by Vaughn M. Greene. If you ever wondered just what those strange statues of Jomon Warriors of ancient Japan were, well– maybe this is your answer! Profusely illustrated with photos of Jomon statuetes & modern comparison, this book is for those who think that nothing is too strange. 6x9 tradepaper, 112pp, photos, diagrams, ISBN 0-80481-311-6, $6.95.

ATLANTIS AND LEMURIA

by Rudolf Steiner. A reprint of a rare 1923 book that gives a good overview of the metaphysical concepts of the time of the remote past especially the concept of "Root" races. It is worth it just for the turn of the century concepts of Atlantis, though the Lemuria material seems pretty strange. The Occult Reich used much of the Root Race material, while Steiner himself had to escape Nazi Germany to Switzerland. 6x9 tradepaper, 144 pp, $9.95.

UFO PHOTOGRAPHS AROUND THE WORLD

VOL. I. Whether you are a die-hard skeptic or "contactee" (or maybe somewhere inbetween), you will be fascinated by this series of photographs taken all over the world. Are they all just clever hoaxes? Many may be terrestrial craft of unknown origin; Hi-domed discs, low-domed discs, many different types of saucer aircraft. 6x9 hardback, 254 pp, ISBN 0-934269-00-9, $14.95.

UFO PHOTOGRAPHS AROUND THE WORLD

VOL. II. More photographs of Unidentified Flying Objects for your own discernment. Flying cylinders, needleshaped UFOs, cigar-shaped airships, more. Many of these craft may be ancient Vimanas still in use. 6x9 Hardback, 286 pp, ISBN 0-934269-01-7, $16.95.

UFO CRASH AT AZTEC

by Wendelle Stevens. Chronicles the highly controversial reported crash of of UFO at a desert plateau near Aztec, New Mexico on May 25, 1948. Illustrated. 6x9 Hardbound, 612 pp, ISBN 0-934269-05-X, $18.95.

THE LOST SHIP OF NOAH

by Charles Berlitz. That an ancient ship exists in the mountains of northeast Turkey has been a controversial subject for thousands of years. Berlitz sums up much of the evidence for the remains of ships from some cataclysmic past. Especially interesting are the details of the recent discovery of the remains of a reed ship and its connection to the ancient reed ships of Peru, Easter Island, Egypt, Sumeria and India. 204 pp, photos, maps, 6x9 hardbound, $17.95.

LOST SCIENCE & ALTERNATE HISTORY

EARTHSONG
by Miram Kaplan. A remarkable book that speaks of the earth as a living being, a crystal, in fact. Ancient wisdom on the earth is shared by an Australian Aborigianl woman, an Incan Priest and a Mayan teacher on the the earth's crystaline matrix. Photos and illustrations, 81 pp, 6x9 tradepaper, $9.95. NEW

THE SACRED THEORY OF THE EARTH
by Thomas Frick.
This is an unusual collection of works on geomancy, sacred geology and geometry, the *Egyptian Book of the Dead*,Nepalese stone carving, ley lines, symbolic gardens, African ethnography, and feng shui. "Sacred Theory" includes an inclined Galactic Light Pond in Bamiyan, Afghanistan; the Great Salt Lake of Utah; American Indian myths; José Argüelles' Psychoatmospheric weather reports; the art of cartography; the internal connections of earth and sky; and the outer-space and geomantic art of Lowry Burgess and Victor Flach. A great book for the person wanting an extensive overview of Archaeo-astronomy. 6x9 tradepaper, photos & illustrations, 264 pp, $12.95.

STONEHENGE ... A closer look
By Bonnie Gaunt
In all the ages since its construction men have built nothing comparable to Stonehenge. Like the Great Pyramid of Gizeh, the solutions to the mystery that surrounds its construction, its architect and its purpose have been sincerely sought by scientists, theologians, archaeologists and historians. Now Stonehenge begins to give up her secrets. The story that it tells not only takes us through 4,000 years of mans history, but far beyond, into the timeless forces of the universe, and into the future of man on this planet. ISBN 0-9602688-0-4
236 Pages 47 Line Drawings
5 Black & white photographs
5 1/4 x 7 3/4 Paperback. $9.95

STONEHENGE SCROLLS
Edited by Donald L. Cyr
For eighteen years Donald L. Cyr has been publishing a fantastic News letter , Stonehenge Scrolls. Now for the first time the best of these newsletters have been published in one volume. Containing photographs never before published and unique theories on the mysteries of Stonehenge. Filled with new ideason ley lines, SCEMB lines, and hidden halos. A definite must for all those interested in ancient megalithic structures.
$12.95 176 pp Index Appendix
Photographs Illustrations
8 1/2 x 11 Tradepaper

THE MAGNIFICENT NUMBERS OF THE GREAT PYRAMID AND STONEHENGE
By Bonnie Gaunt
On the rocky plateau of Gizeh, fifteen miles from Cairo, stands the world's most amazing wonder, the Great Pyramid. It stands silent and serene against the Egyptian sky, yet its very presence is a bold defiance against time, a sacred memorial to the intelligence of its builders and their knowledge of time, space, and the universe. The amazing correlation of its geometry with ancient Stonehenge and with the earth, sun, and moon gives us insight into its purpose and its builder. The story that it tells touches us with wonder, and fills us with a profound respect for its architect.
ISBN 0-9602688-1-2
216 Pages 97 Line drawings
23 Black & white photographs
5 1/4 x 7 3/4 Paperback. $9.95

PROJECT KENYA - THE ADVENTURE

By David Round-Turner

In June 1986 an unusual safari set out from Nairobi. Unusual in the sense that instead of a staying and traveling together, sixty seven photographers were grouped in individual teams, each briefed with a separate theme— to record the essence that is Kenya. Brought together by the National Institute for Exploration they had traveled from the United States of America to Africa to record on film the people, the wildlife, the culture of this unique continent.

Some did it the hard way, a tough slog in the semi-arid north; some in style , staying at comfortable lodges, while others spent much of their time beneath the sparkling waters of the Indian Ocean. All lived their adventure to the full, capturing the excitement, variety, grandeur and sheer pleasure of the Kenya safari experience.

The result is a photographic chronicle from which shines the uninhibited joy of discovery. The unparalleled diversity of Kenya is superbly portrayed with nearly 200 color pictures creating a unique record of this multi-faceted and beautiful country.

The pictures are drawn together with an informative narrative which recounts the emotions and experiences of each team. But the photographs are the pedestal upon which this book stands— a magnificent tribute to dedication, skill and gratitude towards a glorious and hospitable land.

$9.95   NEW RELEASE
133 pp
192 Color Photographs
8 x12 Hardcover

GUIDE TO JAMAICA

including Haiti
by Harry S. Pariser

No other island in the Caribbean conjures up such exotic images as Jamaica and no other guide treats Jamaica with more depth, historical detail, or practical travel information than Guide to Jamaica. The chapter on Jamaican music will have your hair in treadlocks before you know.

$7.95   ISBN 0-918373-8-6
165 pp 4 Color pages
51 Black and white photographs
39 Illustrations 10 Charts18 Maps
Booklist Glossary Index

GUIDE TO THE YUCATAN PENINSULA

by Chicki Mallan

Like a hitchhiking thumb jutting into the Caribbean, the Yucatan Peninsula is an invitation to adventure and discovery. Once isolated from the rest of Mexico, new roads through the rainforest make the Peninsula easily accessible at last.

$11.95 ISBN 0-918373-11-5
300 pp 4 Color pages
154 Black and white photographs 55 Illustrations 53 Maps 68 Charts
Appendix Book list
Vocabulary Index
5 x 7 Tradepaper

INDONESIA HANDBOOK

Fourth Edition
by Bill Dalton

Not only is Indonesia Handbook the most complete and contemporary guide to Indonesia yet prepared, it is a sensitive analysis and description of one of the world's most fascinating human and geographical environments. It is a travel encyclopedia which scans island by island, Indonesia's history,ethanology, art forms, geography, flora and fauna– while making clear how the traveler can move around,eat, sleep, and generally enjoy an utterly unique travel experience. 25,000 copies in print.

$17.95 ISBN 0-918373-12-3
600 pp
5 X7 Tradepaper

SOUTH PACIFIC HANDBOOK
3rd Edition
by David Stanley
Here is paradise explored, photographed and mapped - the first original, comprehensive guide to the history, geography, climate, cultures, and customs of the 19 territories in the South Pacific. No other travel book covers such a phenomenal expanse of the earth's surface.
$13.95   ISBN 0-918373-05-0
588 pp 12 Color pages
195 Black and white photographs
121 Illustrations 35 Charts
138 Maps Booklist Glossary Index
5 x 7 Tradepaer

FINDING FIJI
by David Stanley
Fiji, everyone's favorite South Pacific country, is now easily accessible either as a stopover, or a whole Pacific experience in itself. This guide covers it all - the amazing variety of land and seascapes, customs and climates, sightseeing attractions, hikes, beaches, and how to board a copra boat to the outer islands. Finding Fiji is packed with practical tips, everything you need to know in one portable volume.
$6.95   ISBN 0-918373-03-4
127 pp 20 Color photos
78 Illustrations 26 Maps 3 Charts
Subject and name place index
Vocabulary
5 x 7 Tradepaper

MICRONESIA HANDBOOK
by David Stanley
Apart from atomic blasts at Bikini and Enewetak in the late '40s and early '50s, the vast Pacific area between Hawaii and the Philippines has received little attention. This informative book covers the seven North Pacific territories in detail.
$ 7.95   ISBN 0-918373-06-9
238 pp 8 Color pages 58 Maps
12 Charts 77 Photographs
68 Drawings Index
5 x 7 Tradepaper

JAPAN HANDBOOK
by J.D. Bisignani
Packed with practical money-saving tips on travel, food and accomodation. Japan Handbook is essentially a cultural and anthropological manual on every facet of Japanese life.
$12.95 ISBN 0-9603322-2-7
504 pp 35 Color photographs
200 Black and white photographs
92 Illustrations 29 Charts
112 Maps and town plans
Appendix on Japanese language
Booklist Glossary Index
5 x 7 Tradepaper

JOBS IN JAPAN
by John Wharton
The complete guide to living and working in the land of rising opportunity. Jobs in Japan explains in full detail how to make travel plans, deal with Japanese culture, set up a house, negotiate with employers, get the proper visas, and also bring along your children! There are thousands of long and short-term jobs available. There is no special education, experience, or speaking of language required. Work with friendly people and earn upto US $30,000-$40,000 in teaching, writing, entertainment, and business.
$9.95 ISBN 0-911285-00-8
264 pp Illustrated
5 x 8 Tradepaper

TEACHING TACTICS
For Japan's English Classroom's
by John Wharton
Teaching Tactics, a supplement to Jobs in Japan, explains in detail- how a native-speaker can teach English conversations, current English-teaching methods, popular classroom activities, types of Japanese students and their needs, the history of English and how it works, and Names and addresses of hundreds of English schools in Japan - most of which are now hiring!
$6.95 ISBN 0-911285-02-4
144 pp
5 x 8 Tradepaper

MAUI HANDBOOK

J.D. Bisignani

MAUI HANDBOOK

by J.D. Bisignani

Boasting historic Lahina, sensitively planned Kaanapali resort, power center Haleakala, and preciptous Hana Road, Hawaii's Maui is one of the most enchanting and popular islands in all of Oceania. Ramble the rollicking streets where missionaries landed in search of souls, and whalers berthed in search of trade and pleasure. Scan the sun-dappled waters for the spout of humpback whales as they play and mate in their winter sanctuary. Luxuriate on glistening beaches. Swim and snorkel in reef-protected waters or dive into the mysterious world of a half-submerged volcano crater. Challenge the surf at world-famous beaches. No 'fool-'round" advice is offered on Maui's full range of accommodations, eateries, rental cars, tours, charters, shopping, and transportation. Maui Handbook also contains practical money-saving tips, plus a comprehensive introduction to island ways, geography, and history.

$10.95 ISBN 0-918373-02-6
235 pp 6 Color photographs
50 Black and white photographs
62 Illustrations 27 Maps
13 Charts Booklist Glossary Index
5 x 7 Tradepaper

BLUEPRINT FOR PARADISE

By Ross Norgrove

Do you dream of living on a tropical island paradise? Blueprint for Paradise clearly and concisely explains how to make that dream a reality. Derived from his own and others' experiences, Norgrove covers: choosing an island, owning your own island, designing a house for tropical island living, transportation, getting settled, and successfully facing the natural elements. Breathtaking illustrations complete this remarkable guide.

$14.95 ISBN 0-918373-15-8
202 pp Illustrations
5 x 7 1/4 Tradepaper

NEW ZEALAND HANDBOOK

by Jane King

New Zealand is nature's improbable masterpiece, a world of beauty and wonder jammed into three unforgettable islands. Whether you want to run wild white-water rapids, ski the slopes of a smodering volcano, cast a fly-rod in an icy stream, or have a bet on "the trots," New Zealand's got it all. And New Zealanders have a rare knack for making visitors feel right at home. New Zealand Handbook introduces you to the people, places history, and culture of this exraordinary land. This information-packed guide leads you to reasonably priced accommodations, restaurants, enertainment, and outdoor adventure - all the best that only New Zealand can offer.

$13.95 ISBN 0-918373-13-1
512 pp 22 Color photographs
99 Black and white photographs
146 Illustrations 82 Maps
5 x 7 Tradepaper

Blueprint
for
Paradise

How to live on a tropic island

Ross Norgrove

HAWAII HANDBOOK

by J.D. Bisignani

Hawaii Handbook is an easy-to-use guide that leads you through this oasis of excitement and enchantment. It offers a comprehensive introduction to Hawaii's geography, vibrant social history, arts, and events. The travel sections inform you of the best sights, lodging, food, entertainment, and services. J.D. Bisignani has discovered bargains on excursions, cruises, car rentals, and airfares. Take advantage of these insider's shopping tips - hot bargains, souvenirs, fine arts, and Hawaiian handicrafts. Compare the finest resorts and bargain condos t budget motels and campgrounds. You'll even discover how to recognize and gather your own food from tree to sea!

$15.95 ISBN 0-918373-14-X
650 pp Highly detailed maps
Graphs and charts
Exceptional illustrations
Color and Black and white photographs
5 x 7 Tradepaper

Please PRINT all information.

Please do not send cash.

We accept **telephone credit card orders**. Call anytime 815 253 6390

We give a 10 % discount when you order 3 or more items.

BACKORDERS:
We will backorder forthcoming and out-of stock titles unless otherwise requested.

RETAILERS:
Standard discounts available Call or write for more information

SHIPPING CHARGES:
UNITED STATES

Postal Bookrate:
$1.25 for 1st book
50¢ each additional book.
Maximum you pay $5.00

United Parcel Service (UPS):
$2.75 for 1st book
50¢ each additional item
Maximum you pay $6.00

Airmail:
$4.00 per item
Sorry , No C.O.D.
Residents of Illinois add 7% sales tax.

SHIPPING CHARGES:
CANADA

Postal Bookrate:
$1.50 for 1st book
50¢ each additional book.
Maximum you pay $5.00

United Parcel Service (UPS):
$2.75 for 1st book
75$ each additional book
Maximum you pay $6.50
YOU MUST include your phone number for UPS
Airmail:
$4.00 per item

SPECIAL PAYMENT NOTICE
for **Canadian** customers
1. Remittance MUST BE $US
2. Canadian Postal Money Orders Accepted
3. Other Checks MUST BE drawn on a US Bank

SHIPPING CHARGES:
ALL OTHER COUNTRIES
Surface Delivery:
$2.75 1st item
$1.00 each addition item

SPECIAL PAYMENT NOTICE
for **International** customers
1. Remittance MUST BE $US
2. Checks MUST BE drawn on a US Bank

Name _____

Address _____

City _____ State ____ Postal Code _____

Telephone Home ( )_____ Work ( )_____

Full Book Title	Price	Qty	$ Amount

Adventures Unlimited Press
Post Office Box 22
Stelle, IL USA 60919-9989
815 253 6390

Item Totals	
Less Discount	
Shipping	
Sales Tax	
Total Remit	

MasterCard **VISA**

[ ] **Charge It**   Credit Card Number _____
Expiration _____ Today's Date_____
Signature _____

[ ] **Check Enclosed** (Drawn on US Bank in $US)

Write for a free catalogue of books and expeditions.

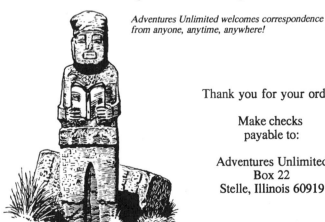

Adventures Unlimited welcomes correspondence from anyone, anytime, anywhere!

Thank you for your order!

Make checks payable to:

Adventures Unlimited
Box 22
Stelle, Illinois 60919